Enhanced sensitivity radio telescopes are producing dramatic results in a wide range of areas in astronomy. An international conference was held in Jodrell Bank, University of Manchester to take stock of these advances. This timely volume presents the review articles presented by a host of world experts who gathered at this meeting.

The articles show how high sensitivity is leading to a much improved understanding and breakthroughs in radio spectral line analysis, radio continuum observations of galaxies, cosmology, pulsars, and radio emission from stars. They also review the new and enhanced instruments now available. Finally, we are given a glimpse of the exciting telescopes being planned for the future.

This volume provides graduate students and researchers with an up-to-date and wide-ranging review of the new and future research possible with high-sensitivity radio telescopes.

CAMBRIDGE CONTEMPORARY ASTROPHYSICS

High-Sensitivity Radio Astronomy

CAMBRIDGE CONTEMPORARY ASTROPHYSICS

Series editors
José Franco, Steven Kahn, Andrew King and Barry Madore

Also available in this series

Gravitational Dynamics, *edited by O. Lahav. E. Terlevich and R. J. Terlevich*

High-Sensitivity Radio Astronomy

Proceedings of a meeting
held at Jodrell Bank, University of Manchester
January 22-26, 1996

Edited by
N. Jackson
R. J. Davis
*Jodrell Bank, Nuffield Radio Astronomy Laboratories,
University of Manchester*

PUBLISHED BY THE PRESS SYNDICATE OF THE UNIVERSITY OF CAMBRIDGE
The Pitt Building, Trumpington Street, Cambridge CB2 1RP United Kingdom

CAMBRIDGE UNIVERSITY PRESS
The Edinburgh Building, Cambridge CB2 2RU, United Kingdom
40 West 20th Street, New York, NY 1001–4211, USA
10 Stamford Road, Oakleigh, Melbourne 3166, Australia

© Cambridge University Press 1997

This book is in copyright. Subject to statutory exception
and to the provisions of relevant collective licensing agreements,
no reproduction of any part may take place without
the written permission of Cambridge University Press.

First published 1997

Printed in the United Kingdom at the University Press, Cambridge

Set in Computer Modern

A catalogue record for this book is available from the British Library

Library of Congress Cataloguing in Publication data

ISBN 0 521 57350 5 hardback

Contents

Participants	xi
Preface	xv
Foreword	xvii
Acknowledgements	xviii

Planets, stars and pulsars

Longitudinal asymmetry of Jupiter's radiation belts after the encounter with comet D/Shoemaker-Levy 9
  Richard Strom, Imke de Pater and Floris van der Tak 2

Pulsars at the highest radio frequencies
  Richard Wielebinski 6

Present and future pulsar searches
  F. Camilo 14

The study of stellar radio emission with very large aperture radio telescopes
  E.R. Seaquist 23

High-resolution MERLIN maps of the radio emission from Nova Cygni 1992 and Nova Cassiopeiae 1993
  Stewart Eyres, Richard Davis, Mike Bode and Huw Lloyd 33

Orbital motion and a model for HM Sge
  H.T. Kenny, A.R. Taylor, S. Kwok, S.P.S. Eyres and R.J. Davis .. 37

Variable sources and jets in Cepheus A
  V.A. Hughes 42

A radio survey of southern X-ray binary stars
  R.E. Spencer, A.K. Tzioumis, L.R. Ball, S.J. Newell and V. Migenes 46

MERLIN astrometry of radio stars
  S.T. Garrington, R.J. Davis, L.V. Morrison and R.W. Argyle 50

High-sensitivity radio observations of the 8.665-GHz ^{3}He$^+$ hyperfine line emission from planetary nebulae
  D.S. Balser, T.M. Bania, R.T. Rood and T.L. Wilson 53

High-resolution observations of young planetary nebulae
  M. Bryce, A. Pedlar, T. Muxlow, P. Thomasson, G. Mellema, J. Meaburn and I. Bains 57

Continuum and polarization mapping of diffuse Galactic radio sources
  J.R. Dickel 61

High-sensitivity observations of circumstellar masers
  R.J. Cohen 65

Megamasers
  Willem Baan 73

Weak radio spectral lines

H I and CO in the high-latitude cloud MBM7
  Y.C. Minh 84

A large-scale, sensitive CO survey of the Mon OB1 region
  Richard Oliver, M.R.W. Masheder and Patrick Thaddeus 88

Quest for the 3-cm spectral limit: high-sensitivity measurements of ^{3}He$^+$ emission from Galactic H II regions
  T.M. Bania, D.S. Balser, R.T. Rood and T.L. Wilson 93

High-reliability galactic H I line observations
  P. Kalberla, D. Hartmann, U. Mebold and G. Westphalen 97

Radio H I and optical absorption-line studies of interstellar gas
  B. Bates, C.R. Shaw, S.N. Kemp, F.P. Keenan, R.D. Davies and R.S. Roger 101

Galaxies and cosmology

High-resolution imaging of H I in galaxies
  J.M. van der Hulst 106

Radio emission from galactic disks
  Rainer Beck 117

Weak radio emission from active galactic nuclei
  Marek Kukula, J.S. Dunlop and A. Pedlar 129

High-sensitivity decimetre observations of Seyfert nuclei
  A. Pedlar, N.G. Hamilton and M. Kukula 139

Neutral hydrogen in Seyfert galaxies
  C.G. Mundell 143

SNR and ionized gas in M82
  T.W.B. Muxlow, A. Pedlar, K.A. Wills, P.N. Wilkinson and D.J. Axon 149

Atomic hydrogen absorption in radio-loud sources
  John E. Conway 153

VLBI observations of low-power radio galaxies
  G. Giovannini, W.D. Cotton, L. Feretti, L. Lara and T. Venturi .. 157

5 GHz EVN polarization of 3C286
  H.S. Sanghera, D.R. Jiang, D. Dallacasa, R.T. Schilizzi, E. Lüdke and W.D. Cotton 161

Polarization imaging with MERLIN at K-band to resolve Faraday effects in CSS jets
  E. Lüdke, H.S. Sanghera and W.D. Cotton 165

The cosmological evolution of radio sources
  James Dunlop 167

Determining the Hubble constant
  Michael Rowan-Robinson 177

Radio objects at redshifts $300 \geq z \geq 10$
  V.K. Dubrovich 189

New-generation telescopes

The Millimetre Array projects
  Roy S. Booth 194

Submillimetre-wave technology for extragalactic spectral-line astronomy
  Stafford Withington 203

Very deep continuum observations at submillimetre wavelengths
  E.I. Robson 207

Cold dust in galaxies at 1.2-mm wavelength
  R. Wielebinski 211

The Lovell Telescope upgrade
  R.J. Davis 213

The Giant Metrewave Radiotelescope
  *G. Swarup. S. Ananthakrishnan, C. R. Subrahmanya, A. P. Rao,
  V. K. Kulkarni and V. K. Kapahi* 217

Status and plans for the Effelsberg 100-m telescope
  W. Reich 225

Renovating the Nançay radio telescope: the FORT project
  W. van Driel, J. Pezzani and E. Gerard 229

High-sensitivity and high-dynamic-range observations with the WSRT
  A.G. de Bruyn 233

MERLIN Phase 3
  R.J. Cohen 238

The VLA upgrade project
  R.A. Perley 242

A high-sensitivity second-generation space VLBI mission
  J.S. Ulvestad, L.I. Gurvits and R.P. Linfield 252

Giga-bit VLBI storage system for the next generation
  *J. Nakajima, H. Kiuchi, Y. Chikada, M. Miyoshi, N. Kawaguchi, H. Kobayashi
  and Y. Murata* 256

The concept of the square kilometre array interferometer
  Robert Braun 260

Alternative array configuration and antenna elements for the square kilometre array
  interferometer
  G. Swarup 269

A proposed large radio telescope of new design
  T.H. Legg 274

Further site survey for the next-generation large radio telescope in Guizhou
  B. Peng, R. N. An, Y. Qiu, Y. Nie, B. Zhu, X. Xu and R. Strom .. 278

Technical aspects of the square kilometre array interferometer
  A. van Ardenne and F. Smits 282

Argus: a next-generation omnidirectional radio telescope
  R.S. Dixon 287

Afterword

Towards high sensitivity and high resolution
  R. Hanbury Brown 292

Author index   299
Object index   300
Subject index   302

Participants

Scott Aaron	Brandeis University, USA: sea@quasar.astr.brandeis.edu
Arnold van Ardenne	NFRA, Netherlands: ardenne@nfra.nl
Bryan Anderson	NRAL, UK: ba@jb.man.ac.uk
Habib Asareh	NRAL, UK: ha@jb.man.ac.uk
Max Avruch	MIT, USA:
Dave Axon	NRAL, UK: axon@stsci.edu
Willem Baan	Arecibo, USA: wbaan@naic.edu
John Baker	British National Space Centre, UK
Dana Balser	NRAO, USA: dbalser@nrao.edu
Thomas Bania	Boston, USA: bania@buast4.bu.edu
Gareth Banks	University of Wales, UK: gdb@astro.cf.ac.uk
Brian Bates	Belfast, UK: B.Bates@qub.ac.uk
Tony Beasley	NRAO, USA: tbeasley@aoc.nrao.edu
Rainer Beck	MPI Bonn, Germany: r.beck@mpifr-bonn.mpg.de
Roy Booth	Onsala, Sweden: roy@oso.chalmers.se
Peter Boyce	University of Wales, UK: gdb@astro.cf.ac.uk
Robert Braun	NFRA, Netherlands: rbraun@nfra.nl
John Brooks	ATNF Epping, Australia: jbrooks@atnf.csiro.au
R Hanbury Brown	Andover, Hampshire, UK:
Ian Browne	NRAL, UK: iwb@jb.man.ac.uk
Ger de Bruyn	NFRA, Netherlands: ger@nfra.nl
Myfanwy Bryce	Manchester, UK: mbryce@jb.man.ac.uk
Harvey Butcher	NFRA, Netherlands: hbutcher@nfra.nl
Fernando Camilo	NRAL, UK: fc@jb.man.ac.uk
Jim Cohen	NRAL, UK: rjc@jb.man.ac.uk
John Conway	Onsala, Sweden: jconway@oso.chalmers.se
Antonio da Costa	IST Lisbon, Portugal: dsc25_08@beta.ist.utl.pt
Rod Davies	NRAL, UK: rdd@jb.man.ac.uk
Richard Davis	NRAL, UK: rjd@jb.man.ac.uk
Ketan Desai	NRL Washington, USA: ketan@moon.nrl.navy.mil
Peter Dewdney	DRAO, Canada: ped@drao.nrc.ca
John Dickel	Illinois, USA: johnd@sirius.astro.uiuc.edu
Simon Dicker	NRAL, UK: srd@jb.man.ac.uk
Robert Dixon	Ohio State, USA: Bob_Dixon@osu.edu
John Dreher	SETI Inst., USA: john_dreher@phoenix.seti-inst.edu
Wim van Driel	Paris Observatory, France:
Viktor Dubrovich	SAO, Russia: dubr@sao.stavropol.su
James Dunlop	Edinburgh, UK: jsd@roe.ac.uk
Darrel Emerson	NRAO Tucson, USA: demerson@nrao.edu
Stewart Eyres	NRAL, UK: spe@astro.keele.ac.uk
Chris Fassnacht	Caltech, USA: cdf@astro.caltech.edu
Michael Garrett	NRAL, UK: mag@jb.man.ac.uk
Simon Garrington	NRAL, UK: stg@jb.man.ac.uk
Gabriele Giovannini	Bologna, Italy: ggiovannini@astbo1.bo.cnr.it
Leonid Gurvits	NFRA, Netherlands: lgurvits@nfra.nl
Christian Henkel	MPI Bonn, Germany: p220hen@mpifr-bonn.mpg.de
David Henstock	NRAL, UK: drh@jb.man.ac.uk
Anthony Holloway	University of Manchester, UK: ajh@ast.man.ac.uk

Participants

Vic Hughes	Queen's University, Canada: hughes@qucdn.queensu.ca
Thijs van der Hulst	Groningen, Netherlands: vdhulst@astro.rug.nl
Busaba Hutawarakorn	NRAL, UK: bh@jb.man.ac.uk
Julia Hutchison	University of Central Lancashire, UK: jmh@gal.star.uclan.ac.uk
Carole Jackson	RGO Cambridge, UK: cjackson@ast.cam.ac.uk
Neal Jackson	NRAL, UK: njj@jb.man.ac.uk
Peter Kalberla	Bonn, Germany: pkalberl@astro.uni-bonn.de
Ken Kellermann	NRAO Charlottesville, USA: kkellerm@nrao.edu
Harold Kenny	Queen's University, Canada:
Leon Koopmans	Groningen, Netherlands: leon@astro.rug.nl
Marek Kukula	ROE Edinburgh, UK: m.kukula@roe.ac.uk
Andrzej Kus	Torun, Poland: ajk@astro.uni.torun.pl
Patrick Leahy	NRAL, UK: jpl@jb.man.ac.uk
Tom Legg	NRC Ottawa, Canada: legg@hiaras.hia.nrc.ca
Felix Lockman	NRAO Greenbank, USA: jlockman@nrao.edu
Sir Bernard Lovell	NRAL, UK:
Andrew Lyne	NRAL, UK: agl@jb.man.ac.uk
Geoff Macdonald	University of Kent, UK: ghm@star.ukc.ac.uk
Richard McMahon	IoA Cambridge, UK: rgm@ast.cam.ac.uk
Maria Marchã	University of Lisbon, Portugal: mmarcha@milkyway.cc.ul.pt
Chris Martin	NRAL, UK: cem@jb.man.ac.uk
Mike Masheder	University of Bristol, UK: mike.masheder@bristol.ac.uk
Young Minh	Korea Astronomy Observatory, Korea: minh@hanul.issa.re.kr
Ian Morison	NRAL, UK: im@jb.man.ac.uk
Carole Mundell	NRAL, UK: cgm@jb.man.ac.uk
Tom Muxlow	NRAL, UK: twbm@jb.man.ac.uk
Junichi Nakajima	CRL Kashima, Japan: nakaji@crl.go.jp
Sunita Nair	NRAL, UK: sunita@jb.man.ac.uk
Richard Oliver	Bristol, UK: r.j.oliver@bristol.ac.uk
Yuri Parijskij	SAO Pulkovo, Russia: par@homesao.spb.su
Alan Pedlar	NRAL, UK: ap@jb.man.ac.uk
Bo Peng	Beijing, China: pb@bao01.bao.ac.cn
Richard Perley	NRAO Socorro, USA: rperley@nrao.edu
Richard Porcas	MPI Bonn, Germany: porcas@mpifr-bonn.mpg.de
Marc Price	Parkes, Australia: rmprice@atnf.csiro.au
Anita Richards	NRAL, UK: amsr@jb.man.ac.uk
Maria Rioja	NFRA/JIVE, Netherlands: rioja@nfra.nl
Ian Robson	JAC Hawaii, USA: eir@jach.hawaii.edu
Michael Rowan-Robinson	Imperial College London, UK: m.rrobinson@ic.ac.uk
Hardip Sanghera	NFRA/JIVE, Netherlands: hss@nfra.nl
Bob Sault	ATNF Epping, Australia: rsault@atnf.csiro.au
Richard Schilizzi	NFRA/JIVE, Netherlands: rts@nfra.nl
Ernie Seaquist	Toronto, Canada: seaquist@astro.utoronto.ca
Carl Shaw	Queen's University Belfast, UK: c.shaw@qub.ac.uk
Andrew Slavin	Liverpool John Moores University, UK: ajs@starul.livjm.ac.uk
Graham Smith	NRAL, UK: fgs@jb.man.ac.uk
F. Smits	NFRA, Netherlands: carolien@nfra.nl
Jonathan Smoker	NRAL, UK: jvs@jb.man.ac.uk
Lister Staveley-Smith	ATNF, Australia: lstavele@atnf.csiro.au
Wolfgang Steffen	University of Manchester, UK: wsteffen@ast.man.ac.uk

Participants

Richard Strom	NFRA, Netherlands: strom@nfra.nl
Ralph Spencer	NRAL, UK: res@jb.man.ac.uk
Dave Stannard	NRAL, UK: ds@jb.man.ac.uk
Govind Swarup	TIFR Pune, India: gswarup@gmrt.ernet.in
Russ Taylor	University of Calgary, Canada: russ@bear.ras.ucalgary.ca
Andrew Thean	NRAL, UK: ahct@jb.man.ac.uk
Peter Thomasson	NRAL, UK: pt@jb.man.ac.uk
Tasso Tzoumis	ATNF Epping, Australia: atzoumi@atnf.csiro.au
J Ulvestad	JPL Pasadena, USA: 6031::jsu
François Viallefond	Observatoire de Paris, France: viallefond@obspm.fr
Dennis Walsh	NRAL, UK: dw@jb.man.ac.uk
Kelvin Wellington	ATNF Epping, Australia: kwelling@atnf.csiro.au
Richard Wielebinski	MPI Bonn, Germany: p022rwi@mpifr-bonn.mpg.de
Peter Wilkinson	NRAL, UK: pnw@jb.man.ac.uk
Karen Wills	NRAL, UK: kaw@jb.man.ac.uk
Stafford Withington	MRAO Cambridge, UK: stafford@mrao.cam.ac.uk
Shengyin Wu	Beijing Observatory, China: wsy@bao01.bao.ac.cn
G Zalamanski	Besançon, France: zala@obs-besancon.fr

Preface

The Conference on "High-Sensitivity Radio Astronomy" was held at Jodrell Bank, University of Manchester, on 22-26 January 1996, to review and discuss developments concerned with high-sensitivity radio astronomy, both now and in the future, and brought together 116 astronomers from 17 countries.

Radio astronomy has come a long way since the aerials of Jansky and Reber, and the discovery of a few bright radio sources in the sky. The first such sources, Cas A and Cyg A, were later identified as two quite different objects; a Galactic supernova remnant and an extragalactic object respectively. New eras in astronomy generally were opened up with the identification of the radio sources 3C48 and 3C273 with quasars, and a few years later with the discovery of pulsars. Radio sources are still prominent in lists of the highest-redshift objects known, and radio selection is at the present time the most straightforward way to select galaxies at very high redshift without observable central quasars.

New areas of radio astronomy are being opened up, comparable to the advances of thirty years ago, with the advent of higher sensitivity interferometers and giant single telescopes. The major growth industry in recent years has been the investigation of thermal sources, in particular the development of stellar radio astronomy. Interferometers such as MERLIN and the VLA now have sufficient sensitivity to map tens of radio stars and investigate phenomena such as stellar winds; starburst objects are now accessible too, a particular highlight being the new images of M82 (Muxlow *et al.*, this conference).

In extragalactic astronomy the major areas include the study of "radio-quiet" (though definitely not radio-silent) objects. Detailed maps can now be made of Seyfert galaxies with resolution well-matched to that of the HST and detailed physics drawn from radio-optical comparisons. Moreover, this work will undoubtely be extended in the future to Seyferts' high redshift cousins, the radio-quiet quasars, which have hitherto been inaccessible. Further research will also be possible in gravitational lens studies, with followup of fainter objects being possible and also extended monitoring campaigns to determine H_o.

Finally, a large range of new telescopes is being developed together with upgrades to existing observatories. Both single-telescope and interferometer proposals are summarised in this book, culminating in designs for a "Square Kilometre Array Interferometer" (SKAI). Such an instrument would be unique in that it would allow us to see the majority of the universe's gas – cold gas at high redshift – rather than being restricted to the high-redshift emission-line objects we have so far tended to observe. This will require a high degree of cooperation (and funding), but one can expect that conferences in fifty years' time will consist of discussions of science from such an array rather than technical details of its construction.

Neal Jackson, Richard Davis
Jodrell Bank
July 1996

Foreword

The Conference on High Sensitivity Radio Astronomy was convened to commemorate the 50th anniversary of radio astronomy at Jodrell Bank. It is a convenient title to obscure the vagueness of those days. The term "radio astronomy" as a description of the new techniques for observing the universe in the radio wave region of the spectum did not emerge in the literature until late in the 1940s. Of more significance is that when I arrived at Jodrell Bank with trailers of ex-army radar equipment in December 1945 there was no intention of initiating an astronomical observatory — indeed, my permission to work at Jodrell Bank in order to escape the electrical interference of the City of Manchester was for two weeks only.

Before the war I had been a conventional cosmic ray physicist using geiger counters and cloud chambers and the hope was to use the war-time radar techniques to study large cosmic ray air showers. However, this was not to be. The serendipitous nature of scientific discovery prevailed then, as it has done throughout the subsequent 50 years. Although the cosmic ray proposal figured largely in the reason for building the 218 ft transit telescope and then the 250 ft MkI, the cosmic ray observations have never been seriously attempted at Jodrell Bank nor, as far as I know, elsewhere.

In those early days at Jodrell Bank very few of the titles of the papers given in this symposium would have made sense. The discoveries of Jansky and Reber were not thought to be of much astronomical interest and the phenomena were believed to originate solely in intergalactic space. Even after the discovery of the localised sources of radio emission the instinctive belief that these were objects in the galaxy prevailed. The recognition of their extragalactic nature belongs to the decade of the 1950s and as Hanbury Brown related in his introductory talk the high redshift objects known as quasars were not discovered until the early 1960s.

Fifty years ago talk of radio galaxies, quasars, pulsars, the cosmic microwave background, spectral lines, masers, even of magnetic fields and the title of this conference would nearly all have lain in the realms of fiction. They were the substance of this commemorative meeting and now we must dream of the talking points of the centenary celebrations in 2045 AD.

Sir Bernard Lovell

Acknowledgements

A large number of people contributed to making this conference a success. We would particularly like to thank Janet Eaton, who organised much of the logistics and worked tirelessly before and during the conference. Thanks are also due to the other secretarial staff at Jodrell, to the rest of the LOC – Ian Browne, Ian Morison, Carole Mundell, and Alan Pedlar – and to the students who helped with the organisation. The conference was supported by the University of Manchester.

Planets, stars and pulsars

Longitudinal asymmetry of Jupiter's radiation belts after the encounter with comet D/Shoemaker-Levy 9

By RICHARD G. STROM[1], IMKE DE PATER[2] AND FLORIS VAN DER TAK[2]†

[1]Netherlands Foundation for Research in Astronomy, P.O. Box 2, 7990 AA Dwingeloo, The Netherlands

[2]Astronomy Department, 601 Campbell Hall, University of California, Berkeley CA 94720, USA

Jupiter's synchrotron radiation was observed extensively before, during and after comet Shoemaker-Levy 9 crashed into the planet. The total flux density showed a dramatic increase at all wavelengths. Here we describe the time evolution of the brightness distribution of the radio emission as observed by the VLA and the WSRT telescopes at 21 cm. Preliminary reduction and analysis of the images reveal, roughly, the following phenomena: **a)** The increase in Jupiter's flux density is, overall, dominated by an enhancement in the magnetic equatorial emission at longitudes $\lambda_{III} \approx 160° - 260°$, the so-called *active sector*. **b)** There is *no* obvious connection with the impact sites or times. **c)** The radiation peak which brightens is often displaced inwards, closer to Jupiter, suggestive of radial diffusion of the high-energy electrons to regions of greater magnetic field strength. **d)** High resolution VLA images show that the impact-induced asymmetry between the peaks lasts for less than 4 days, a period similar to that expected from the energy-dependent drift of synchrotron-radiating particles. **e)** Most intriguing is the apparent discrepancy in the time evolution of the ratios between the radiation peaks in high and low resolution images. At low resolution the induced asymmetry persists much longer, and continues to vary daily. In these observations one is also sensitive to higher latitude material. We are probably seeing the consequences of continued pitch-angle scattering in these images.

1. Introduction

The discovery of radio emission from Jupiter can be considered one of the highlights of the heroic days of radio astronomy. The sequence of events leading up to Burke and Franklin's momentous discovery was as humorous as it was unexpected (see Franklin 1983). For they had the good fortune to be observing near 20 MHz, where an ionospheric emission process (still not entirely understood, but almost certainly involving cyclotron radiation) produces highly anisotropic decametric emission with peak bursts at the 10^6 Jy level (for a review, see Carr, Desch & Alexander 1983). While the emission which concerns us here is of a completely different nature, and dominates the decimetric wavelengths, the decametre discovery put Jupiter on the map, as it were, as a radio source.

Jupiter's decimetre radiation is the result of synchrotron emission from highly energetic electrons trapped in the Jovian magnetic field: the equivalent of the Earth's Van Allen belts. (At the shorter wavelengths, below about 6 cm, thermal emission from the planet's disk becomes dominant. This emission has been removed, where necessary, so all the results presented here refer only to the nonthermal component of the emission.) Early indications of the brightness distribution of this emission came from Berge (1966) and Branson (1968). For a review of what is now known about the emission from the radiation belts, see de Pater (1990).

† Present address: Leiden Observatory, P.O. Box 9513, 2300 RA Leiden, The Netherlands

In addition to short-term changes in the radio emission, produced by Jupiter's nearly ten-hour rotation period, gradual long-term variations in the total flux density have also been observed. The emission has been monitored at wavelengths near 12 cm with a variety of telescopes for many years. Over a period of nearly two and a half decades, the flux density (normalized to the standard distance of 4.04 AU) has drifted up and down between 4 and 5 Jy on a timescale of years (Klein, Thompson & Bolton 1989; Bolton et al. 1989). This should be kept in mind when considering the changes observed during the encounter with D/Shoemaker-Levy 9 (SL9), although we note that the changes we observed were of a larger magnitude at these wavelengths, and occurred over a much shorter timescale (days rather than years).

2. Observations

Jupiter was observed by many radio telescopes during and around the period of the SL9 impacts (de Pater et al. 1995). This extensive monitoring enabled us to follow the spectral evolution of the synchrotron emission while the flux density increased dramatically at all wavelengths, from 92 to 6 cm. At 20 and 90 cm, we used both the VLA and the Westerbork Synthesis Radio Telescope (WSRT), and here we report on the results of imaging at the shorter wavelength with those instruments.

The VLA, by virtue of its good instantaneous u,v-plane coverage (baselines covering a range of orientations), is generally able to generate two-dimensional images from even short observations. This means that for Jupiter, which changes in the course of a Jovian day of just under 10 hours, we are able to study these changes fully, limited only by the inherent resolution of the array which can be as good as $\approx 3''$ arc. In our data reduction we have made a series of maps at different Jovian longitudes, and used the deconvolution algorithm CLEAN (Högbom 1974).

The WSRT, being an east-west (one-dimensional) array, has to build up the different orientations required for a two-dimensional map by means of the earth's rotation. Earth-rotation synthesis assumes source nonvariability, which is clearly invalid in the case of Jupiter. Consequently, in our analysis, we have used the data to construct a series of one-dimensional maps to study variations in the peak brightness with Jovian longitude, with a resolution of about $20''$ arc.

3. Results

The nonthermal emission from Jupiter's radiation belts showed a dramatic increase in July, 1994, starting at about the time of impact of the first SL9 fragment. The total flux density increased roughly linearly at all wavelengths during the week of the remaining collisions. The increase ranged from 15% to 40% at the various observing frequencies in a systematic way, so that the effect was an overall hardening of the radio spectrum. After the last impact the radio emission declined in intensity, at a rate which increased with wavelength. At 70–90 cm, the flux density dropped back to near its pre-impact level within just a few days. At 21 cm, after about 6 weeks the decrease was only about half way to recovering its normal level. These global changes have been detailed by de Pater et al. (1995).

Although changes in Jupiter's flux density have been observed in the past, as noted above, such a sharp increase in intensity is unprecedented. Apart from its obvious temporal link to the SL9 impacts, both in terms of the commencement of the flux density increase and the inception of its decline after the last collision, there is no evidence linking changes in the brightness distribution to individual fragments or their locations.

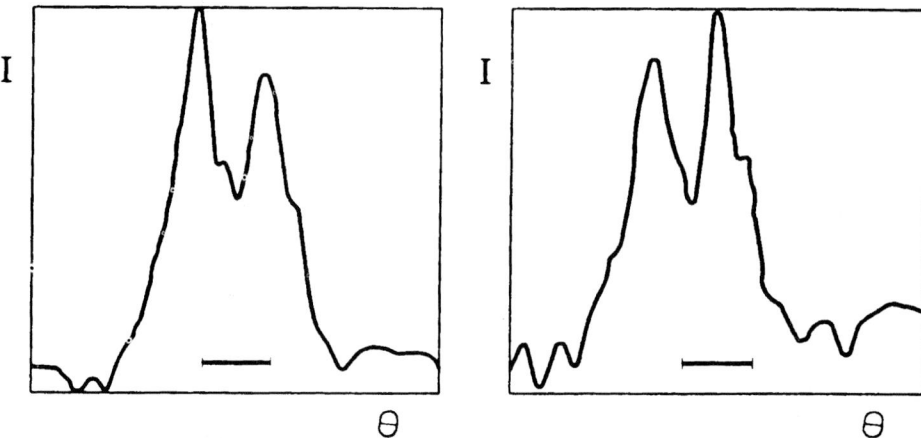

FIGURE 1. The east-west Jovian brightness distribution at 21 cm, as observed with the WSRT near the same longitude. The bar indicates Jupiter's diameter. (*Left*) On 15 July, 1994, at $\lambda_{III} \simeq 274°$; (*Right*) on 20 July, 1994, at $\lambda_{III} \simeq 288°$.

The VLA maps show an overall brightening, especially in the so-called active sector ($\lambda_{III} \simeq 160° - 260°$). Moreover, the bright peak tends to be displaced inwards, closer to Jupiter, suggesting that the radiating particles may have diffused inwards to regions of greater magnetic field strength. The asymmetry in the brightness peaks, brought about by the enhancement of the active sector, persists for only a few days, consistent with the drift time of the radiating particles.

In our analysis of the WSRT one-dimensional brightness distributions, we noticed that the relative intensity of the eastern and western peaks at different Jovian longitudes showed an unusual asymmetry after the last impact, which changed in an irregular way, and persisted for at least a month. This behaviour had not been seen before, and in particular is not present in our observations from just before the encounter with SL9. The effect is illustrated in Fig. 1, which shows strip scans made with the WSRT at similar longitudes one day before the first impact (15 July), and after most of the collisions had occurred (20 July): the inversion in the strongest peak can be clearly seen.

Since this effect was not obviously present in the VLA observations, and considering that the WSRT beam is much more extended in the north-south direction, we suspected that it might be related to emission well-displaced from the Jovian equator. The high resolution VLA snapshot maps do indeed show that the high latitude secondary peaks first imaged by de Pater & Jaffe (1984) were brighter after the impacts. These peaks are probably caused by a population of small-pitch-angle electrons which cross the equator at a fairly large distance from the planet ($\approx 2.5\,R_J$). We tentatively conclude that it is emission at high Jovian latitudes which is mainly responsible for the unusual peak ratios seen in the WSRT observations.

4. Conclusions

The increase in the flux density of Jupiter observed at all centimetre wavelengths during the encounter with the comet D/Shoemaker-Levy 9 reflects a brightening of the radiation belts generally, but the increase did not affect all structural features equally. Similarly, the decline observed, which showed a strong wavelength dependence, affected some parts of the radiation belts more strongly than others. The impact-induced asymmetry between the magnetic-equatorial radiation peaks persists for only a few days at 21 cm, a time scale consistent with the global drift of electrons around Jupiter. During this readjustment of the radiating electrons, the change in total emission at 21 cm was negligible.

The larger-scale asymmetry at 21 cm, which we believe is produced by emission regions at higher latitudes and which was especially evident in the WSRT observations, persisted for weeks rather than days. Since we are not even sure what the cause of the high-latitude radiation peaks is, it would be speculation to try to explain the slow return to normality, although it is likely to be related to the behaviour of electrons of small pitch angle. Interestingly, some efforts to model the increase in total flux density have concentrated upon pitch-angle scattering of the already existing electrons. Hopefully as our understanding of the processes involved improves, we will also turn up clues regarding the origin of the high-latitude emission.

The Westerbork Synthesis Radio Telescope is operated by the Foundation for Research in Astronomy with the financial support of the Netherlands Organization for Scientific Research (NWO). The VLA is part of the National Radio Astronomy Observatory, which is operated by Associated Universities, Inc., under a cooperative agreement with the National Science Foundation (NSF). This research was supported in part by NSF grants 22122 and 26284, and National Aeronautics and Space Administration grant NAGW-3917 to the University of California, Berkeley.

REFERENCES

BERGE, G.L. 1966. *ApJ* **146**, 767

BOLTON, S.J., GULKIS, S., KLEIN, M.J., DE PATER, I. & THOMPSON, T.J. 1989. *J. Geophys. Res.* **94**, 121

BRANSON, N.J.B.A. 1968. *MNRAS* **139**, 155

CARR, T.D., DESCH, M.D. & ALEXANDER, J.K. 1983. In *Physics of the Jovian magnetosphere.* (ed. A.J. Dessler), p. 226, Cambridge University Press.

DE PATER, I. 1990. *ARA&A* **28**, 347

DE PATER, I. & JAFFE, W.J. 1984. *ApJS* **54**, 405–419.

DE PATER, I., HEILES, C., WONG, M., MADDALENA, R.J., BIRD, M.K., FUNKE, O., NEIDHOEFER, J., PRICE, R. M., KESTEVEN, M., CALABRETTA, M., KLEIN, M.J., GULKIS, S., BOLTON, S.J., FOSTER, R.S., SUKUMAR, S., STROM, R.G., LEPOOLE, R.S., SPOELSTRA, T., RIBISON, M., HUNSTEAD, R.W., CAMPBELL-WILSON, D., YE, T., DULK, G., LEBLANC, Y., GALOPEAU, P., GERARD, E. & LECACHEUX, A. 1995. *Science* **268**, 1879

FRANKLIN, K. L. 1983. In *Serendipitous discoveries in radio astronomy* (ed. K. I. Kellermann & B. Sheets), p. 252, NRAO.

HÖGBOM, J. A. 1974. *A&AS* **15**, 417

KLEIN, M.J., THOMPSON, T.J. & BOLTON, S.J. 1989. In *Time variable phenomena in the Jovian System.* (ed. M.J.S. Belton, R.A. West & J. Rahe), p. 151, NASA.

Pulsars at the highest radio frequencies

By RICHARD WIELEBINSKI

Max-Planck-Institut für Radioastronomie, Auf dem Hügel 69, 53121 Bonn, Germany

Pulsars were discovered in Cambridge at 81.5 MHz, and the early years of research on these objects concentrated on their low-frequency characteristics. A technical reason for this development was the fact that pulsar spectra are very steep, while at the same time the sensitivity of the available radio telescopes was dropping rapidly at higher frequencies.

From theoretical considerations it was clear soon after the discovery that we were dealing with the radio emissions from the magnetosphere of a rotating (not pulsating) neutron star. Higher-frequency radio signals are considered to come from the deeper layers of the magnetosphere and hence are possibly more reliable signatures of the real emission mechanism.

The 100-m radio telescope of the Max-Planck-Institut für Radioastronomie in Effelsberg has been for the last 25 years the most sensitive single radio telescope in the short-cm to long-mm wavelength range. The advent of HEMT amplifiers allowed the construction of sensitive receivers for the 10–50 GHz frequency range. Pulsar observations could now be extended into this hitherto inaccessible frequency range.

---

1. Introduction

The discovery of pulsars by Hewish et al. (1968) startled not only the astronomical world. The original discovery paper contained information on many of the pulsar characteristics like dispersion, time variation etc. Soon after the discovery announcement all the radio telescopes around the earth studied pulsars. Many further parameters of pulsars were established, such as the exact period, its derivative, the polarization and spectrum. A spectrum derived from five frequency studies (Robinson et al. 1968) showed that the spectral index became very steep at high radio frequencies, much steeper than what was normally observed for other types of radio sources. In fact the spectral index of CP 1919 dropped from $\alpha \sim -1.5$ in the MHz range to $\alpha \sim -3.0$ in the GHz range. The large single telescopes at Jodrell Bank, Parkes, and Arecibo collected an enormous set of data at frequencies between 100 MHz and 1400 MHz.

The volume of data on pulsars at high radio frequencies is rather small. The earliest publications in 1968 gave some information on pulsars up to the frequency of 2.7 GHz. The important milestone in high-frequency pulsar research was the completion of the 100-m radio telescope of the MPIfR. In fact the first publication from this new instrument (Wielebinski et al. 1972) described the detection of six pulsars at 2.8-cm wavelength (10.6 GHz). The first detections of pulsars at 22.7 GHz were reported by Bartel et al. (1977), again using the 100-m radio telescope. Finally with the new sensitive HEMT receiver system placed in the 100-m radio telescope it was possible to detect pulsars at 8-mm wavelength (Wielebinski et al. 1993). In this paper the characteristics of pulsars at high radio frequencies (short cm and mm wavelengths) will be described. These high-frequency characteristics give important hints about the emission mechanism of pulsars.

2. The instrumental requirements

At lower radio frequencies, where the bulk of pulsar data was obtained, the signals are strong and detection is relatively easy. In fact some historical reviews suggest that pulsars were obvious in the early continuum surveys of the Milky Way but went undetected because they were rejected as interference. Also the time constants of receivers were made longer so as to integrate the cosmic noise. The detection by Hewish et al. (1968) was made with an antenna optimised for short time constants needed for studies of interplanetary scintillations. All antennas with large collecting areas could discover numerous pulsars in the 80–400 MHz frequency range, first with fast pen recorders, later with digital searching methods. Searches for pulsars at 1.4 GHz are only recent and have contributed to our knowledge about pulsars with high dispersion. Pulsar searches at 4.75 GHz have now started.

Pulsar searches and later pulsar studies profited enormously from the fast developments of digital techniques. Since pulse periods were found to be highly accurate it was possible to integrate the signal synchronously with the pulsar period and hence reach very sensitive detection levels. This "stacking" of data is of particular significance at the high radio frequencies where the pulse energy drops rapidly. Integrations over 100000 pulses are no rarity.

3. Pulse shapes

Pulsars are observed to have a variety of pulse shapes. The shapes are believed to be due to a combination of the pulsar emission beams and the way these cut the observer's line of sight. Observations of pulsars at high frequencies published by Sieber et al. (1975) showed that it is necessary to follow the development of a pulse shape to the highest frequencies before a classification is possible. A suggestion by Cordes (1986), that the pulse width is directly related to the height of the emission region above the surface of the neutron star, gained much support from recent high-frequency pulsar studies. A larger sample of pulse shapes from multi-frequency observations was published by Izvekova et al. (1994). A new catalogue of integrated pulse shapes by Seiradakis et al. (1995) supplemented by mm-wavelength observations of Kramer et al. (1996) gives the complete state of information to date on pulse shapes at four frequencies. We know pulse shapes of 183 pulsars at 1.4 GHz, 46 pulsars at 4.75 GHz, 24 pulsars at 10.6 GHz and 8 objects at 33 GHz.

The fact that average pulse profiles can be separated into individual sub-pulse components has been known for a long time (e.g. Manchester & Taylor 1977). A new technique of identifying seperate components was developed recently by Kramer et al. (1994). This technique was applied by Kramer (1994) to a sample of pulsars at three frequencies allowing an exact description of pulse-shape evolution. This technique combined with pulse time alignment is of great importance in high-frequency pulsar studies. Since the signals from pulsars at the high frequencies are very weak, the prediction of the pulse position in pulse phase, based on lower-frequency observations, is of great importance.

4. Pulse energy, pulse spectrum

The significant changes in pulse intensities and pulse shapes which occur on many time scales make the determination of pulse flux or pulse energy rather unreliable. Repeated observations are necessary with equal weight being given to those observations when a pulse is not detected. This is of particular significance at high frequencies where the signal-to-noise is a problem. Some details of the problems encountered can be found in Taylor et al. (1993) and Lorimer et al. (1995). Principally pulse energy should be given which can be derived from the integration below the average pulse shape in Joules $m^{-2} Hz^{-1}$. Alternatively some observers give the "equivalent pulse flux" which is defined

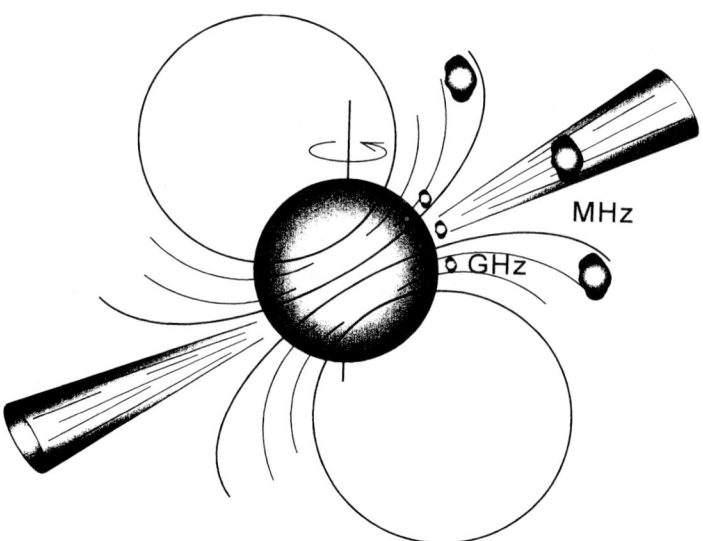

FIGURE 1. In the radius-to-frequency mapping concept the higher frequency emission comes from regions near the surface of the neutron star.

by the integrated energy under the pulse divided by the pulsar period in Jansky s^{-1}. An important aspect of all these different definitions is that observers use common programmes so that intercomparison of pulse fluxes from recent observations in overlapping frequencies shows consistent results.

The average spectral index derived for lower radio frequencies has been given as $\alpha = -1.5$ (Manchester & Taylor 1981) and $\alpha = -1.6$ by Lorimer et al. (1995). Once some higher-frequency flux values were included the spectra steepened to $\alpha = -1.8$ (Sieber 1973; Taylor et al. 1993). A comprehensive study of the spectra of 45 pulsars was made by Malofeev et al. (1994) based on the high-frequency observations described by Seiradakis et al. (1995). Two types of pulsar spectra are evident from this work. In 20 objects a single power-law spectrum is observed with a spectral index $-1.2 < \alpha < -2.6$ with an average $\alpha = -1.85$. The second type of pulsar spectrum found in 25 pulsars can best be fitted with two power laws where the spectral indices α_1, α_2 are $-0.2 < \alpha_1 < -1.5$ and $-1.3 < \alpha_2 < -4.1$ with $\langle \alpha_1 \rangle = -1.1$ and $\langle \alpha_2 \rangle = -2.7$. It may be a coincidence that can be noted that this gives the average spectrum $\langle \alpha \rangle = -1.9$. The extrapolation of the pulsar flux values to mm wavelengths suggested that detections could be feasible in some flat-spectrum objects only after many hours of integration. It was therefore quite a surprise that some pulsars could be easily detected at mm wavelengths after only some minutes of observing time (Wielebinski et al. 1993). This indicated a dramatic change of the pulsar spectrum for several objects, a fact that was studied in more detail recently by Kramer et al. (1996). These authors proposed that a spectral index turn-up (as expected from the known spectrum of the Crab pulsar) did indeed exist.

SEIRADAKIS et al. (1995) 6cm 46 Profiles

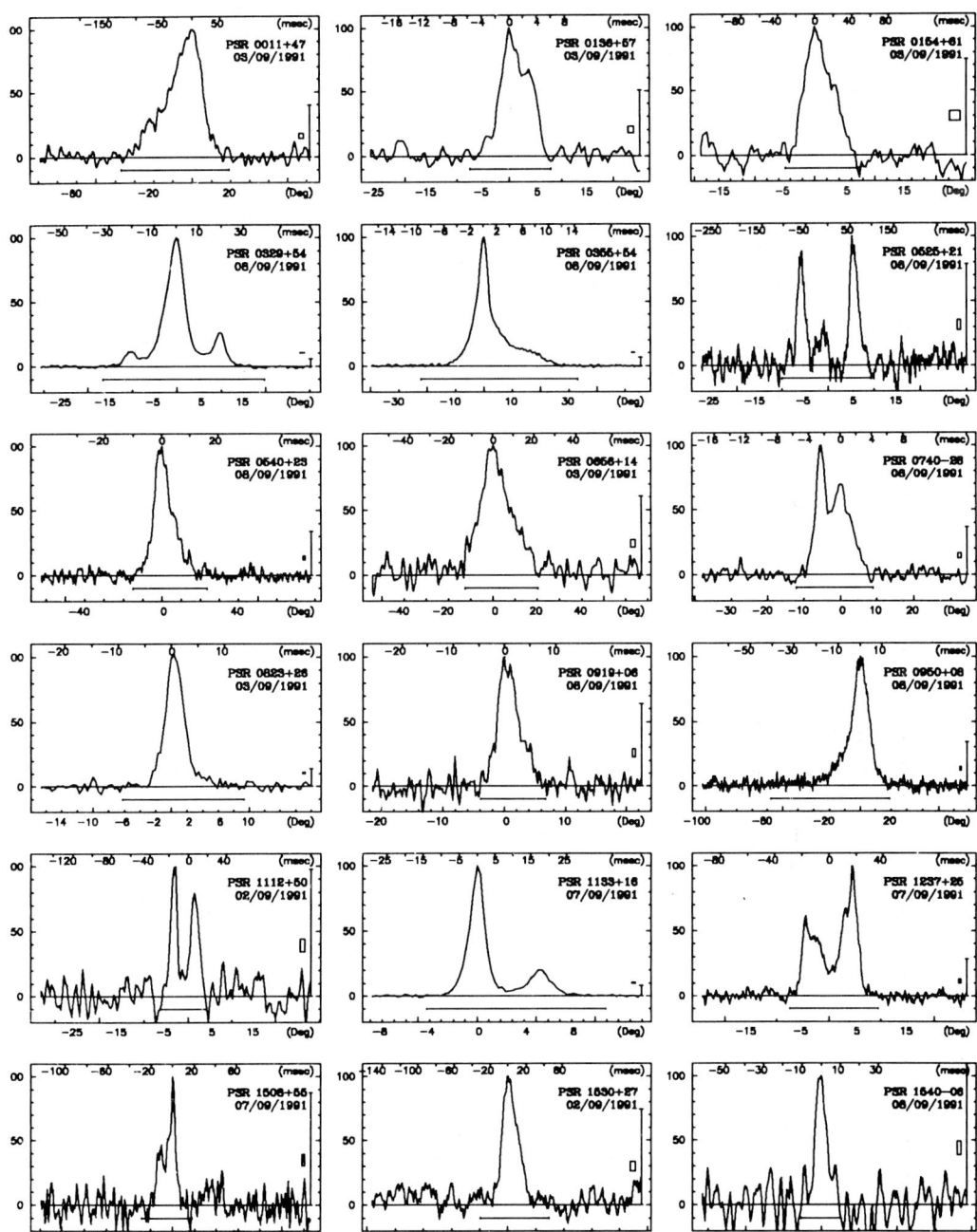

FIGURE 2. After nearly 25 years of Effelsberg observation 46 pulse shapes have been observed at λ6 cm wavelength.

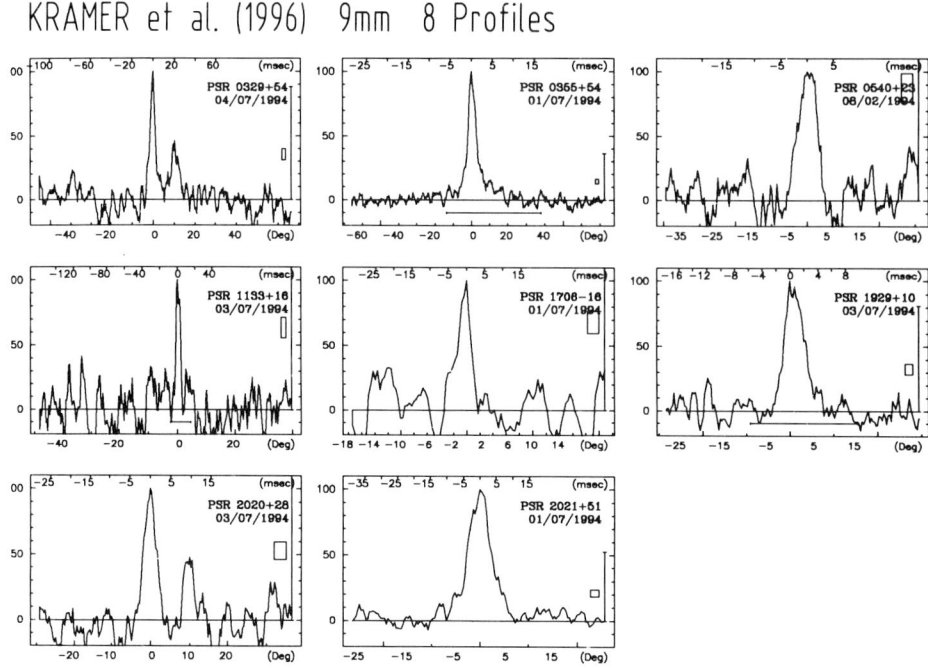

FIGURE 3. The present set of eight 9 mm detections.

5. Pulsar polarization

In all radio polarization measurements the general rule is: the higher the observing frequency the higher the polarization. This is the result of the depolarization by Faraday effects through the interstellar medium. Observations of polarized Galactic emission, polarization of radio sources (supernova remnants, normal galaxies, radio galaxies etc.) follow this general rule. Pulsar polarization at metre wavelength range was known to be high, sometimes up to 100% (e.g. Manchester 1971). It was also realized that pulsar polarization falls (Manchester et al. 1973) to higher frequencies but the exact depolarization development was not known due to lack of any data.

In a series of papers using the 100-m Effelsberg radio telescope (Morris et al. 1981; Xilouris et al. 1991; Xilouris et al. 1995) the integrated polarization characteristics of pulsars were presented at 1.7 GHz, 2.75 GHz, 8.7 GHz and 10.55 GHz. In a more recent paper (Xilouris et al. 1996) the observations were extended to 32 GHz. A very specific development of polarization with frequency can be deduced from these data.

The polarization which is high across most of the pulse shape at low frequencies remains but only in single sub-pulse components. At the highest radio frequencies only a very narrow sub-pulse is higly polarized. This explains why the earlier pulse polarization observation gave low values. At 32 GHz for most of the pulsars the polarization falls to zero!

PSR B2021+51

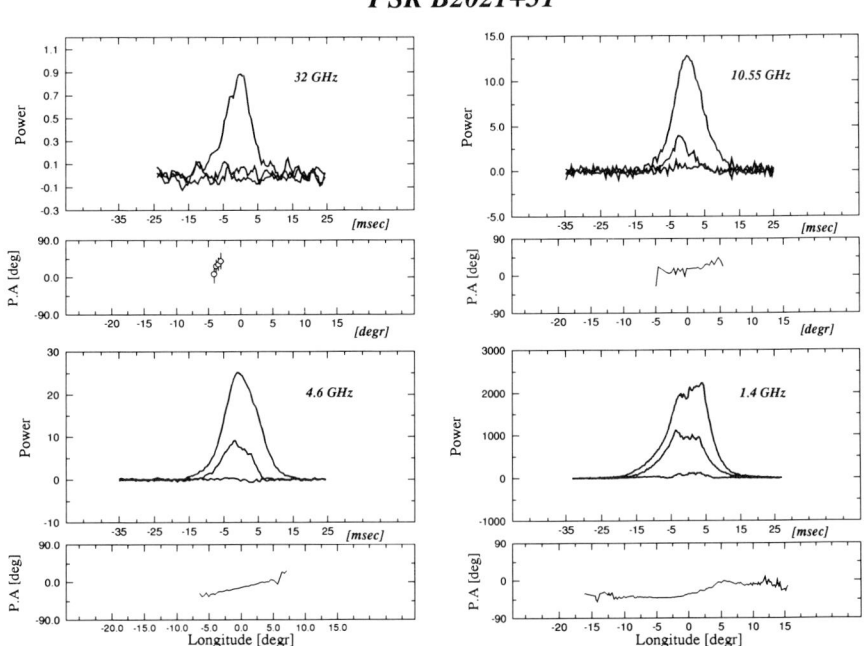

FIGURE 4. The frequency dependence of the integrated polarization of PSR 2021+51

6. Towards new horizons

The data described so far were a summary of nearly 25 years of pulsar observations with the 100-m radio telescope of the MPIfR. The next 25 years could be posssibly just as exciting. To begin with the 100-m telescope will be continually upgraded. Also the Green Bank Telescope should be coming on line with many new possibilities.

An example of such new possibilities is the recent addition of a new 4.75 GHz two-horn, four-channel HEMT receiver system in the secondary focus of the 100-m telescope. The system noise is below 30 K and the instantenous bandwidth is 500 MHz. The band is split into eight channels each 60 MHz wide. There are two such filter banks for each polarization. The channels can be added with predetermined dispersion correction or they can be used in a search mode. In the past 20 years of pulsar observing with the 100-m telescope only 46 pulse shapes could be obtained at this frequency, limited by sensitivity. With the new system more than 80 weaker pulsars were studied in only 60 hours of observing time! The weakest pulsars detected were at the level of $40\,\mu\text{Jy s}^{-1}$. Studies of pulsars as weak as $10\,\mu\text{Jy s}^{-1}$ seems feasible. A search for highly dispersed pulsars towards the galactic centre has been started.

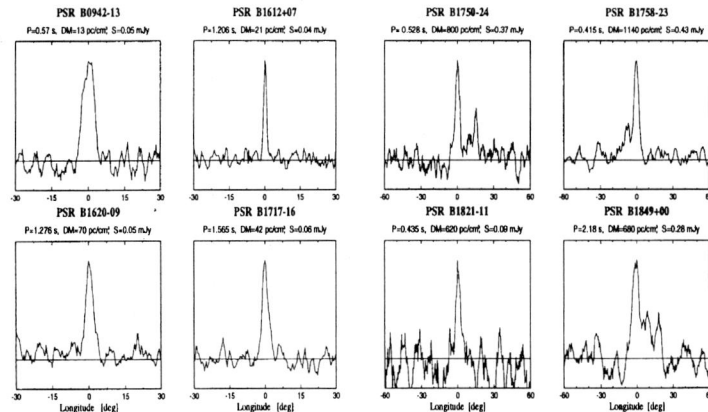

FIGURE 5. The new Effelsberg pulsar survey at 6 cm wavelength. Left: the weakest pulsars detected with a new HEMT receiver ($S = 40 - 60\,\mu\mathrm{K\,s^{-1}}$); right: the four pulsars with the highest dispersion measure ($DM = 620 - 1140\,\mathrm{pc\,cm^{-3}}$)

I would like to thank the many observers at the MPIfR who contributed their data freely to this talk. In particular I thank M. Kramer, A. Jessner, J. Kijak and K. Xilouris.

REFERENCES

BARTEL, N., SIEBER, W. & WIELEBINSKI, R. 1977. *A&A* **55**, 319

CORDES, J.M. 1986. *Astrophys. J.* **311**, 183

HEWISH, A., BELL, S.J., PILKINGTON, J.D.H., SCOTT, P.F. & COLLINS, R.A. 1968. *Nature* **217**, 709

IZVEKOVA, V.A., JESSNER, A., KUZMIN, A.D., MALOFEEV, V.M., SIEBER, W. & WIELEBINSKI, R. 1994. *A&AS* **105**, 235

KRAMER, M. 1994. *A&AS* **107**, 527

KRAMER, M., WIELEBINSKI, R., JESSNER, A., GIL, J.A. & SEIRADAKIS, J.H. 1994. *A&AS* **107**, 515

KRAMER, M., XILOURIS, K.M., JESSNER, A., WIELEBINSKI, R. & TIMOFEEV, M. 1996. *A&A* **306**, 867

LORIMER, D.R., YATES, J.A., LYNE, A.G. & GOULD, D.M. 1995. *MNRAS* **273**, 411

MALOFEEV, V.M., GIL, J.A., JESSNER, A., MALOV, I.F., SEIRADAKIS, J.H., SIEBER, W. & WIELEBINSKI, R. 1994. *A&A* **285**, 201

MANCHESTER, R.N. 1971. *ApJS* **23**, 283

MANCHESTER, R.N. & TAYLOR, J.H. 1977. *Pulsars*. Freeman & Co., San Francisco

MANCHESTER, R.N., TAYLOR, J.H. 1981. *ApJ* **86**, 1953

MANCHESTER, R.N., TAYLOR, J.H. & HUGUENIN, G.R. 1973. *ApJ* **179**, L7

MORRIS, D., GRAHAM, D.A., SIEBER, W., BARTEL, N. & THOMASSON, P. 1981. *A&A* **46**, 421

ROBINSON, B.J., COOPER, B.F.C., GARDNER, F.F., WIELEBINSKI, R. & LANDECKER T.L. 1968. *Nature* **218**, 1143

SEIRADAKIS, J.H., GIL, J.A., GRAHAM, D.A., JESSNER, A., KRAMER, M., SIEBER, W. & WIELEBINSKI, R. 1995. *A&A Suppl.* **111**, 205

SIEBER, W. 1973. *A&A* **28**, 237

SIEBER, W., REINECKE, R. & WIELEBINSKI, R. 1975. *A&A* **38**, 169

TAYLOR, J.H., MANCHESTER, R.N. & LYNE, A.G. 1993. *ApJS* **88**, 529

WIELEBINSKI, R., SIEBER, W., GRAHAM, D.A., HESSE, H. & SCHÖNHARDT, R.E. 1972. *Nature* **240**, 131
WIELEBINSKI, R., JESSNER, A., KRAMER, M. & GIL, J.A. 1993. *A&A* **272**, L13
XILOURIS, K.M., RANKIN, J.M., SEIRADAKIS, J.H. & SIEBER, W. 1991. *A&A* **241**, 87
XILOURIS, K.M., SEIRADAKIS, J.H., GIL, J., SIEBER, W. & WIELEBINSKI, R. 1995. *A&A* **293**, 153
XILOURIS, K.M., KRAMER, M., JESSNER, A., WIELEBINSKI, R. & TIMOFEEV, M. 1996. *A&A* (in press)

Present and future pulsar searches

By FERNANDO CAMILO

University of Manchester, NRAL, Jodrell Bank, Macclesfield, Cheshire SK11 9DL, UK

We describe recent pulsar searches, concentrating on undirected surveys for millisecond pulsars. We present survey parameters and regions, and tabulate parameters for 47 binary and recycled pulsars in the disk of the Galaxy, including 34 discovered in the last five years in the surveys detailed. We briefly discuss some future search projects.

1. Introduction

In 1974, seven years after the discovery of the first radio pulsar (Hewish et al. 1968), the Crab pulsar was by far the most rapidly spinning neutron star known, with period $P = 33$ ms. Pulsars had typical periods of ~ 0.5 s and were isolated objects. In the first large-scale survey to search systematically in period and dispersion measure (DM), introducing search techniques that remain in use today with few alterations, Hulse & Taylor (1975a,b) discovered the first binary pulsar, PSR B1913+16, with $P = 59$ ms and spindown rate $\dot{P}$ two orders of magnitude smaller than for most pulsars. The unusual spin characteristics of the pulsar were soon linked to its binary evolution (Smarr & Blandford 1976) and the original binary pulsar is in retrospect the first "recycled" pulsar, where the neutron star was spun-up by matter accreted from its evolved companion.

By 1981 two more binary pulsars, B0655+64 (Damashek, Taylor, & Hulse 1978) and B0820+02 (Manchester et al. 1978), had been discovered, when Backer et al. (1982) unveiled the 1.56-ms isolated pulsar B1937+21 in a directed search of a steep-spectrum polarized point source. PSR B1953+29, the second millisecond pulsar, was discovered in a directed search of an unrelated γ-ray error box (Boriakoff, Buccheri, & Fauci 1983). Until the late 1980s searches for millisecond pulsars covering large areas of sky were impractical to carry out due to the computational resources needed, and discoveries trickled in slowly: PSR B1855+09 was discovered in 290 deg^2 of sky at Arecibo (Stokes et al. 1986), and the eclipsing pulsar B1957+20 (Fruchter, Stinebring, & Taylor 1988) was discovered in the second such undirected survey. Meanwhile, prompted by the existence of many low-mass X-ray binary (LMXB) systems in globular clusters (GC) and the suggestion (Alpar et al. 1982) that LMXBs were linked to millisecond pulsar formation, Lyne et al. (1987) found the first millisecond pulsar in a GC. Extremely successful searches of many clusters followed, with 33 pulsars now known in GCs, of which 25 have $P < 30$ ms. (For an excellent review see Kulkarni & Anderson 1996.) The availability of powerful supercomputers and affordable workstations, and the realization that millisecond pulsars in the disk of the Galaxy have large scale heights, inspired the vast search efforts of the past five years, which we now describe.

2. Recent and continuing surveys

2.1. Arecibo

In 1990, as the Arecibo telescope lay partially disabled, Wolszczan (1990) surveyed whatever part of the sky drifted through the telescope beam for rapidly-rotating pulsars. His discovery of the nearby ($d < 1$ kpc) pulsars B1257+12 and B1534+12 in 150 deg^2 of sky at

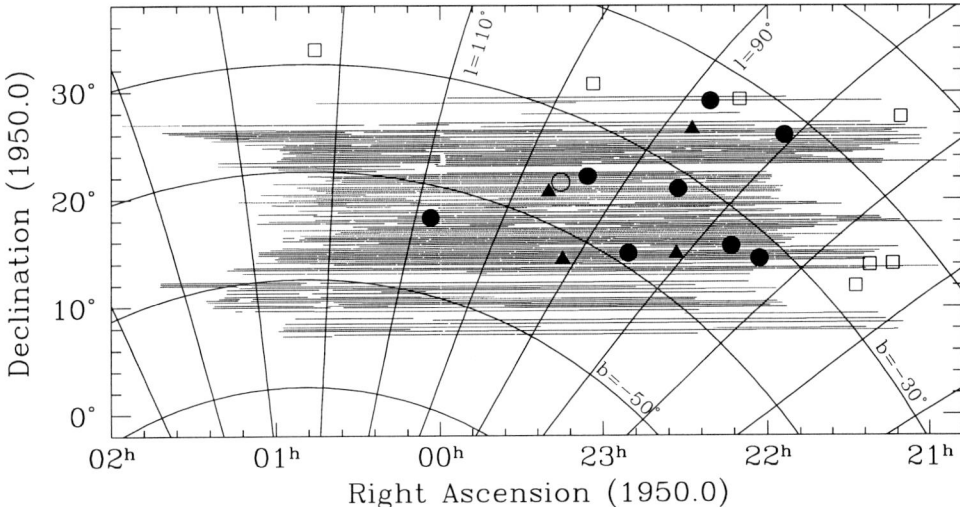

FIGURE 1. Beam areas searched by the Princeton–Arecibo high-latitude millisecond pulsar survey (Camilo, Nice, & Taylor 1996). Open symbols denote previously known pulsars, and filled symbols represent new discoveries (triangles are fast pulsars and circles are slow pulsars).

high Galactic latitudes suggested that large numbers of millisecond pulsars awaited discovery in essentially all directions (Johnston & Bailes 1991). His data acquisition setup used an autocorrelation spectrometer dividing a bandwidth of 10 MHz into 128 channels, at a centre frequency of 430 MHz, with total power samples averaged every $516\,\mu$s. This setup was also being used at the time to complete a survey of $260\,\deg^2$ of sky at $|b| \leq 8°$ that resulted in the discovery of PSRs B1957+20 and J2019+2425, and the redetection of PSRs B1855+09 and B1937+21 (Nice, Fruchter, & Taylor 1995).

In 1991 three groups started systematic drift surveys of high Galactic latitude regions at Arecibo. The Princeton group was assigned the "box" in the right ascension range $21^h30^m \leq \mathrm{RA} \leq 01^h00^m$ (Figure 1), while Caltech surveyed in the $08^h00^m \leq \mathrm{RA} \leq 13^h00^m$ range, and Penn State/NRL took the $13^h00^m \leq \mathrm{RA} \leq 18^h00^m$ box. These surveys used a filterbank-based backend, with search parameters summarized in Table 1 — the integration time of 32.8 s is a good match to the $40/\cos\delta$ s transit time between half-power points of the $10'$ beam. Overall $1750\,\deg^2$ were searched in this manner until early 1994. Together with the original high-latitude search, a total of $1900\,\deg^2$ were surveyed free from RFI (Camilo, Nice, & Taylor 1996, Foster et al. 1995, Ray et al. 1996, Ray et al. 1995, Thorsett et al. 1993b), resulting in the detection of 6 millisecond pulsars ($P < 30$ ms), as well as PSRs B1534+12 and J2235+1506.

Table 1 lists a limiting flux density of $S_{\min} = 0.5\,\mathrm{mJy}$ for these Arecibo surveys. It must be noted that the actual flux-density threshold for detection of a pulsar depends in a complex way on the location of the pulsar relative to the beam centre, period, dispersion, pulse shape, sky temperature, zenith angle of telescope, and other factors, and any serious inferences about the underlying pulsar population must take into account these factors affecting detectability, many details of which can be found in Camilo, Nice, & Taylor (1996). The 0.5-mJy limit can be regarded as a figure of merit, useful for comparison with other surveys, and is appropriate only for ideal conditions, for long-period pulsars located at the centre of the telescope beam pattern, for observations at zenith, and at high Galactic latitudes with sky background temperature $T \sim 25$ K.

In early 1994 the pace of pulsar searching at Arecibo increased, enabled by a telescope

TABLE 1. Four surveys for millisecond pulsars.

	Parkes	Arecibo	Green Bank	Jodrell Bank
Survey Region	$\delta < 0°$	$-2° < \delta < +38°$	$\delta > 0°$	$\delta > +35°$
Telescope	64-m	305-m	43-m	76-m
Frequency (MHz)	436	430	370	408
Integration Time (s)	157	32.8	134	314
Bandwidth (MHz)	2×32	2×8	2×40 (2×20)	2×8
Channel Width (kHz)	125	250	78	125
Sampling Rate (kHz)	3.3	4	3.9	3.3
System Temp. (K)	≥ 50	≥ 70	≥ 55	≥ 50
$S_{\min}$ (mJy)	≥ 3	$\gtrsim 0.5$	≥ 8	≥ 4
Total Area (deg^2)	20,600	13,400	20,600	8800
Area Surveyed	$\sim 20,000$	$\sim 4400 + 260$	15,900 (1500)	~ 4000
Fast Pulsars Found	17	13+4	2	1

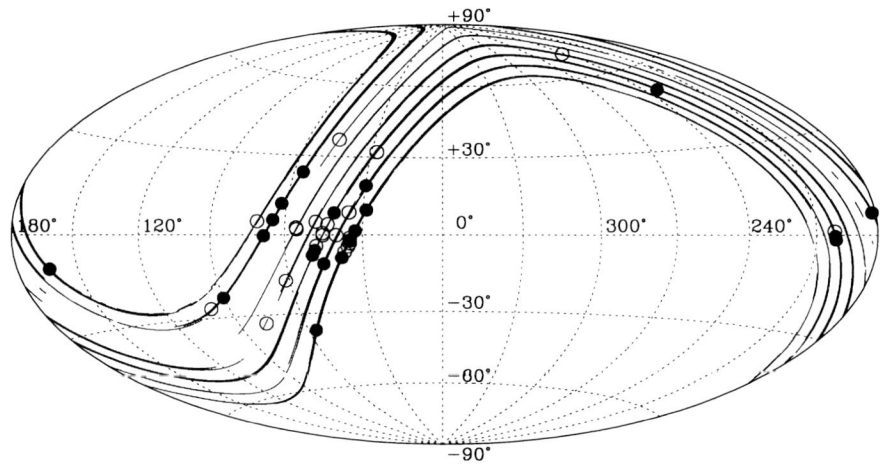

FIGURE 2. Sky coverage to date of the Princeton–Arecibo declination-strip millisecond pulsar survey (Camilo et al. 1996). Filled circles denote newly discovered pulsars; open circles represent previously known pulsars detected in the survey.

upgrading project now nearly completed, and collaborations at Cornell/Berkeley and STScI/Arecibo were also assigned regions to survey. Each of the 5 groups now searches in 8 interspersed declination strips, 24^h-long in RA and one degree wide. As the telescope beam width at 430 MHz is about $10'$ each group has further subdivided their degree-wide strips into 6 rows.

To date three groups have reported results, searching a total of about 2500 deg^2 and discovering 5 millisecond pulsars (Camilo et al. 1996b, Foster et al. 1995, Ray et al. 1996). This surface density of millisecond pulsars is apparently smaller than for the high-latitude surveys: partly this is due to the lower average sensitivity of these surveys, due to both higher average zenith angles and sky background temperatures, while partly it is an artifact, as some of the areas included in the 2500 deg^2 had been previously searched in

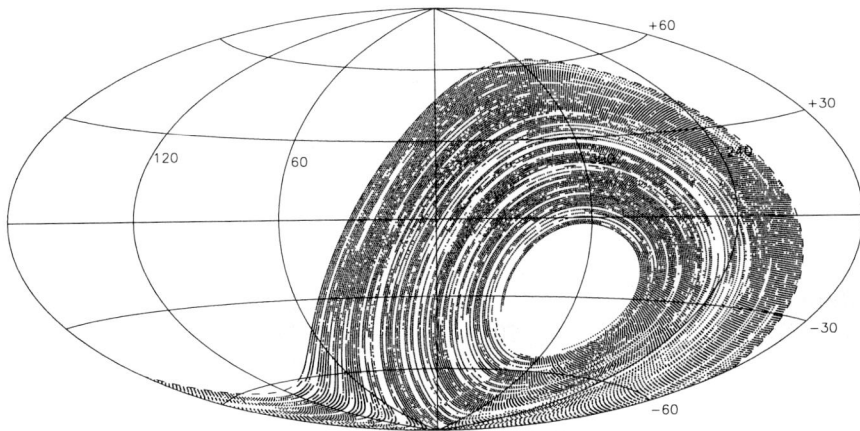

FIGURE 3. Partial coverage of the Parkes southern sky survey (Manchester et al. 1996).

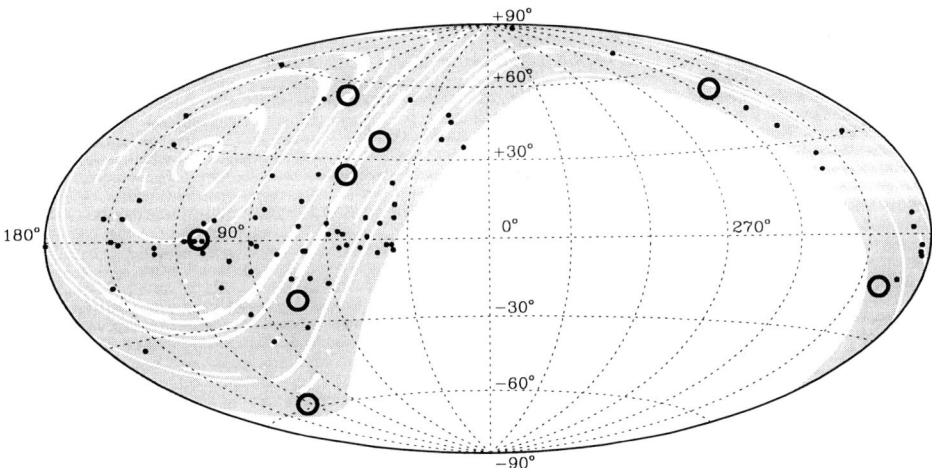

FIGURE 4. Areas of sky searched by the Princeton–NRAO millisecond pulsar northern sky survey (Sayer, Nice, & Taylor 1996). Dots represent known pulsars detected in the survey; open circles are newly discovered pulsars.

the high-latitude surveys. Again, serious statistical inferences require consultation of the original references. The areas surveyed by the Princeton group are shown in Figure 2.

In addition to the undirected surveys, a recent Arecibo search of unidentified EGRET γ-ray error boxes resulted in the discovery of the millisecond pulsar J0751+1807, not related to the γ-ray source (Lundgren, Zepka, & Cordes 1995).

2.2. Parkes

The recently completed ATNF/Jodrell Bank/Bologna survey of the southern sky at the Parkes telescope was enormously successful, discovering 17 millisecond pulsars (all those with $\delta < 0°$ in Table 2), or about half of all known in the disk of the Galaxy. Carried out over 1991–1995, this collaboration searched approximately 44,000 different telescope beam pointings. Some 28,000 of those are shown in Figure 3, reproduced from Manchester et al. (1996), where details concerning observing system and data reduction, as well as the first 2/3 of the processed beam pointings, are given. Some observing details are summarized in Table 1. Notice the higher sensitivity of this survey to intermediate DMs, compared to the Arecibo surveys: neglecting interstellar scattering, a pulse is broadened by interstellar dispersion and by instrumental sampling approximately $(t_{\rm DM}^2 + t_{\rm s}^2)^{1/2}$, where at ~ 430 MHz, $t_{\rm DM} \simeq 102 {\rm DM} \delta\nu\,\mu$s with $\delta\nu$ the channel width in MHz. From Table 1 we see that the Parkes and Arecibo surveys have identical time resolution at DM $\simeq 10\,{\rm cm}^{-3}\,$pc, while the former is relatively more sensitive for larger DMs.

The most spectacular pulsar discovered in the Parkes survey was PSR J0437−4715. Intrinsically this binary millisecond pulsar is quite ordinary, but its distance of only 150 pc makes it very bright, with average flux density near 430 MHz of 600 mJy, compared with more typical flux densities for millisecond pulsars of ~ 10 mJy! Being so bright and close, it promises to be a superb object for studies of millisecond pulsar emission mechanisms, binary companions (e.g., Danziger, Baade, & DellaValle 1993), and timing stability.

2.3. Green Bank

The discovery of PSR J0437−4715 less than half-way through the Parkes survey prompted other observers to try to find similar objects in the northern hemisphere. As only 60% of the northern sky is visible from Arecibo, and the rate of coverage is slow, Sayer, Nice, & Taylor (1996) undertook a relatively fast survey for bright millisecond pulsars with the Green Bank 140-foot telescope. This survey is now completed, with sky coverage reproduced in Figure 4 and some survey parameters listed in Table 1. In this survey the mildly relativistic binary pulsar J1518+4904 was discovered, as was the intermediate-mass binary millisecond pulsar J1022+1001, independently found at Arecibo.

This survey and the completed Parkes survey suggest that objects as bright as the nearby PSR J0437−4715 are very rare indeed. The non-detection of weaker millisecond pulsars appears at first glance of Table 1 to come as a surprise. However, a direct comparison of sensitivities for the Green Bank and Parkes surveys in the region of interest (e.g., $P \simeq 4$ ms and DM $\simeq 20\,{\rm cm}^{-3}\,$pc), is complicated by the simplified treatment for the Parkes survey of the reduction in sensitivity at short periods caused by the finite number of detected harmonics (Manchester et al. 1996).

2.4. Jodrell Bank

In parallel with the Parkes survey, a similar data-acquisition system has been used at Jodrell Bank to survey the sky northwards of the Arecibo declination range (Table 1). To date approximately half of the survey region has been analyzed, with the discovery of the millisecond pulsar J1012+5307 (Nicastro et al. 1995). This binary system is remarkable in that the pulsar companion is extremely bright by millisecond pulsar/white dwarf companion standards, enabling detailed photometry and spectroscopy to be carried out.

Also at Jodrell Bank, in a search of a steep-spectrum, highly-polarized compact source, Navarro et al. (1995) discovered a very luminous binary millisecond pulsar, J0218+4232.

2.5. Table of Field Binary and Recycled Pulsars

In Table 2 we present some parameters for all binary and recycled pulsars known in the disk of the Galaxy. For this purpose we regard isolated pulsars as recycled if they have a small enough period and low surface magnetic dipole field strength, as inferred from their $\dot{P}$. Somewhat arbitrarily we settle on all isolated pulsars with periods $P < 60$ ms, with the exception of PSRs B0531+21, B0540−69, and B1951+32, all clearly young pulsars associated with supernova remnants. Although we chose to concentrate on these pulsars, there may well be a large group of isolated pulsars with larger periods that have been recycled as well (see, e.g., Deshpande, Ramachandran, & Srinivasan 1995).

TABLE 2. Parameters of binary and recycled pulsars known in the disk of the Galaxy.

PSR	P (ms)	$\dot{P}$ (10^{-18})	P_b (d)	$a_1 \sin i/c$ (s)	e	m_2 ($M_\odot$)	d (kpc)	$|z|$ (kpc)	Refs.
\multicolumn{10}{c}{Circular Orbit Pulsars ($e \lesssim 0.01$)}
J0034−0534	1.88	0.0051	1.589	1.438	<0.00002	0.17	1.0	0.9	(1,2)
J0218+4232	2.32	0.08	2.029	1.984	<0.00002	0.20	>5.8	>1.8	(3)
J0437−4715	5.76	0.057	5.741	3.367	0.000019	0.17	0.1	0.1	(4,5)
J0613−0200	3.06	0.0096	1.199	1.091	<0.00002	0.15	2.2	0.4	(6,7)
J0621+1002	28.85	<0.08	8.319	12.032	0.0025	0.54	1.9	0.1	(8)
J0751+1807	3.48	0.008	0.263	0.397	<0.0001	0.15	2.0	0.8	(9)
J1012+5307	5.26	0.015	0.605	0.582	<0.00002	0.13	0.5	0.4	(10,11)
J1022+1001	16.45	0.04	7.805	16.765	0.000098	0.87	0.6	0.5	(8)
J1045−4509	7.47	0.017	4.084	3.015	0.000024	0.19	3.3	0.7	(1,7)
J1455−3330	7.99	0.024	76.175	32.362	0.00017	0.30	0.7	0.3	(6,2)
J1603−7202	14.84	0.014	6.309	6.881	<0.00002	0.33	1.6	0.4	(12)
J1640+2224	3.16	<0.003	175.461	55.330	0.00080	0.30	1.1	0.7	(13,14)
J1643−1224	4.62	0.018	147.017	25.073	0.00051	0.14	>4.9	>1.8	(6,7)
J1709+23	4.63						1.9	1.0	(13)
J1713+0747	4.57	0.0085	67.825	32.342	0.000075	0.33	1.0	0.4	(15,16)
J1804−2717	9.34	0.042	11.129	7.281	0.000035	0.24	1.2	0.1	(12)
B1855+09	5.36	0.018	12.327	9.231	0.000022	0.26	1.0	0.1	(17)
J1911−1114	3.63	0.013	2.717	1.763	<0.00001	0.14	1.6	0.3	(12)
B1953+29	6.13	0.030	117.349	31.413	0.00033	0.21	5.4	0.0	(18)
B1957+20	1.61	0.017	0.382	0.0892	<0.00004	0.03	1.5	0.1	(19)
J2019+2425	3.93	0.0070	76.512	38.768	0.00011	0.37	0.9	0.1	(20,21)
J2033+17	5.95		56.2	20.07	<0.05	0.22	1.4	0.3	(22)
J2051−0827	4.51	0.013	0.099	0.045	<0.0003	0.03	1.3	0.6	(23)
J2129−57	3.73		6.633	3.500		0.16	>2.6	>1.8	(12)
J2145−0750	16.05	0.030	6.839	10.164	0.000019	0.51	0.5	0.3	(1,2)
J2229+2643	2.98	0.0019	93.016	18.913	0.00025	0.15	1.4	0.6	(24)
J2317+1439	3.45	0.0024	2.459	2.314	0.000001	0.21	1.9	1.3	(25,24)
B0655+64	195.67	0.69	1.029	4.126	<0.00003	0.81	0.5	0.2	(18,26)
B0820+02	864.87	104.3	1232.395	162.145	0.012	0.23	1.4	0.5	(18,26)
J1803−2712	334.42	17.3	406.781	58.940	0.00051	0.17	3.6	0.2	(18)
B1831−00	520.95	10.7	1.811	0.723	<0.004	0.08	2.6	0.2	(18,26)
\multicolumn{10}{c}{Other Binary Pulsars ($e \gtrsim 0.25$)}
J0045−7319	926.28	4486.	51.169	174.254	0.808	8.8	57.		(27)
B1259−63	47.76	2280.	1236.81	1296.00	0.870	4.19	4.6	0.1	(28)
J1518+4904	40.93	<0.1	8.634	20.044	0.249	1.01	0.7	0.6	(29)
B1534+12	37.90	2.43	0.421	3.729	0.274	1.34	0.7	0.5	(30,18)
B1820−11	279.83	1376.	357.762	200.673	0.795	0.80	6.3	0.1	(18,26)
B1913+16	59.03	8.63	0.323	2.342	0.617	1.39	7.1	0.3	(18)
B2303+46	1066.37	569.1	12.340	32.688	0.658	1.46	4.2	0.9	(18,26)
\multicolumn{10}{c}{Other Recycled Pulsars}
J0712−6820	5.49	0.017	Isolated				1.0	0.4	(31)
J1024−0719	5.16	0.018	Isolated				0.3	0.2	(31)
B1257+12	6.22	0.11	Planets				0.6	0.6	(32,33)
J1730−2304	8.12	0.020	Isolated				0.5	0.1	(6,2)
J1744−1134	4.07	0.0079	Isolated				0.2	0.0	(31)
B1937+21	1.56	0.11	Isolated				3.6	0.0	(17)
J2124−3358	4.93	0.020	Isolated				0.2	0.2	(31)
J2235+1506	59.77	0.16	Isolated				1.2	0.7	(25,24)
J2322+2057	4.81	0.0097	Isolated				0.8	0.5	(20,21)

(1) Bailes et al. 1994; (2) Camilo et al. 1996a; (3) Navarro et al. 1995; (4) Johnston et al. 1993; (5) Bell et al. 1995a; (6) Lorimer et al. 1995b; (7) Bell et al. 1996; (8) Camilo et al. 1996b; (9) Lundgren, Zepka, & Cordes 1995; (10) Nicastro et al. 1995; (11) Lorimer et al. 1995a; (12) Lorimer et al. 1996; (13) Foster et al. 1995; (14) Wolszczan (priv. comm.); (15) Foster, Wolszczan & Camilo 1993; (16) Camilo, Foster, & Wolszczan 1994; (17) Kaspi, Taylor, & Ryba 1994; (18) Taylor, Manchester, & Lyne 1993; (19) Arzoumanian, Fruchter, & Taylor 1994; (20) Nice, Taylor & Fruchter 1993; (21) Nice & Taylor 1995; (22) Ray et al. 1996; (23) Stappers et al. 1996; (24) Camilo, Nice, & Taylor 1996; (25) Camilo, Nice, & Taylor 1993; (26) Arzoumanian 1995; (27) Kaspi et al. 1994; (28) Manchester et al. 1995; (29) Nice, Sayer, & Taylor 1996; (30) Wolszczan 1991; (31) Bailes et al. 1996; (32) Wolszczan & Frail 1992; (33) Wolszczan 1994: References.

The table lists observed parameters, although for most millisecond pulsars the intrinsic $\dot{P}$ is significantly different from the observed one (Camilo, Thorsett, & Kulkarni 1994). The companion masses, m_2, are derived from the pulsar mass function for an assumed pulsar mass of $1.4\,M_\odot$ (Thorsett et al. 1993a), and for the "median" inclination angle of $i = 60°$, with the exception of PSRs B1855+09, B1534+12, and B1913+16, for which they are measured with the help of relativistic orbital effects, and PSR J0045−7319, from optical measurements (Bell et al. 1995b). Distances are obtained from the measured DM and the Taylor & Cordes (1993) model for the Galactic distribution of free electrons.

Thirty-four of the 47 pulsars in the table were discovered in the last five years in the pulsar surveys summarized in this paper: PSRs B1257+12 and B1534+12, and all those with names based on the J2000 equinox, with the exceptions of PSRs J1803−2712 and J0045−7319. For these pulsars we provide the discovery reference and the most up-to-date source of parameters, while for pulsars discovered in other surveys we generally provide only the most recent publication.

Of the 35 millisecond pulsars in Table 2 (about 5% of all known pulsars), 22 (3/5) are in nearly circular orbits with companions of mass $0.15 \lesssim m_2 \lesssim 0.45\,M_\odot$, presumably helium white dwarfs. Seven (1/5) are isolated objects. Three have intermediate-mass companions ($m_2 \geq 0.45\,M_\odot$), probably carbon-oxygen white dwarfs, 2 are in eclipsing systems with companions of very low mass ($m_2 \lesssim 0.05\,M_\odot$), and one has at least three planetary-mass companions. Four of these 35 pulsars yield time-of-arrival (TOA) uncertainties of less than $1\,\mu s$ with present telescopes (scaling from the TOA uncertainties obtained at Jodrell Bank for several of the pulsars discovered at Parkes it appears plausible that the NRAO Green Bank Telescope will increase this number).

3. Future surveys

The completion of the present Arecibo and Jodrell Bank surveys are likely to result in the discovery of about 10–20 more millisecond pulsars within the next two years. At Arecibo, the next stage will involve searching the same region of sky already covered with better frequency and time resolution (of order 60 kHz-wide channels and sampling times under $100\,\mu s$). Once a substantial area is surveyed at this resolution we may know more about the limiting period of neutron stars (we could of course be unlucky enough that the right combination of magnetic field strengths and accretion rates never occurs in interacting binary systems so that the break-up period for neutron stars is never realized).

We conclude with the outline of a survey to begin in late 1996 at the Parkes telescope. Despite the success of the recent millisecond pulsar surveys, they probe mostly the neighbourhood of the Sun (only 4 of the 31 millisecond pulsars in Table 2 discovered in undirected surveys have dispersion-derived distances in excess of 2 kpc). To learn about the true Galactic distribution of these objects we need to probe deeper. Higher-frequency surveys do that, reducing the effects of interstellar scattering. The Johnston et al. (1992) and Clifton et al. (1992) surveys were extremely successful, discovering numerous young pulsars near the Galactic plane (Table 3). However, they were not very sensitive to millisecond pulsars due to integration and sampling times, and channel widths used. The Parkes multi-beam survey (Table 3) aims to substantially improve on all these fronts. Its ~ 0.2 mJy flux density limit at 1400 MHz corresponds to about 2–4 mJy at ~ 400 MHz for typical spectral indices, i.e., the same sensitivity limit as the Parkes southern sky survey (§2.2). The narrow channels and rapid sampling of the new survey maintain sensitivity to fast pulsars to greater distances than before: at $\mathrm{DM} = 38\,\mathrm{cm}^{-3}\,\mathrm{pc}$ the new survey has $(t_{\mathrm{DM}}^2 + t_{\mathrm{s}}^2)^{1/2} = 416\,\mu s$, the same as the southern sky survey at $\mathrm{DM} = 20\,\mathrm{cm}^{-3}\,\mathrm{pc}$. Even

TABLE 3. Two 21-cm surveys at Parkes.

Type	Conventional (Johnston et al. 1992)	Multi-beam
Number of beams	1	13
Integration Time (min)	2.5	~ 30
Bandwidth (MHz)	2×320	2×288
Channel Width (MHz)	5	3
Sampling Rate (kHz)	0.83	4
$S_{\rm sys}$ (Jy)	70	40
$S_{\rm min}$ (mJy)	~ 1	~ 0.2
Survey Region	$270° \leq l \leq 20°, |b| \leq 4°$ (800 deg^2)	—
Pulsars Found (Detected)	46 (100)	—

at $DM = 100 \, {\rm cm}^{-3} \, {\rm pc}$ a pulse detected in the new survey will be smeared by only 0.9 ms. We expect to discover well over 100 pulsars, many of them young and fast.

I thank Matthew Bailes, Jon Bell, Dunc Lorimer, Andrew Lyne, Dick Manchester, Ron Sayer, and Ben Stappers for generously allowing me to use some unpublished data.

REFERENCES

ALPAR, M.A., CHENG, A.F., RUDERMAN, M.A., & SHAHAM, J. 1982. *Nature* **300**, 728

ARZOUMANIAN, Z. 1995. PhD thesis, Princeton University

ARZOUMANIAN, Z., FRUCHTER, A.S., & TAYLOR, J.H. 1994. *ApJ* **426**, L85

BACKER, D.C., KULKARNI, S.R., HEILES, C., DAVIS, M.M., & GOSS, W.M. 1982. *Nature* **300**, 615

BAILES, M. ET AL. 1994. *ApJ* **425**, L41

BAILES, M. ET AL. 1996, ApJ. in preparation

BELL, J.F., BAILES, M., MANCHESTER, R.N., LYNE, A.G., & CAMILO, F. 1996. in preparation

BELL, J.F., BAILES, M., MANCHESTER, R.N., WEISBERG, J.M., & LYNE, A.G. 1995a. *ApJ* **440**, L81

BELL, J.F., BESSELL, M.S., STAPPERS, B., BAILES, M., & KASPI, V. 1995b. *ApJ* **447**, L117

BORIAKOFF, V., BUCCHERI, R., & FAUCI, F. 1983. *Nature* **304**, 417

CAMILO, F., BAILES, M., LYNE, A.G., MANCHESTER, R.N., & BELL, J.F. 1996a. in preparation

CAMILO, F., FOSTER, R.S., & WOLSZCZAN, A. 1994. *ApJ* **437**, L39

CAMILO, F., NICE, D.J., SHRAUNER, J.A., & TAYLOR, J.H. 1996b. *ApJ* In press

CAMILO, F., NICE, D.J., & TAYLOR, J.H. 1993. *ApJ* **412**, L37

CAMILO, F., NICE, D.J., & TAYLOR, J.H. 1996. *ApJ* **461**, 812

CAMILO, F., THORSETT, S.E., & KULKARNI, S.R. 1994. *ApJ* **421**, L15

CLIFTON, T.R., LYNE, A.G., JONES, A.W., MCKENNA, J., & ASHWORTH, M. 1992. *MNRAS* **254**, 177

DAMASHEK, M., TAYLOR, J.H., & HULSE, R.A. 1978. *ApJ* **225**, L31

DANZIGER, I.J., BAADE, D., & DELLAVALLE, M. 1993. *A&A* **276**, 382

DESHPANDE, A.A., RAMACHANDRAN, R., & SRINIVASAN, G. 1995. *J. Astrophys. Astr.* **16**, 69

FOSTER, R.S., CADWELL, B.J., WOLSZCZAN, A., & ANDERSON, S.B. 1995. *ApJ* **454**, 826

FOSTER, R.S., WOLSZCZAN, A., & CAMILO, F. 1993. *ApJ* **410**, L91

FRUCHTER, A.S., STINEBRING, D.R., & TAYLOR, J.H. 1988. *Nature* **333**, 237

HEWISH, A., BELL, S.J., PILKINGTON, J. D.H., SCOTT, P.F., & COLLINS, R.A. 1968. *Nature* **217**, 709

HULSE, R.A. & TAYLOR, J.H. 1975a. *ApJ* **201**, L55

HULSE, R.A. & TAYLOR, J.H. 1975b. *ApJ* **195**, L51

JOHNSTON, S. & BAILES, M. 1991. *MNRAS*, **252**, 277

JOHNSTON, S. ET AL. 1993. *Nature* **361**, 613

JOHNSTON, S., LYNE, A.G., MANCHESTER, R.N., KNIFFEN, D.A., D'AMICO, N., LIM, J., & ASHWORTH, M. 1992. *MNRAS* **255**, 401

KASPI, V.M., JOHNSTON, S., BELL, J.F., MANCHESTER, R.N., BAILES, M., BESSELL, M., LYNE, A.G., & D'AMICO, N. 1994. *ApJ* **423**, L43

KASPI, V.M., TAYLOR, J.H., & RYBA, M. 1994. *ApJ* **428**, 713

KULKARNI, S.R. & ANDERSON, S.B. 1996, in Dynamical Evolution of Star Clusters – Confrontation of Theory and Observations: IAU Symposium 174, Available at http://astro.caltech.edu/~srk/gciau.ps

LORIMER, D.R., LYNE, A.G., BAILES, M., MANCHESTER, R.N., D'AMICO, N., CAMILO, F., STAPPERS, B.W., & JOHNSTON, S. 1996. *MNRAS* in preparation

LORIMER, D.R., FESTIN, L., LYNE, A.G. & NICASTRO, L. 1995a. *Nature* **376**, 393

LORIMER, D.R. ET AL. 1995b. *ApJ* **439**, 933

LUNDGREN, S.C., ZEPKA, A.F., & CORDES, J.M. 1995. *ApJ* **453**, 419

LYNE, A.G., BRINKLOW, A., MIDDLEDITCH, J., KULKARNI, S.R., BACKER, D.C., & CLIFTON, T.R. 1987. *Nature* **328**, 399

MANCHESTER, R.N., JOHNSTON, S., LYNE, A.G., D'AMICO, N., BAILES, M., & NICASTRO, N. 1995. *ApJ* **445**, L137

MANCHESTER, R.N. ET AL. 1996, *MNRAS* In press

MANCHESTER, R.N., LYNE, A.G., TAYLOR, J.H., DURDIN, J.M., LARGE, M.I., & LITTLE, A.G. 1978. *MNRAS* **185**, 409

NAVARRO, J., deBRUYN, G., FRAIL, D., KULKARNI, S.R., & LYNE, A.G. 1995. *ApJ* **455**, L55

NICASTRO, L., LYNE, A.G., LORIMER, D.R., HARRISON, P.A., BAILES, M., & SKIDMORE, B.D. 1995. *MNRAS* **273**, L68

NICE, D.J., FRUCHTER, A.S., & TAYLOR, J.H. 1995. *ApJ* **449**, 156

NICE, D.J., SAYER, R.W., & TAYLOR, J.H. 1996. in preparation

NICE, D.J. & TAYLOR, J.H. 1995. *ApJ* **441**, 429

NICE, D.J., TAYLOR, J.H., & FRUCHTER, A.S. 1993. *ApJ* **402**, L49

RAY, P.S. ET AL. 1995. *ApJ* **443**, 265

RAY, P.S., THORSETT, S.E., JENET, F.A., KERKWIJK, M. H.V., KULKARNI, S.R., PRINCE, T.A., SANDHU, J.S., & NICE, D.J. 1996. *ApJ* submitted

SAYER, R.W., NICE, D.J., & TAYLOR, J.H. 1996. *ApJ* submitted

SMARR, L.L. & BLANDFORD, R. 1976. *ApJ* **207**, 574

STAPPERS, B.W. ET AL. 1996. *ApJ* submitted

STOKES, G.H., SEGELSTEIN, D.J., TAYLOR, J.H., & DEWEY, R.J. 1986. *ApJ* **311**, 694

TAYLOR, J.H. & CORDES, J.M. 1993. *ApJ* **411**, 674

TAYLOR, J.H., MANCHESTER, R.N., & LYNE, A.G. 1993. *ApJS* **88**, 529

THORSETT, S.E., ARZOUMANIAN, Z., MCKINNON, M., & TAYLOR, J. 1993a. *ApJ* **405**, L29

THORSETT, S.E., DEICH, W. T.S., KULKARNI, S.R., NAVARRO, J., & VASISHT, G. 1993b. *ApJ* **416**, 182

WOLSZCZAN, A. 1990. IAU circular 5073

WOLSZCZAN, A. 1991. *Nature* **350**, 688

WOLSZCZAN, A. 1994. *Science* **264**, 538

WOLSZCZAN, A. & FRAIL, D.A. 1992. *Nature* **355**, 145

The study of stellar radio emission with very large aperture radio telescopes

By E. R. SEAQUIST[1]

[1]Department of Astronomy, University of Toronto, Toronto, ON M5S 3H8, Canada

We consider the impact of a future large radio telescope of aperture approximately 1 km^2 on research in stellar radio astronomy, and the impact of this research field on the design of a large telescope. We note that approximately 10^6 stars should be detectable in the continuum alone, and we briefly examine examples of the way several topics would be revolutionized. The needs of stellar radio astronomy are governed by the compact and variable nature of the sources. For effective use in stellar radio astronomy, the telescope should operate at centimetre wavelengths, with maximum baselines exceeding 300 km, and with good u-v plane coverage permitting snapshots to be made with high fidelity.

1. Introduction

We examine the potential impact of a new-generation large-aperture radio telescope operating at centimetre wavelengths on the detection and study of stellar radio emission. We also examine the impact of the scientific demands on the basic design features of the telescope, namely aperture, frequency coverage, angular resolution (or the filling factor), field of view, and $u - v$ plane coverage. The aperture will be assumed to be 10^6 m^2, equivalent to that for the Square Kilometre Array Interferometer (SKAI) concept (see Braun, this conference), but we will refer to such a telescope by the more generic term "Large Radio Telescope" (LRT). By filling factor, we mean $f = A_e/B_m^2$, where A_e is the effective aperture and B_m is the maximum baseline. Note that for $A_e = 1$ km^2, $B_m = f^{-1/2}$, and the angular resolution is given by $\theta = 2.1(\frac{\lambda}{\text{cm}})f^{\frac{1}{2}}$ arcsec.

Current detection rates of continuum stellar sources, largely with the VLA (e.g. Seaquist 1993) and some simple assumptions about the number counts show that an LRT should be able to detect about 10^6 objects. Note that radio detections of stars generally refer to their circumstellar envelopes, not the optical photospheres or chromospheres. This is because of the small angular sizes of stars, and "long-wavelength penalty" associated with the Rayleigh-Jeans law. However, this will change significantly with an LRT since stellar photospheres would be detectable in some instances. The actual capacity for achieving these detection rates and undertaking useful studies will depend on the design features referred to above.

2. Sensitivity and confusion

Figure 1 shows an estimate of the sensitivity and confusion levels for an LRT with different filling factors based on a 12-hour integration. In order to arrive at figures for the sensitivity of the telescope, we scaled the sensitivity of the VLA by the aperture ratio, using figures based on the proposed VLA upgrade (Bastian & Bridle 1995). The curves for confusion by extragalactic sources are based on a measurement of the rms confusion with the VLA scaled to other resolutions and frequencies (Condon 1974). A spectral index ($S \sim \nu^\alpha$) of $\alpha = -0.7$ is assumed for the confusing sources and a slope of 2.0 is used for the slope of the differential source count in the relevant flux range. Note that confusion appears negligible at all frequencies for $f < 10^{-5}$, corresponding to

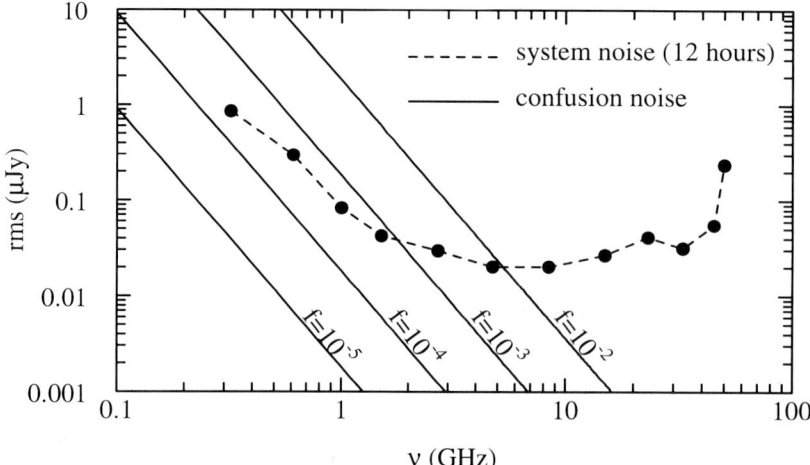

FIGURE 1. Plot of the anticipated sensitivity (1σ) of a large radio telescope of collecting aperture of 1 km^2 (LRT). The sensitivities are based on an integration time of 12 hours, and are obtained by scaling the sensitivity of an upgraded VLA by the appropriate aperture ratio. Also shown is the expected confusion noise produced by extragalactic sources.

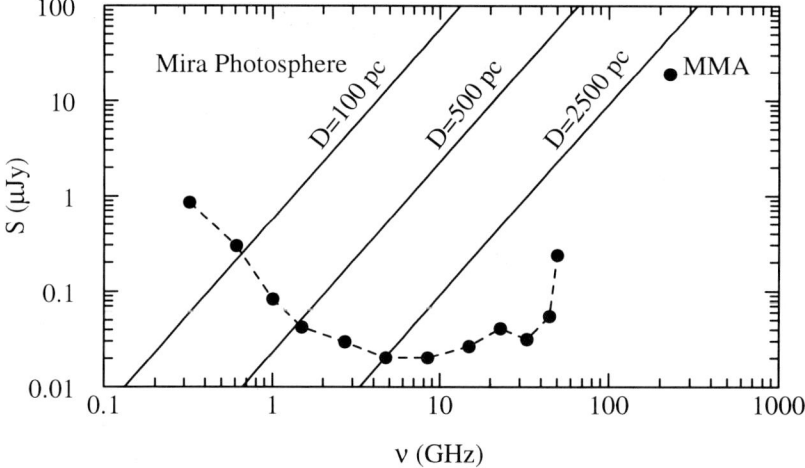

FIGURE 2. The flux density expected for a typical Mira variable as a function of frequency and distance compared with the sensitivity curve in Figure 1. Also shown is the expected sensitivity of the MMA for the same integration time at a wavelength of approximately 1 mm.

$B_m > 300$ km, with shorter maximum baselines permissible at cm wavelengths. Factors not discussed here, but which may lead to significant degradation, are interference and confusion power in the array sidelobes, the latter especially in the galactic plane (where most stars will be observed). These effects may impose severe problems for achieving the theoretical sensitivity.

Solar-type chromospheres and coronae
Red giant photospheres, chromospheres, and winds
Wolf-Rayet and OB star winds
Symbiotic star (SS) envelopes
Classical Novae ejecta

TABLE 1. Examples of thermally emitting stars

3. Thermal radio-emitting stars

3.1. Impact on frequency coverage

Some examples of thermally-emitting stars are listed in Table 1. A common feature of these objects is their positive spectral index ($\alpha > 0$), which means they would be optimally detected at short-cm wavelengths. For example, Figure 2 shows the flux densities expected from the photosphere of a Mira variable at various distances together with the sensitivity curve in Figure 1. Presently, isolated Miras can be detected at radio wavelengths out to distances of only a few hundred pc, but this will be increased to more than 3 kpc at short-cm wavelengths making possible the detection of more than 1000 Miras. The impact will be directly on the understanding of the mass-loss mechanism in red giants. Recent work by Reid & Menten (1995) shows from observations of nearby Miras at 1.3 cm that the radio-emitting photospheres are twice the size of their optical photospheres and that they do not pulsate. Thus the shocks associated with pulsation dissipate within two stellar radii, the region where mass loss is driven. Since mm observations should penetrate this region, simultaneous cm and mm continuum observations should lead to a better understanding of the driving mechanism. Note in Figure 2 that an LRT operating at 1-cm wavelength would be more sensitive than the proposed Millimetre Array (MMA) to photospheric emission from Miras.

Another example is the shells of classical and recurrent novae. Figure 3(a) shows the light curves at different frequencies based on a uniform slab model for the expanding shell and a representative value for the shell mass, showing once again the importance of cm observations in detecting nova emission sensitively and promptly. Figure 3(b) shows the maximum flux vs. frequency compared to the projected sensitivity of an LRT for an integration time of 12 hours. In principle every classical nova in the Galaxy would be detectable, even at levels far below the peak flux, thus increasing the potential detection rate from about $1\,\text{yr}^{-1}$ to about $30\,\text{yr}^{-1}$. Radio measurements provide the most accurate estimates of the shell mass and kinetic energy (e.g. Seaquist 1993), and imaging in the radio provides a unique way to study shock heating and nonthermal particle production produced by the collision between the fast wind and the nova ejecta (Bode & Lloyd 1995).

3.2. Impact on the resolution or filling factor

The limiting brightness temperature of a synthesis telescope is given approximately by

$$T_B = 4.6 \times 10^{-4} \left(\frac{S}{\mu\text{Jy}}\right) \left(\frac{B_m}{\text{km}}\right)^2 \text{K} \tag{3.1}$$

For $S = 0.02\,\mu\text{Jy}$, which would be typical at cm wavelengths, $B_m = 1000\,\text{km}$ ($f = 10^{-6}$) yields a 1σ brightness sensitivity of 10 K, adequate to map exceedingly faint features of free-free emitting sources at a resolution of 2 milliarcsec at 1 cm. This is roughly the same brightness temperature level as currently reached by the VLA, but at thirty times the angular resolution. The mapping capability is shown in Figure 4 which is a plot

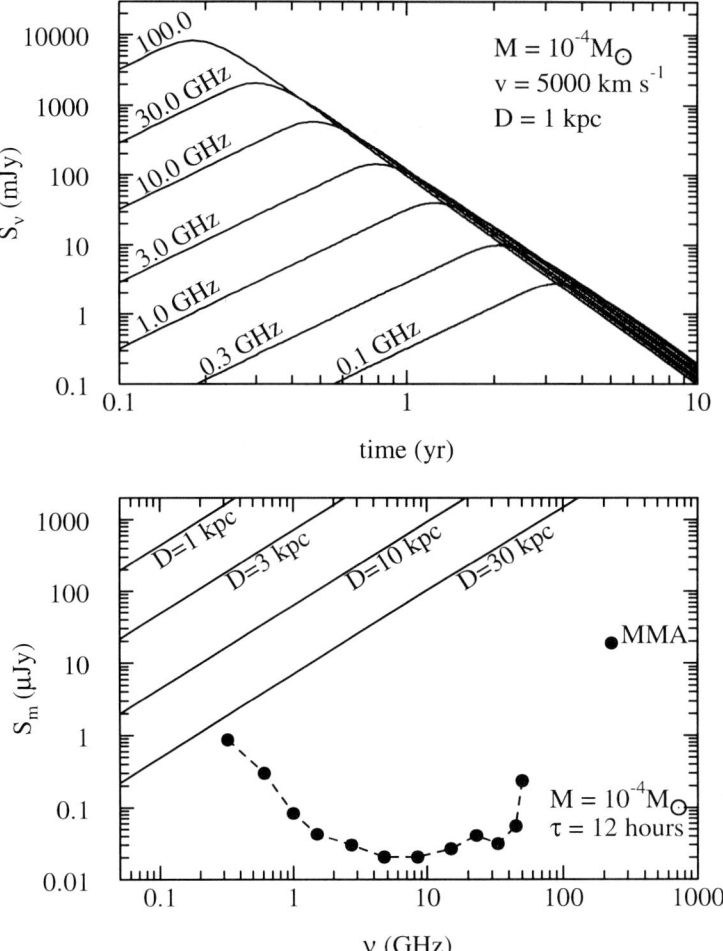

FIGURE 3. (a) Theoretical radio light curves for a classical nova based on a simple model consisting of free-free emission from an expanding uniform slab. (b) The maximum flux density in the nova light curve vs. frequency and distance compared with the sensitivity curve in Figure 1. Also shown is the expected sensitivity for the MMA

of linear resolution vs. distance D at two frequencies for $B_m = 1000$ km, compared with the linear diameters of some typical thermal sources. Consider as an example the origin of nonthermal emission in the thermally emitting winds of WR and OB stars. There are about a dozen cases known, with the WR system HD 193793 being a notable example. There are at least two different possibilities for the origin and acceleration of nonthermal particles, including turbulence in the wind and interaction between the winds of binary companions (Williams 1995). The LRT could map the winds of essentially all WR stars within 10 kpc at several frequencies, permitting the position and structure of the nonthermal region to be pinpointed in relation to the wind structure.

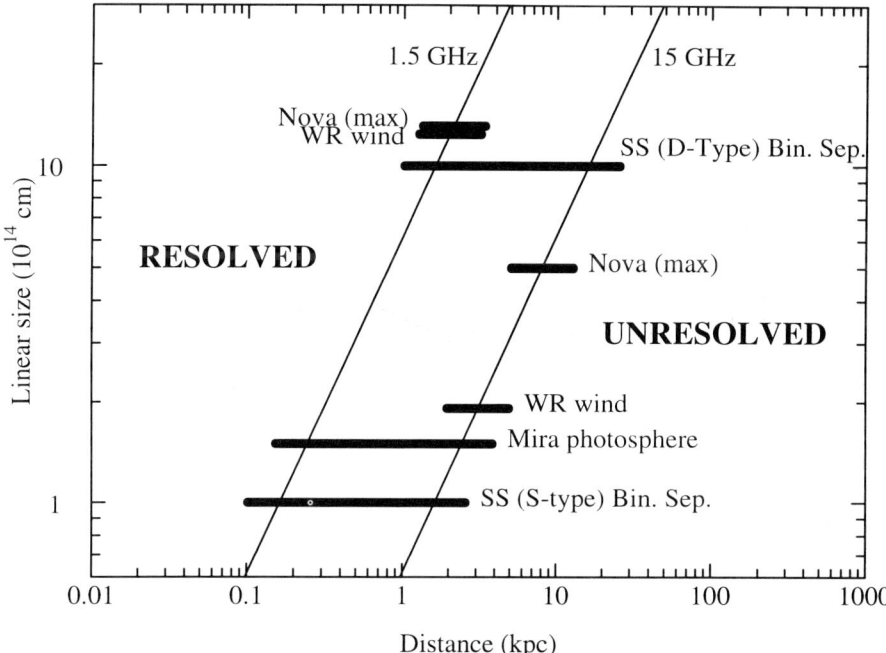

FIGURE 4. The linear resolution vs. distance at 1.5 GHz and 15 GHz for an LRT with $B_m = 1000$ km, compared with the linear sizes of a number of thermally emitting stars. Where the source is due to free-free emission the dimension refers to the optically thick photosphere. SS (S-type) and SS (D-type) refer to typical binary separations for the two IR classes of symbiotic stars.

4. Nonthermal radio-emitting stars

4.1. Impact on frequency coverage

Table 2 shows a number of examples of stars exhibiting nonthermal radio emission. Note that most of these types of objects also have relatively flat or rising spectra, relating to optical depth effects and the energy spectral index of the gyrosynchrotron emitting electrons. Therefore, with the exception of flare stars it would also be important to have observing capability at cm wavelengths for observations of nonthermally emitting stars. Figure 5 shows a plot of the flux density vs. distance together with the sensitivity in 12 hours of an LRT at 6-cm wavelength for a range of nonthermally emitting stars using typical observed luminosities. It should be borne in mind that these luminosities may range over several orders of magnitude, and that the observed ranges are affected by observational selection. The kink in the curve for flare stars reflects the choice of an integration time of 1 second, and a bandwidth of 1 MHz relevant to the observation of dynamic spectra of individual flares. Below we consider some specific examples of the impact on the study of nonthermal sources.

4.2. Examples, illustrating the impact on the science, field of view, and u-v plane coverage

4.2.1. Solar-type stars

With an aperture of $1\,\text{km}^2$, it would be possible for the first time to explore the solar-stellar connection in the radio. Though such studies are possible in the optical, UV, and X-ray bands, even the nearest stars with solar radio luminosity are too faint to observe

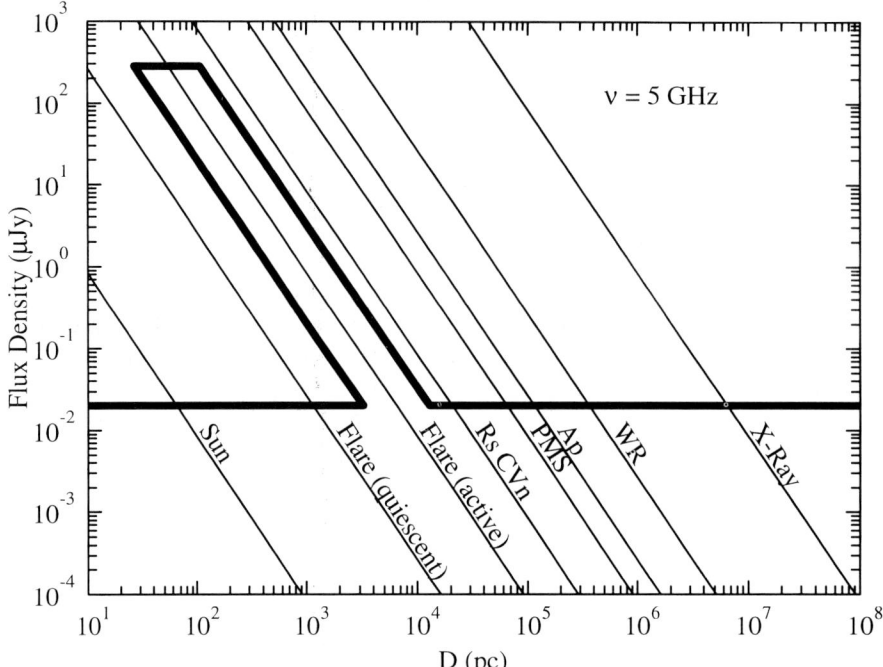

FIGURE 5. The flux density vs. distance at a frequency of 5 GHz for a number of nonthermally emitting stars compared with the sensitivity for an integration time of 12 hours. The luminosities used are typical values, and it should be noted that the luminosities may range over several orders of magnitude. The kink in the sensitivity curve for flare stars is relevant to the measurement of the dynamic spectrum of a flare event on a flare star at a frequency and time resolution of 1 MHz and 1 second respectively.

Solar-type stars
Flare stars
RS CVn and Algol binaries
Pre-main sequence stars
Extragalactic Supernovae
X-Ray Binaries

TABLE 2. Examples of nonthermal radio emission

---

at present. Figure 5 shows that a quiescent solar-type star could be detected to D=50 pc with an LRT. The corresponding volume contains more than 2000 G-dwarfs, and thus an entire new field of stellar investigation would be opened up.

4.2.2. *Pre-main sequence stars*

Pre-main sequence (PMS) stars may be sources of either thermal or nonthermal emission. In the former case, they constitute the so called class 0 sources which not only exhibit f-f emission from ionized gas, but also sub-mm continuum from circumstellar disks and large scale molecular outflows (e.g. André 1995). They are believed to be the earliest stages of star formation. A good example is VLA 1623 associated with a molecular outflow in the ρ Oph molecular cloud. PMS stars later evolve into T Tauri variables, which in their later stages exhibit nonthermal (gyrosynchrotron) emission, probably relating to diminished amounts of circumstellar material and to dynamo action stimulated by rapid rotation and convective envelopes. Several studies show that the nearby ρ Oph cloud, for example, contains dozens of such cm-emitting sources (e.g. Leous et al. 1991) distributed over an area approaching one square degree. Figure 5 shows that such stars, even with lower luminosities than those now observed, will be detectable in molecular clouds throughout the entire Galaxy. With the LRT observations of these and their massive counterparts - the Herbig Ae and Be stars, could be used to map star formation throughout the entire Galaxy.

This capability would exist only if such a large-aperture telescope also had a large field of view and good instantaneous u-v plane coverage (ie snapshot capability). These characteristics would be necessary to survey large areas in molecular cloud regions, and to provide images of crowded fields within a short observing time, particularly since PMS stars are variable. Note that mosaicing techniques may not work well in this case because of the short-term variability of the sources.

4.2.3. X-ray binaries

Interest in X-ray binaries is enjoying a revival because new X-ray and gamma-ray data have led to the discovery of superluminal radio jets associated with the hard X-ray sources GRS 1915+105 and GRO J1655-40. These are sometimes referred to as microquasars because of the similarity in form to their large scale counterparts in quasars. Figure 5 shows that luminous X-ray binaries could be studied and mapped in all Local Group galaxies with essentially the same sensitivity and nearly the same linear resolution as the most distant candidates are now studied in our own Galaxy.

Mirabel & Rodriguez (1995) have suggested that jet speed and X-ray hardness correlate, and have divided the sources into binaries containing neutron stars (softer X-ray spectrum and non-relativistic jet) and black holes (harder X-ray spectrum and relativistic jet). An LRT would provide significantly enlarged samples of such objects to confirm (or deny) this correlation, which will then yield new insights into the driving mechanisms for both stellar and quasar jets. Moreover, the sensitivity to low brightness (10 K) permits mapping at levels as much as 10 orders of magnitude lower than the theoretical maximum brightness of the compact core ($10^{11} - 10^{12}$ K). With good imaging capability, it will be possible to probe the interaction of fossil jets with the circumstellar and interstellar media surrounding X-ray binaries, yielding information on the history of mass loss from the system.

4.2.4. Extragalactic supernovae

Studies of extragalactic supernovae (SN) at radio wavelengths are strongly hampered by lack of sensitivity, particularly when compared to optical wavelengths. A general rule of thumb is that radio emission can be detected for supernovae brighter than about visual magnitude 12, or out to the distance of the Virgo Cluster (Weiler, this conference). This situation would be dramatically changed by an LRT, permitting essentially all optically-detected supernovae to be detected in the radio. Not only would the sample size be greatly enlarged, but as Figure 6 shows, a typical type II SN could be detected out to a redshift of unity or even larger. Since the distances to some radio emitting SN can be independently determined by the VLBI measurement of their expansion, a luminosity calibration is possible, and thus radio studies of SN could provide a new cosmological probe, providing an entirely independent method of measuring H_0 and Ω_0. An important point for emphasis is that radio studies are unaffected by extinction within the host galaxy, and that an intervening medium may in fact be viewed as a benefit by producing 21-cm absorption against the SN.

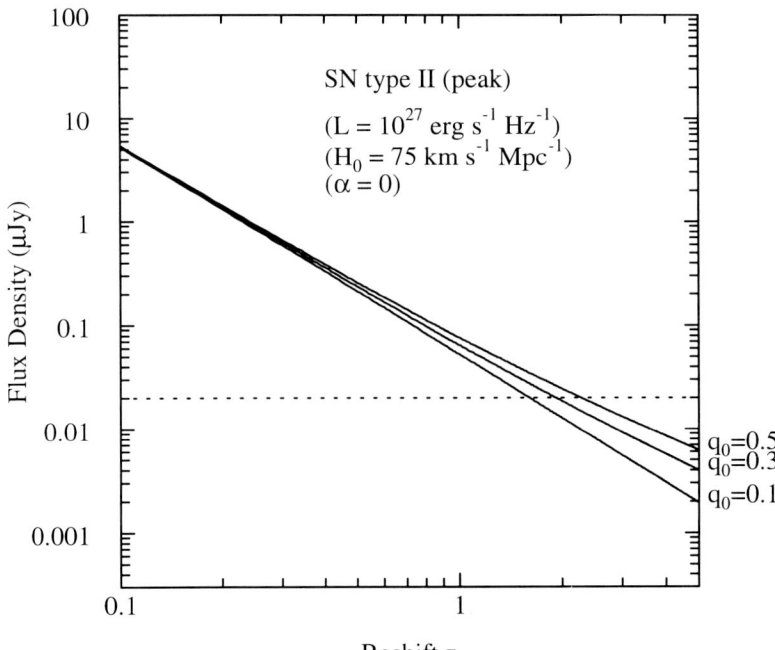

FIGURE 6. The expected 5 GHz flux density of a typical type II supernova as a function of the redshift of the host galaxy and the deceleration parameter q_0. The horizontal line represents the approximately 1σ sensitivity at 5 GHz for an LRT based on Figure 1. The radio spectrum is assumed to be flat ($\alpha = 0$).

The nonthermal emission in SN is produced by the shock interaction between the expanding ejecta and the circumstellar material which is the remnant of previous mass loss. Thus the radio light curve provides important information on the history of the mass-loss process for the progenitor. The potential increase from 1 every 1-2 years to 10-20 per year would lead to an enormous acceleration in knowledge of the SN progenitor. Note that due to the effects of free-free absorption in the expanding ejecta, the radio spectra of SN are flat or slowly rising with frequency. In addition, like the case of a nova shell, the shorter-wavelength emission is more prompt than the longer-wavelength emission, facilitating collaborative studies in shorter wave bands. These facts argue again for operation of any new large array at cm wavelengths.

4.2.5. *The impact of time variability*

Stellar sources are by their nature rapidly variable. Since such variations will be readily detectable with a large collecting aperture, there are significant impacts on the required u-v plane coverage. The following analysis yields an indication of the shortest detectable time variation measurable with an LRT as a function of distance and brightness temperature. We model the variation as a source of brightness temperature T_B which switches on and off in time interval τ, governed by the light-crossing time. The minimum observable time is set by equating this to the integration time. The result is that for a detection at the 1σ level, the minimum time is given by

$$\tau = 2.12 \times 10^{-3} \left(\frac{\lambda}{\text{cm}}\right)^{\frac{4}{5}} \left(\frac{D}{\text{pc}}\right)^{\frac{4}{5}} \left(\frac{T_B}{10^{10}\text{K}}\right)^{-\frac{2}{5}} \text{sec} \qquad (4.2)$$

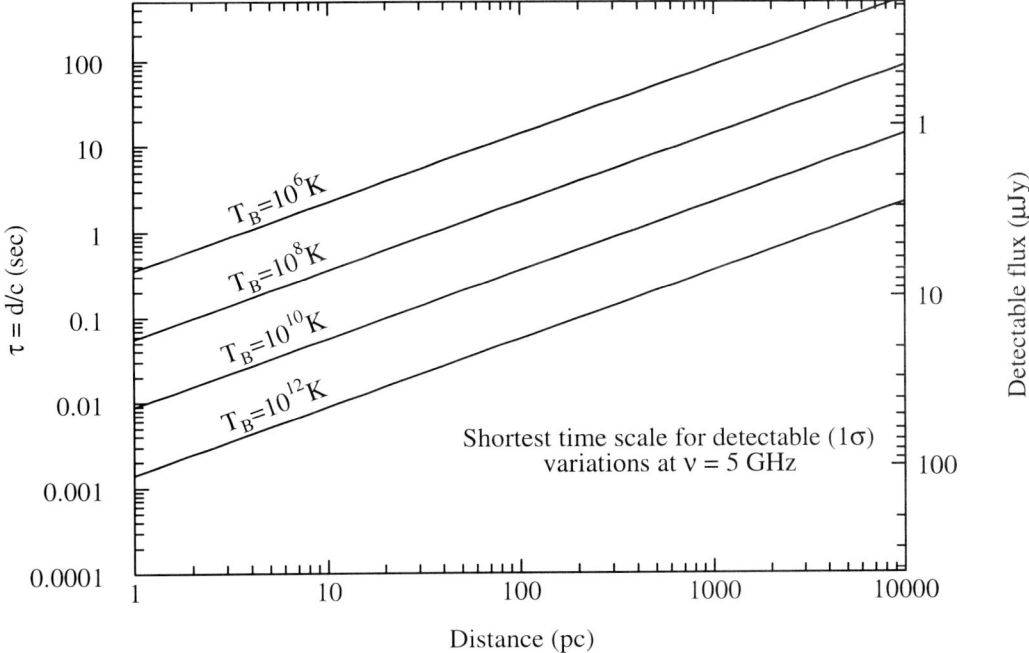

FIGURE 7. The shortest detectable time variation for a source as a function of the distance and brightness temperature at 5 GHz. The right hand scale refers to the minimum detectable flux variation.

Figure 7 shows τ as a function of D for $\nu = 5\,\text{GHz}$ and various brightness temperatures usually associated with stellar nonthermal sources. It is clear that variation on time scales of minutes are detectable in stellar sources throughout the entire galaxy. A typical example might be a field containing say 100 PMS sources. Whether or not it is intended to study the variations in such sources, they would produce unpredictable sidelobes in synthesis maps and could severely limit the dynamic range. Therefore maps must be made by taking short snapshots, perhaps minutes in length, and deep integrations made by combining data in the image plane instead of the u-v plane. Thus good instantaneous u-v plane coverage, requiring perhaps 20 or more individual elements, is required for an LRT array.

5. Spectral line sources

Thus far we have made no reference to spectral line sources of stellar emission. At cm wavelengths, these are primarily stellar OH and H_2O masers (OH/IR stars) and radio recombination line sources (stars with massive ionized envelopes). We confine our coverage of this topic to a comment on the impact of maser line emission on the field of view and u-v plane coverage. Detected OH/IR stars now number in the thousands, and their distribution and kinematics are being used to probe the gravitational potential of the Galaxy, both near the centre and in the bulge region (e.g. Sevenster et al. 1995). To date, the selection of OH/IR candidates is based on their brightness and colours at IRAS wavelengths. However, sensitive blind surveys will probably reveal many more such objects. For example, Blommaert et al. (1994) used the VLA to survey seven fields along the galactic plane, and found that 35 of the 44 detections were previously unknown.

The density of objects in this survey corresponds to roughly 10 per square degree. With an increase of nearly three orders of magnitude over the current VLA sensitivity, an LRT would be expected to detect many more, depending on the luminosity function at the low luminosity end. Wide fields and good u-v plane coverage would again optimize the detection rate. In addition, the light curves of the OH components are needed to obtain "phase lag distances". Such measurements for large numbers of stars would be feasible only with large fields of view. Finally, we note that the Magellanic Clouds could be searched for OH emission to luminosity levels well below that now detectable for sources in the Galaxy.

6. Further remarks on the design of an LRT Array

The current vision for an LRT is the SKAI (Braun, this conference). Implementation schemes call either for a large number (30) of array elements with each element comprising an adaptive array of smaller elements and receivers per band, or a smaller number of large apertures with one receiver per band. Operation at short wavelengths required for effective use in stellar astronomy favours a relatively small number of large apertures since expensive receivers are needed. Large apertures could include inexpensive designs such as Arecibo-type or Adaptive Reflector type (Legg, this conference). However, the u-v plane coverage may not be as good and adaptive beam formation may be more difficult. In this case, it will be important to achieve good u-v plane coverage using bandwidth synthesis methods.

I am grateful for discussions with Jim Condon, Kurt Weiler, and Philippe André all of whom provided me with important insights for this contribution. I thank S. Scott for assistance in preparing the manuscript, and the Natural Sciences and Engineering Council of Canada and the National Research Council of Canada for their support.

REFERENCES

ANDRÉ, P. 1995 Proceedings of Conference *Radio Emission from the Stars and the Sun*, held in Barcelona, Spain 3-7 July, 1995

BASTIAN, T.S. & BRIDLE, A.H. 1995 The VLA Development Plan, Proceedings of a Science Workshop held in Socorro, NM, 13-15 January 1995

BLOMMAERT, J.A.D.L., VAN LANGEVELDE H.J., & MICHIELS, W.F.P. 1994. *A&A* **287**, 479

BODE, M.F. & LLOYD, H.M. 1995. Proceedings of Conference *Radio Emission from the Stars and the Sun*, held in Barcelona, Spain 3-7 July, 1995

CONDON, J.J. 1974. *ApJ* **188**, 279

LEOUS, J.A., FEIGELSON, E.D., ANDRÉ, P. & MONTMERLE, T. 1991. *ApJ* **379**, 683

MIRABEL, I.F. & RODRIGUEZ, L.F. 1995 to be published in *Proceedings of the 17th Texas Symposium on Relativistic Astrophysics and Cosmology*, ed. H. Bohringer et al., New York Academy of Science.

REID, M.J. & MENTEN, K.M. 1995. Proceedings of Conference *Radio Emission from the Stars and the Sun*, held in Barcelona, Spain 3-7 July, 1995

SEAQUIST, E.R. 1993. Radio Emission from Stars, in *Rep. Prog. Phys.* **56**, 1145

SEVENSTER, M.N., DEJONGHE, H. & HABING, H.J. 1995. *A&A* **299**, 689

WILLIAMS, P.M. 1995. Proceedings of Conference *Radio Emission from the Stars and the Sun*, held in Barcelona, Spain 3-7 July, 1995

High resolution MERLIN† maps of the radio emission from Nova Cygni 1992 and Nova Cassiopeiae 1993

By STEWART EYRES[1], RICHARD DAVIS[2], MIKE BODE[3] AND HUW LLOYD[3]

[1] Astrophysics Group, Physics Department, Keele University, Keele, ST5 5BG, UK

[2] University of Manchester, NRAL, Jodrell Bank, Macclesfield, Cheshire, SK11 9DL, UK

[3] Astrophysics Group, Liverpool John Moores University, Byrom Street, Liverpool, L3 3AF, UK

1. Introduction

Classical novae at outburst are the most luminous members of the class of binary stars known as cataclysmic variables. The central system is an interacting binary consisting of a white dwarf (WD) primary, and a main–sequence dwarf secondary, which fills its Roche lobe. Matter is transferred via an accretion disc around the WD, and after a period of $10^5 - 10^4$ yr an explosive outburst caused by a thermonuclear runaway on the surface of the WD component can occur. This produces the observed large increase in visual magnitude, and is accompanied by the ejection of between 10^{-4} and $10^{-5} M_\odot$ of material at velocities of order $1000\,\mathrm{km\,s^{-1}}$ (Starrfield 1989).

Radio emission in classical novae is believed to originate from the ionised ejecta. As a result, radio observations allow the evolution of that ejecta to be traced. The emitting region is compact during the initial, optically-thick rise phase. This means that MERLIN is ideally suited to observing this phase, as the combination of resolution and sensitivity allows imaging at this time. Prior to the outburst of Nova Cygni 1992, the radio emitting shells of classical novae had remained unresolved in this stage of evolution.

2. Observations

2.1. *Nova Cygni 1992*

Nova Cygni 1992 (V1974 Cygni) was discovered at a visual magnitude of 6.8 on 1992 February 19 (Collins 1992), rising to magnitude 4.4 two days later. This made it one of the brightest novae in recent years, and the subsequent evolution of the optical light curve indicated a time to decline by two magnitudes, $t_2 \sim 20$ d (see for example Dolzan & Mikuz 1992), classifying it as a fast nova (Payne-Gaposhkin 1957). The presence of enhanced Ne line emission suggests that the primary star is an ONeMg WD (Shore *et al.* 1992).

Radio observations were acquired early in the nova's evolution, with the first detection being made by Hjellming (1992) on 1992 March 30 using the VLA. These observations indicated that the nova had an optically thick, thermal spectrum. The nova was first resolved in the radio 80 d after outburst, using MERLIN, the first time that this proved possible during the optically thick expansion phase of a classical nova (Pavelin *et al.* 1993). Here we present maps from 11 epochs of observations of Nova Cygni 1992 with MERLIN between 1992 and 1994. These are shown in Figs. 1 & 2, and are discussed in more detail by Eyres *et al.* (1996).

Our observations reveal a number of new and unexpected features. The emission

† MERLIN is a national facility operated by the University of Manchester on behalf of the Particle Physics and Astronomy Research Council

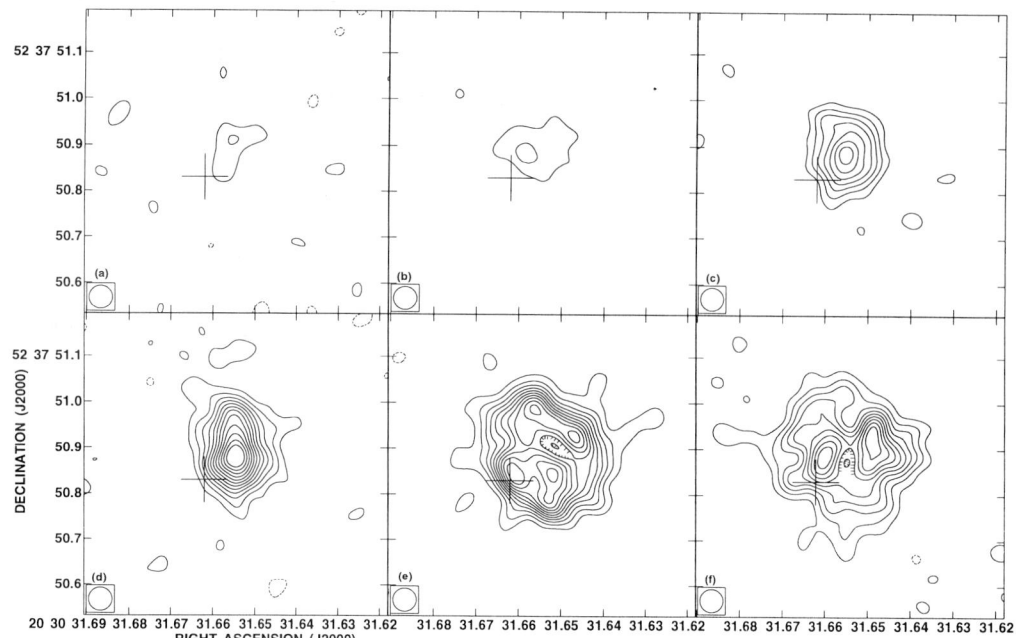

FIGURE 1. Radio maps at 6 cm of Nova Cygni 1992 obtained using MERLIN at epochs (a) 1992 May 9; (b) 1992 June 12; (c) 1992 July 22; (d) 1992 August 9; (e) 1992 November 15; (f) 1992 December 30. The contours are −3, 3, 6, 9, 12, 15, 18, 21, 24, 30, and 33 times the rms noise of 50 μJy beam^{-1}. The hairs emphasize where the contours are lower than those surrounding them. The beam size (full width half maximum) is 50 mas, as shown at the bottom left of each map. The cross is the CAMC position of $\alpha = 20^h 30^m 31^s.662$, $\delta = +52°37'50''.8$, and the size of the cross indicates the 1σ positional error of 50 mas, as given in Pavelin et al.(1993).

distribution is clearly not circularly symmetric, and there is evidence that the direction of the preferred expansion axis varies through 90° with time. Most of the observations were made when the emission was completely or partially optically thick, so that the true electron temperatures could be directly measured. The peak brightness temperatures exceeded 40 000 K at early times, which, although relatively high, are not inconsistent with thermal emission.

The data taken at 18 cm represent the first resolved images of a classical nova at this wavelength. A mixture of optically thick and optically thin material, at a range of temperatures, apparent in the variable peaked structure and the complex structure at 6 cm, is required to explain the light curve at 18 cm. Brightness temperatures of $\sim$ 10 000 K have been measured, and the peak-brightness-temperature curve appears to follow the extrapolation of the curve at 6 cm. The emission in Nova Cygni 1992 is consistent with thermal emission mechanisms.

From the results described here it is clear why simple isothermal, spherically-symmetric models of the evolution of nova remnants have failed to match the observed radio light curves of Nova Cygni 1992, particularly at high frequencies (e.g. Ivison et al. 1993). The structure and evolution of the remnant are not as simple as these models assume.

2.2. Nova Cassiopeiae 1993

Nova Cassiopeiae 1993 (V705 Cassiopeiae) was discovered at a visual magnitude of 6.5 on 1993 December 7 (Kanatsu 1993), rising to 5.6 mag by 1993 December 14 (Hurst 1993), which we take as the maximum, and defines our origin of time. The light curve

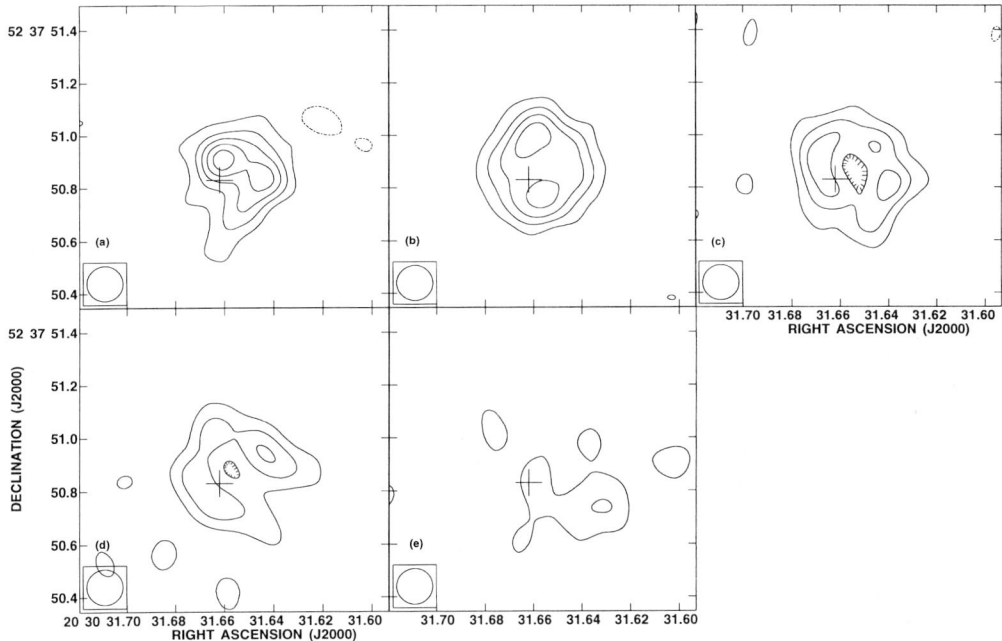

FIGURE 2. Radio maps at 18 cm of Nova Cygni 1992 obtained using MERLIN at epoch (a) 1993 April 15; (b) 1993 May 22; (c) 1993 September 30; (d) 1993 November 22; (e) 1994 January 1. The contours are $-3, 3, 6, 9, 12$ and 15 times the rms noise of 50 μJy beam^{-1} (for (a) and (b)) or 60 μJy beam^{-1} (for (c), (d) and (e)) . The hairs emphasize where the contours are lower than those surrounding them. The beam size (full width half maximum) is 135 mas, as shown at the bottom left of each map. The cross is as Fig. 1.

dropped by 2 mag from the peak in $\sim$ 45 days, identifying the nova as being in the 'moderately-fast' speed class. A slow decline followed, reaching V $\sim$ 11 on 1994 February 18 (Midtskogen 1994) and then undergoing a sharp, DQ Her–type dip, to 16.2 mag on 1994 March 11 (Szentaskó 1994). By 1994 May, the light had recovered, and has remained stable at V $\sim$ 12.5 since.

We present the first published radio observations of this nova. Two sets of data were taken with MERLIN, on 1995 July 22 (day 593) at 6 cm and 1996 January 23-25 (days 778-80) at 18 cm. The 18 cm data were concatenated to increase the total integration time. At 6 cm (Fig. 3(a)), we have detected extended emission with a total flux of $\sim 2.17 \pm 0.02$ mJy, and a peak of 0.370 ± 0.065 mJy beam^{-1}. This is equivalent to a peak brightness temperature of 10 500 K, commensurate with optically thick thermal emission. Fig. 3(b) shows the later uniformly weighted map at 18 cm. Here, the total flux is 0.66 ± 0.023 mJy, and the peak flux density 0.303 ± 0.046 mJy beam^{-1} in the uniformly weighted map, again giving a brightness temperature of around 10 000 K. The evolution of this nova is slow, so we can use the spectral index of ~ 1.1 between these two images to infer that we are observing thermal emission. This is further supported by the fact that the angular sizes at the two wavelengths are the same, once the effects of differing resolution are taken into account.

3. Conclusion

The MERLIN images presented in Figs. 1, 2, & 3 demonstrate our ability to probe the early spatial evolution of the ejecta from classical novae. Widening our database of MER-

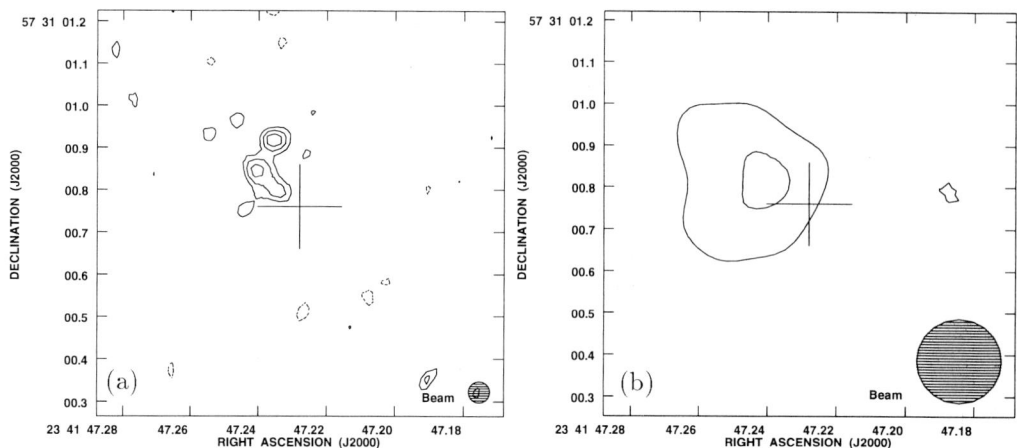

FIGURE 3. Radio maps of Nova Cassiopeiae 1993 obtained using MERLIN at (a) 6 cm on 1995 July 22 and (b) 18 cm in 1996 January. The contours are -3, 3, 4, and 5 times the rms noise of 65 μJy beam^{-1} (for (a)) and -1, 1, 2, 3, and 4 times 100 μJy beam^{-1} (for (b)). The beam sizes (full width half maximum) are 50 mas at 6 cm and 200 mas at 18 cm, as shown at the bottom right of each map. The cross is the CAMC position of $\alpha = 23^{\rm h}41^{\rm m}47^{\rm s}228 \pm 0^{\rm s}021$, $\delta = +57°31'0''76 \pm 0''16$, and the size of the cross indicates the 1σ errors, as given by Argyle and Morrison (1994).

LIN observations to include a number of novae with different characteristics, will lead to important constraints on (i) the outburst mechanism, (ii) post-outburst interactions within the ejecta, and (iii) the mass ejected and energy budget of the outburst.

REFERENCES

ARGYLE, R.W., MORRISON, L.V. 1994. Nova Cassiopeiae 1993. *IAU Circ. No. 5920*

COLLINS, P. 1992. Nova Cygni 1992. *IAU Circ. No. 5454*

DOLZAN, A. & MIKUZ, H. 1992. Nova Cygni 1992. *IAU Circ. No. 5475*

EYRES, S.P.S., DAVIS, R.J., & BODE, M.F. 1996. *MNRAS* **279**, 249

HJELLMING, R.M., 1992. Nova Cygni 1992. *IAU Circ. No. 5502*

HURST, G.M. 1993. *IAU Circ. No. 5905*

IVISON, R.J., HUGHES, D.H., LLOYD, H.M., BANG, M.K. & BODE, M.F. 1993. *MNRAS* **263**, L43

KANATSU, K., 1993. Nova Cassiopeiae 1993. *IAU Circ. No. 5902*

MIDTSKOGEN, O., 1994. Nova Cassiopeiae 1993. *IAU Circ. No. 5939*

PAVELIN, P.E., DAVIS, R.J., MORRISON, L.V., BODE, M.F., IVISON, R.J. 1993. *Nature* **363**, 424

PAYNE-GAPOSHKIN, C. 1957, The Galactic Novae. North-Holland, Amsterdam

PORETTI, E. 1993. Nova Cassiopeiae 1993. *IAU Circ. No. 5912*

SHORE, S.N., STARRFIELD, S.G., AUSTIN, S.J., GONZALEZ-RIESTRA, R., SONNEBORN, G., WAGNER, R.M. 1992. Nova Cygni 1992. *IAU Circ. No. 5523*

STARRFIELD, S. 1989. Thermonuclear processes and the Classical Nova outburst. In Classical Novae, (ed. Bode M. F. & Evans A.), pp. 39–60. Wiley.

SZENTASKÓ, L. 1994. Nova Cassiopeiae 1993. *IAU Circ. No. 5954*

Orbital motion and a model for HM Sge

By H. T. KENNY[1], A. R. TAYLOR[2], S. KWOK[2], S. P. S. EYRES[3], AND R. J. DAVIS[4]

[1]Department of Physics, Canadian Military College, Kingston, Ontario, K7K 5L0, Canada

[2]Department of Physics and Astronomy, University of Calgary, Calgary, AB, T2N 1N4, Canada

[3]Department of Physics, Keele University, Keele, Staffordshire, ST5 5BG, England

[4]University of Manchester, NRAL, Macclesfield, Cheshire, SK11 9DL, England

Observations of HM Sge taken at MERLIN and the VLA are interpreted in terms of a model which includes aspects of the colliding winds model (e.g. Kwok 1988, Girard & Willson 1987) and the STB model (Seaquist, Taylor & Button 1984). The model is in good agreement with the observed radio morphologies and spectra. The model indicates that the observed relative motion of the binary nebular features corresponds to orbital motion in a binary system with a period of 65 ± 10 yr. A binary separation of 20 ± 2 AU and a distance of 250 ± 40 pc are indicated for a total system mass of $2\,M_\odot$. This represents only the second determination of binary separation for a D-type symbiotic system (Willson, Garnavich & Mattei 1981), and to our knowledge is the first direct measurement of stellar orbital motion through radio imagery.

HM Sge is of particular importance among the symbiotic stars. As the most recently erupted symbiotic nova (1975), it is one of the few systems whose nova-like behaviour has been monitored with modern instruments. Radio images revealing extended radio structure were first presented by Kwok (1981), and Kwok, Bignell & Purton (1984). Li (1993) presents the results of a multi-frequency, VLA monitoring programme between 1981 and 1988, and the 22.5-GHz images from this study are shown in figure 1. The morphology in all images is dominated by two peaks of emission, one of which is considerably stronger than the other. As noted by Li, the separation between emission peaks appears to be increasing with time. It may be further noted that the position angle of the line joining the peaks appears to rotate with time.

The observed combination of position-angle rotation and increasing separation is difficult to reproduce in a single model. Of those models capable of doing so, the precessing jet model (e.g. Hack & Paresce, 1993) must be ruled out. This model predicts that ejecta, once expelled, follow a ballistic trajectory, contrary to the observations of Fig. 1.

A model which is capable of explaining the observed radio morphology and evolution of HM Sge is illustrated in figure 2 (Kenny 1995). This model includes aspects of the two primary models currently used to describe nebular radio emission: the STB model (Seaquist, Taylor & Button 1984; Taylor & Seaquist 1984); and the colliding winds model. The terminology used here in describing colliding winds features will be that defined by Kenny (1995): CWc = "concentric" colliding winds (e.g. Kwok, Purton & Fitzgerald 1978; Kwok 1988; Kenny 1995); CWb = "binary" colliding winds (e.g. Girard & Willson 1987; Kenny 1995); and CWo = "orbital" colliding winds (Kenny 1995).

In the model illustrated in figure 2, the hot stellar component (o) acts as the source both of high-energy photons and of a high-velocity wind. The high-energy photons ionize a portion of the wind from the cool stellar component (•) out to the dotted line in the figure, in accordance with the STB model. The high-velocity wind from the hot component interacts with the cool-component wind in various ways as indicated by shading in the figure. The region in which the opposing winds meet is described by the CWb model, and this region would take on a conical shell structure ("interaction

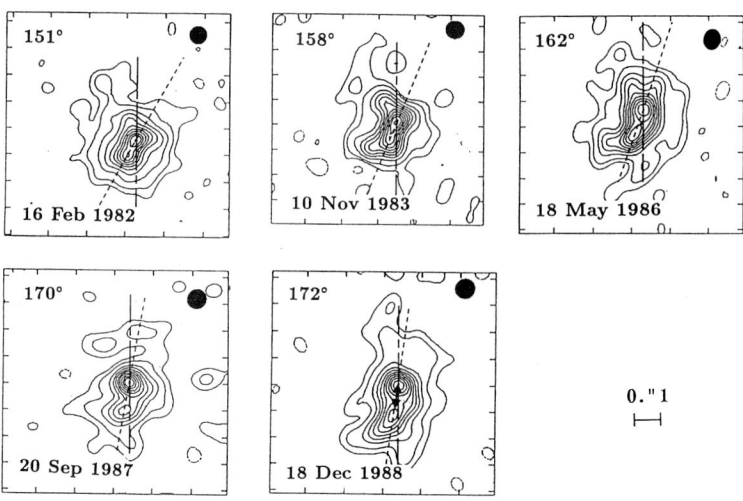

FIGURE 1. HM Sge: VLA images at 22.5 GHz (after Li, 1993). Solid lines are drawn in the NS direction, and dotted lines are drawn joining the major and minor nebular features. Position angles between the lines are indicated in each image. The double headed arrow (↔) in the December 1988 image is used to locate the cool component (north) and the hot component (south) according to the model proposed here.

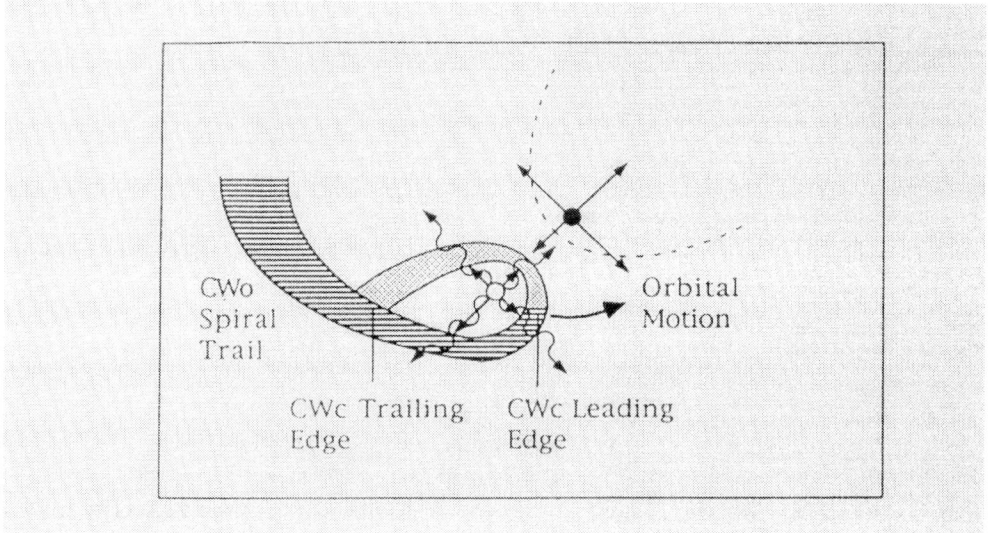

FIGURE 2. Orbital Model of HM Sge. The dotted line indicates the STB ionization boundary.

cone", e.g. Girard and Willson 1987) in the absence of orbital motion. At the flaring end of the interaction cone, the high-velocity wind from the hot component overtakes the "remnant" cool-component wind, blown outward previous to the initiation of the hot-component wind. The structure and dynamics of this region are described by the CWc model. Finally, due to the orbital motion of the stellar components, a trail of compressed nebular material is also produced (CWo spiral trail, see figure 2).

Within the context of this model, it is necessary to inquire which nebular features will be the primary emitters. The answer depends upon the selected system parameters. The model parameters found most suitable in the present analysis indicate that the northern nebular feature (figures 1 and 3) represents the apex of the STB ionization cone, *i.e.* the point of highest density to which ionization penetrates into the cool-component envelope on the line joining the two stars. The emission peak of the interaction cone (CWb) is somewhat weaker, and is unresolved from the STB peak. The southern nebular feature in the model represents the CWc nebula, as labelled in figure 2. The peak emission from this feature occurs at the leading edge where the hot-component wind advances into the previously unencountered cool-component wind. This overall configuration would place the hot component between the 22.5-GHz radio peaks in the images of figure 1, and in the central depression of the 4.9-GHz MERLIN image of figure 3. The central location of the hot component is supported by the work of Eyres et al. (1995), which shows that the peak ultraviolet emission, associated with the hot component, is located in the depression of the MERLIN image.

A primary requirement of the model, as applied to HM Sge, is that the momentum flux of the hot-component wind must be less than that of the cool-component wind. The CWb interaction cone will then sweep over the hot component ("H-facing" interaction) instead of the cool component ("C-facing"). This characteristic is necessary in order to produce nebular emitting components which are: 1) well confined spatially; 2) follow the stellar orbital motion; and 3) achieve agreement with the observed spectrum. The spectral requirement exists since higher values of hot-component momentum flux will result in a CWb interaction cone whose apex remains optically thick to frequencies exceeding the observed optically thin turnover ($\sim 60\,\mathrm{GHz}$: Kenny 1995, Kenny et al. 1996).

Estimates of certain model parameters were based on previously-published studies. Velocities of $10\,\mathrm{km\,s^{-1}}$ and $2000\,\mathrm{km\,s^{-1}}$ were used for the cool-component and hot-component winds respectively (e.g. Wallerstein 1978). For simplicity, a nebular temperature of 10^4 K was assumed throughout, consistent with effective radiative cooling (see Kenny 1995), even though considerably higher values are likely to exist in the inner regions (e.g. Eyres et al. 1995; Mürset et al. 1991). A cool component mass loss rate of $9 \times 10^{-7}\,D_{\mathrm{kpc}}^{3/2}\,[M_\odot \mathrm{yr}^{-1}][\mathrm{km\,s^{-1}}]^{-1}$ was determined by applying a visibility analysis technique (e.g. Kenny, Taylor & Seaquist 1991) to the data presented by Kenny et al. (1993). The remaining parameters necessary to produce model images and spectra, using the radiative transfer solutions of Kenny (1995), are the hot-component mass-loss rate, and the binary separation. To best reproduce the observations (spectrum, morphology, evolution), a value of $9 \times 10^{-7}\,D_{\mathrm{kpc}}^{3/2}\,[M_\odot \mathrm{yr}^{-1}][\mathrm{km\,s^{-1}}]^{-1}$ was found appropriate for the hot-component mass-loss rate. Thereafter, a binary separation of 110 ± 10 D_{kpc} AU was obtained from spectral fitting (Kenny 1995, Kenny et al. 1996).

Model images based on these parameters are shown in figure 3. The 22.5-GHz model image is in good agreement with the observed image. At 4.9 GHz, the primary differences between model and observed images arise from the assumption of spherical symmetry in the model. Despite this obvious weakness, the 4.9-GHz model image does successfully reproduce two ridge-like, optically thick, features bracketing a central depression. Further, while the two model nebular features are of equal intensity at 4.9 GHz, the northern component strongly dominates at 22.5 GHz when the southern feature has become optically thin (as observed). Model spectra (Kenny 1995; Kenny et al. 1996) are in good agreement with the observed spectra.

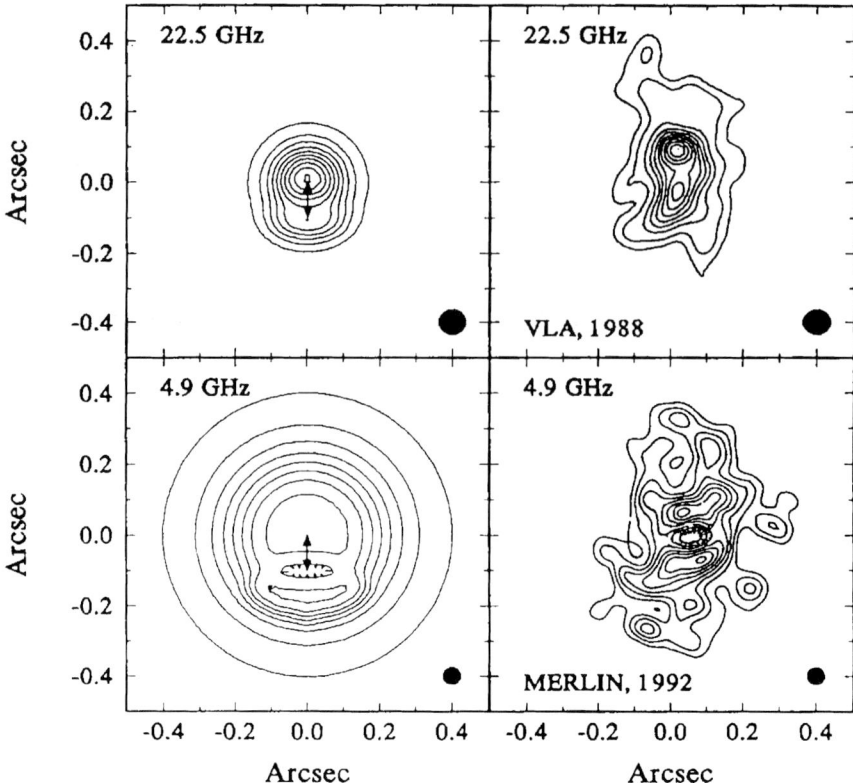

FIGURE 3. Model Images. Observed images are from Li (1993), and Eyres et al. (1995). Double arrows (↔) indicate the locations of the cool component (north) and hot component (south) in the model. Orbital motion not included.

· Orbital solutions for the relative locations of the nebular features, assuming a circular orbit, yield an orbital period of 65 ± 10 yr, which implies a linear binary separation of 20 ± 2 AU for a total system mass of $2\,M_\odot$ (Kenny 1995, Kenny et al. 1996). This result, together with the angular binary separation obtained from spectral fitting, yields a distance of 250 ± 40 pc, considerably closer than values commonly quoted in the literature (1.7 ± 0.8 kpc, Kenny 1995). The difference in location between stellar components and nebular peaks has been taken into account. The fitted inclination of the orbit is $75° \pm 5°$.

Summary

The morphology and evolution of the circumstellar nebula of HM Sge have been studied using high-sensitivity radio images at 22.5 and 4.9 GHz from MERLIN and the VLA. The model which has been developed includes the effects of partial ionization, wind collision and orbital motion. Within the context of this model, the cool stellar component is located in the vicinity of the northern radio peak, while the hot stellar component is located between the two radio peaks (figures 1 and 3). The southern radio peak represents the location at which the high-velocity, hot-component wind overtakes and sweeps up the "remnant" cool-component wind (CWc leading edge, figure 2).

The model requires no new general assumptions about symbiotic stars, but only a consistent synthesis of current models (STB, CWb, CWc), together with a consideration of orbital motion. It seems likely therefore that further members of the class may be describable in similar terms, *e.g.* V1016 Cyg (Kenny 1995); H1-36 (Taylor 1988). Monitoring of the radio morphology of these objects for orbital effects is therefore appropriate.

Significantly, the model presented here facilitates the determination of binary separations and distances from radio images. With the exception of R Aqr (Willson, Garnavich & Mattei 1981; 14-18 AU, eclipsing), HM Sge is the first D-type (dusty) symbiotic system for which a binary separation has been observationally determined. The value obtained, 20 ± 2 AU, is in good agreement with theoretically-predicted values for D-type systems (~ 20 AU, Kenyon 1986, Chapter 6). Distance determinations to D-type symbiotic systems have proven equally difficult, due to strong circumstellar extinction. The fact that HM Sge is found to lie closer (*i.e.* 250 ± 40 pc) than previously suggested may have implications for distance determinations within the class of objects generally.

REFERENCES

Eyres, S.P.S., Kenny, H.T., Cohen, R.J., Lloyd, H.M., Dougherty, S.M., Bode, M.F., & Davis, R.J. 1995. *MNRAS.*, **274**, 317

Girard, T., & Willson, L.A. 1987. *A&A* **183**, 247

Hack, W.J. & Paresce, R. 1993. *PASP* **105**, 1273

Kenny, H.T. 1995. PhD Thesis. University of Calgary.

Kenny, H. T., Taylor, A. R., & Seaquist, E. R. 1991. *ApJ* **366**, 549.

Kenny, H. T. et al. 1996. In preparation.

Kenyon, S. J. 1986. *The Symbiotic Stars.* Cambridge.

Kwok, S. 1981. In *The Nature of Symbiotic Stars* (ed. C. Choisi, and R. Stalio). IAU Colloquium 59, pp. 209-211. Reidel.

Kwok, S. 1988. In *The Symbiotic Phenomenon* (ed. J. Mikolajewska, M. Friedjung, S. Kenyon, and R. Viotti). IAU Colloquium 103, pp. 129-136. Kluwer.

Kwok, S., Bignell, R. C., & Purton, C. R. 1984. *Ap.J.* **279**, 188.

Kwok, S., Purton, C.R., & Fitzgerald, M.P. 1978. *ApJ* **219**, L125.

Li, P. S. 1993 MSc Thesis. University of Calgary.

Mürset, U., Nussbaumer, H.M., Schmid, H.M., & Vogel, M. 1991. *A&A* **248**, 458.

Seaquist, E.R., Taylor, A.R., and Button, S. 1984. *ApJ* **284**, 202.

Taylor, A.R. 1988. In *The Symbiotic Phenomenon* (ed. J. Mikolajewska, M. Friedjung, S. Kenyon, and R. Viotti). IAU Colloquium 103, pp. 77-83. Kluwer.

Taylor, A.R., and Seaquist, E.R. 1984. *ApJ* **286**, 263.

Wallerstein, G. 1978. *PASP* **90**, 36.

Variable sources and jets in Cepheus A

By V. A. HUGHES

Queen's University, Kingston, Ontario K7L 3N6, Canada

VLA Observations of Cepheus A, over the past 14 years, have been re-analyzed. The present data deal with the Sources 7, and interpret them as being the result of a jet travelling to the SE. IR line observations of H_2 and [FeII] are used to identify the 'working surface', or more precisely, the 'working volume' at the head of the jet. The data fit well the jet model by Tenorio-Tagle, Cantó & Różyczka (1988), and lead to an origin of the jet in Source 2. The highly time-dependent Source 9 appears to be part of the jet. A further and much older jet appears to be travelling in a direction to the east. The presence of both red- and blue-shifted CO molecular outflows gives credence to the suggestion that they are due to entrainment by the central jet.

1. Introduction

It is now well established that, in some way, young stellar objects are the source of molecular outflows, and about 150 bipolar ones are now known (e.g. Fukui 1989). Two types are recognized, molecular winds, and higher speed jets. Lines of thought are converging on the idea that the two are intimately related, in so far as it is a collimated jet that causes the entrainment of a molecular wind (Solf 1987, Raga & Cabrit 1993, Hartigan et al. 1993). This solves a number of problems, in particular the fact that the total mass of gas involved is frequently in excess of the mass of the star responsible (e.g. Moriarty-Schieven et al. 1991). Some of the difficulties of associating the two lie in the fact that jets appear to have small percentage ionization and need to be in regions of small extinction in order to be seen, while molecular outflows are in denser and more optically-obscured regions.

Theoretical models have been made which consider the coupling between jet and wind (Stahler 1994, Masson & Chernin 1992, Masson & Chernin 1993), and these can be used to explain the presence of 'interstellar bullets' (Norman & Silk 1979).

Cepheus A is a particular region, which is situated in a molecular cloud with the large density of $\sim 10^6 \, cm^{-3}$. It contains of a number of radio objects, both thermal and non-thermal, one of which has a cross-speed of about $300 \, km \, s^{-1}$ (Hughes 1993), which indicates the presence of a jet. It also has at least two molecular outflows (Rodríguez et al. 1980, Hayashi, et al. 1988, Bally & Lane 1991, Torrelles et al. 1993).

The line of radio sources numbered 7 appears to indicate the presence of a jet, and new IR observations have shown emission from its head (Goetz et al. 1996). This paper describes the fitting of the data to a model for a jet (Tenorio-Tagle et al. 1988), which also indicates a possible mechanism for its source.

2. The Radio and IR Data

Cepheus A has been monitored at the VLA of the National Radio Astronomy Observatory † since 1981; Figure 1 shows the original 6-cm data taken in the B-configuration (Hughes 1984), but convolved to a beamwidth of $2.35 \times 2.35 \, arcsec^2$. The sources are numbered. It was apparent at an early stage that there was variability in some of the

† The National Radio Astronomy Observatory is operated by Associated Universities Inc., under cooperative agreement with the National Science Foundation.

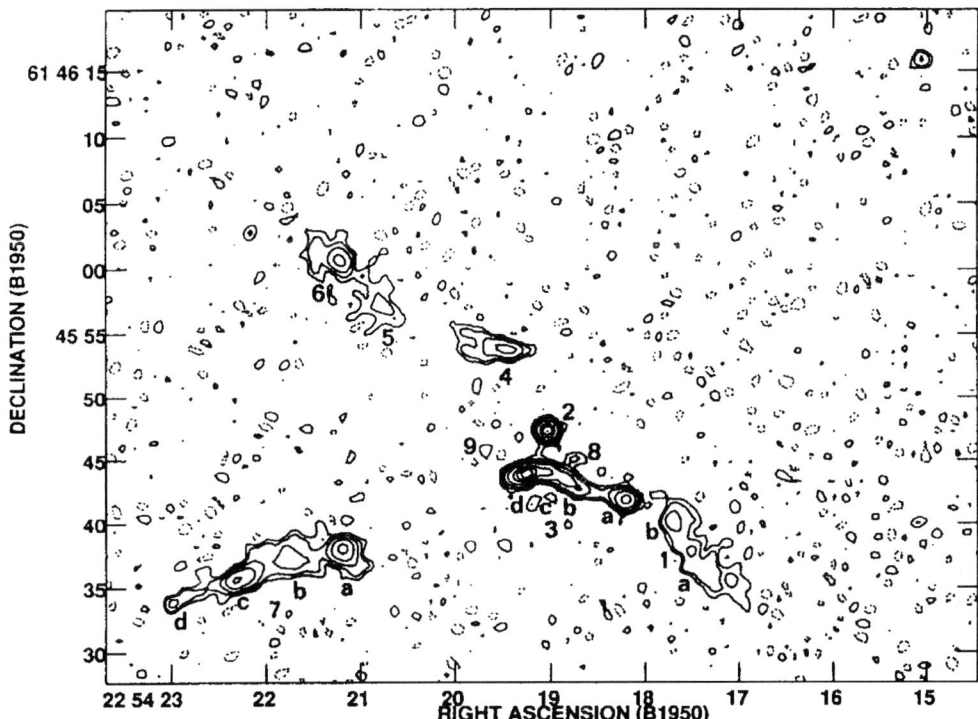

FIGURE 1. 6-cm image of Cepheus A obtained with the VLA in 1981, effective beamwidth 2.35 × 2.35 arcsec^2. Contour levels are at -0.1, 0.1, 0.2, 0.4, 0.8, 1.6, 2.4, 3.2, and 4.0 mJy beam^{-1}.

sources. Source 3(a) has now disappeared, 7(d) has decreased dramatically at 6 cm though less so at 20 cm and has a spectrum which is highly non-thermal. Sources 8 and 9 are highly time-dependent with 9 at its most intense showing a brightness temperature of 10^7 - 10^9 and size < 20 au (Hughes 1991, Hughes, Cohen & Garrington 1996), while 7(c) has two components of which the leading one, 7(c)(ii), has a cross speed of $\sim 300 \,\mathrm{km\,s^{-1}}$ (Hughes 1993). In addition, the structure of some of the sources appears to be variable, Sources 2, 6 and 7(a) showing components that change with time.

A number of molecular lines have been seen, and there are at least two known molecular outflows, chiefly those to the east and SE (Rodríguez et al. 1980, Hayashi, et al. 1988, Bally & Lane 1991, Torrelles et al. 1993), the SE one being in the same direction as the line of objects of Sources 7.

IR observations show the presence of shocks, but there has been little correlation between the 1-2 μm continuum, line and radio data. Recent observations at H$_2$ and FeII have shown good correlation with Source 7(d) (Goetz et al. 1996), and it appears that the leading edge of the shock is defined due to collisional excitation at 7(d) while 7(c) defines the Mach shock; 7(c) and 7(d) are the working surface, or more precisely, the working volume of the shock as previously defined by Blandford & Rees (1974).

The morphology of the line of radio sources suggests the presence of a jet. In sequence from the leading edge is Source 7(d) which is highly non-thermal, 7(c) which contains the moving source and is probably thermal, 7(b) which is non-thermal, and 7(a) which is slightly off the line of sources and appears to contain multiple components. Extrapolating back, Source 9 is highly non-thermal and time-dependent, and Source 2 contains a number

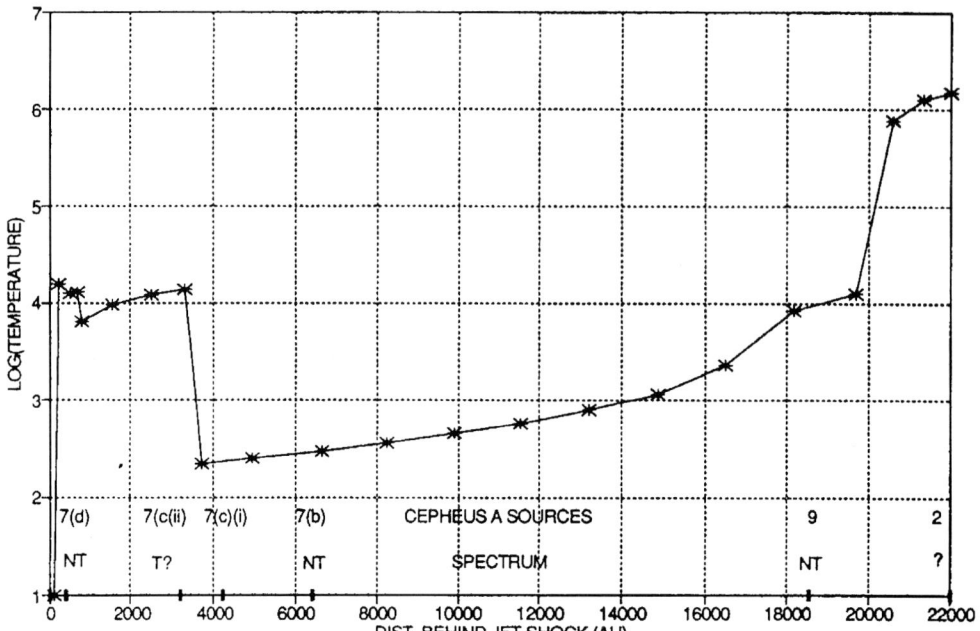

FIGURE 2. Position of sources with respect to the temperature variation along the jet. NT = non-thermal, T = thermal.

of components aligned in a NE-SW direction. We believe that the latter is the source of the jet travelling in the SE direction.

3. The Jet Model

The jet model which fits the line of sources is Figure 3(b) of the paper by Tenorio-Tagle et al. (1988), where the jet is initiated by either the convergence of two streams, as would be produced by the stellar wind from a young star, or to the convergence of winds produced by the breaking up of the accretion disk of a young star due to photoionization from the star (Tenorio-Tagle 1989). The initial velocity of the jets is $\sim 300\,\mathrm{km\,s^{-1}}$, the ambient medium has a density of $10^3\,\mathrm{cm^{-3}}$, temperature 10 K. The age of the jet is 430 years, and the length is 22,000 au, approximately equal to the separation between Sources 2 and 7(d). We have plotted in Figure 2 the variation in temperature along the jet, and also the position of the radio sources. The agreement between position of sources and increases in temperature is excellent for the head of the jet, although not for Source 7(b) and 7(a). However, there is not the expected increase in brightness temperature at Source 9, which could be the result of not including a magnetic field in the model.

It is of interest that Source 7(a) shows multiple components, as does Source 2, and it is possible that it is the source of a further and older jet, and that 7(b) is part of it.

Evidence for the presence of two jets is also to be found in the fact that there are two molecular outflows in the region, the chief one to the SE, and a much older one to the east, both of which are seen in both the red- and blue-shifted CO (Bally & Lane 1991, Moriarty-Schieven et al. 1991). The latter would result if the jets are moving in a direction almost normal to the line of sight, and were producing entrainment of the outflow. Evidence for the former is given by the fact that the IR line emission at the head of the jet has a red shift of 88 km s^{-1} (Goetz et al. 1996) compared with the cross-speed of the Mach shock of 300 km s^{-1}.

The model shows also that the temperature of the jet is far too low to produce ionization, but line emission appears to be due to the collisional excitation of the surrounding gas due to instabilities at the edge of the jet.

4. Conclusions

We have applied the jet model by Tenorio-Tagle et al. (1988) to the line of radio sources in Cepheus A, which results in the following conclusions:

1. The working volume at the head of the jet is clearly seen.
2. The low temperature of the jet leads to an electrically neutral jet; the jet is seen due to collision excitation of the gas in the surrounding medium.
3. The molecular outflows are due to entrainment by the jet.
4. The jet is produced by the breaking up of the accretion disk of a young star, by photoionization by the star (Tenorio-Tagle 1989).

Acknowledgment is made to NRAO for the allocation of observing time, and to the Natural Sciences and Engineering Research Council of Canada for the award of an Operating Grant.

REFERENCES

BALLY, J., LANE, A. P. 1991 in "Astrophysics with Infrared Arrays", ed. R. Elston, ASP Conf. Ser. vol 14
BLANDFORD, R. D., REES, M. J. 1974, *MNRAS* **169**, 396
FUKUI, Y. 1989, "Proceedings of ESO Workshop on Low Mass Star Formation"
GOETZ, J. A., PIPHER, J. L., FORREST, W. J., WATSON, D. M., HUGHES, V. A., GREENHOUSE, M. A., SMITH, H., FISCHER, J., WOODWARD, C. E. 1996. *BAAS* in press.
HARTIGAN, P., MORSE, J. A., HEATHCOTE, S., CECIL, G. 1993. *ApJ* **414**, L121
HAYASHI, S. S., HASEGAWA, T., KAIFU, N. 1988. *ApJ* **332**, 354
HUGHES, V. A. 1984. *ApJ* **276**, 204
HUGHES, V. A. 1991. *ApJ* **393**, 280
HUGHES, V. A. 1993. AJ, **105**, 331
HUGHES, V. A., COHEN, R. J., GARRINGTON, S. 1995. *MNRAS* **272**, 469
MASSON, C. R., CHERNIN, L. M. 1992. *ApJ* **387**, L47
MASSON, C. R., CHERNIN, L. M. 1993. *ApJ* **414**, 230
MORIARTY-SCHIEVEN, G. H., SNELL, R. L., HUGHES, V. A. 1991. *ApJ* **374**, 169
NORMAN, C., SILK, J. 1979. *ApJ* **228**, 197
RAGA, A. S., CABRIT, S. 1993. *A&A* **278**, 267
RÓDRIGUEZ, L. F., HO, P. T. P., MORAN, J. M. 1980. *ApJ* **240**, L149
SOLF, J, 1987. *A&A* **184**, 322
STAHLER, S. W. 1994. *ApJ* **422**, 616
TORRELLES, J. M., VERDES-MONTENEGRO, L., HO, P. T. P., RODRÍGUEZ, L. F., CANTÓ, J. 1993. *ApJ* **410**, 202
TENORIO-TAGLE, G. 1989. "Lecture Notes in Physics," **350**, 264
TENORIO-TAGLE, G., CANTÓ, J., RÓŻYCZKA, M. 1988. *A&A* **202**, 256

A radio survey of southern X-ray binary stars

By R. E. SPENCER[1], A. K. TZIOUMIS[2],
L. R. BALL[3], S. J. NEWELL[1], AND V. MIGENES[4]

[1] NRAL, University of Manchester, UK.
[2] ATNF, Sydney, NSW, Australia
[3] University of Sydney, NSW, Australia.
[4] ISAS, Kanagawa, Japan.

The Australia Telescope Compact Array was used in 1995 to survey the positions of X-ray binary stars south of $-30°$ declination. Detection sensitivities of under 1 mJy were achieved for each of the 91 fields observed at 6- and 3-cm wavelength. The sources were selected from the van Paradijs (1995) catalogue. Eight objects were detected within the inner imaged field at flux densities of a few mJy. A further 9 detections of objects further than 1 arcmin from the field centres were also detected.

1. Introduction

Some of the most astrophysically-interesting objects are to be found among the radio emitting X-ray binary stars (REXRB). The class includes such well-studied objects as SS433 with its moving optical lines and radio jets (Margon 1984), Cygnus X-3, famous for its dramatic radio outbursts (Johnston et al. 1986), Sco X-1 with its peculiar proper motion and association with background sources (Fomalont and Geldzahler 1991), and Circinus X-1 with its unusual X-ray properties and curved radio jets (e.g. Stewart et al. 1993). The recent discoveries of relativistic ejection of radio knots in the X-ray transients GRS 1915+105 (Mirabel and Rodriguez 1994) and GRO 1655-40 (Tingay et al. 1995) well illustrate the extreme nature of some of these objects, and undoubtedly more remain to be discovered.

X-ray binaries are semi-detached binary stars in which matter is transferred from a more or less normal star onto a neutron star or black hole. X-ray satellites have detected large numbers of these objects (193 in a recent catalogue by van Paradijs 1995). However only a small fraction are known to have radio emission (e.g. Hjellming 1988) and most have been found by radio observations from the northern hemisphere. For example, Nelson & Spencer (1988) surveyed a sample of X-ray binaries and cataclysmic variables with the broad-band interferometer (BBI) at Jodrell Bank and found 8 radio emitters out of 50 X-ray binaries at flux levels in excess of around 2 mJy at 6-cm wavelength. In all around 8 high-mass systems (HMXRB) and 16 low-mass systems (LMXRB) are known to be radio emitters.

Many of these objects are weak in the radio ($\sim$ 1 mJy) and are often variable. Several observations have been made by the VLA but the southern sky has not been so well covered. Duldig et al. (1979) surveyed 94 southern X-ray sources using the Parkes 64-m telescope at 2 cm, finding 11 radio detections at > 14 mJy, though several of these sources were extended and others possibly confused, leaving one definite detection (Circ X-1). In order to find more objects a more sensitive search over a large number of fields was required. There are 91 objects in the van Paradijs catalogue with declinations less than $-30°$, 36 HMXRB and 55 LMXRB, and new observations have been undertaken on this sample.

TABLE 1. The Detections

Source	Other Name	Type	Flux density mJy
0050−727	SMC X-3	HM	54.4
0103−762		HM	1.0
0540−697	LMC X-1	HM	83.6
0921−630	V395 Car	LM	2.4
1323−619		LM	4.7
1516−569	Circ X-1	LM	8.0
1642−455	GX 340+0	LM	18.6
1659−487	GX 339-4	LM	5.3
1722−363		HM	115.0

2. The Observations

The uncertainty in X-ray positions means that a relatively large area must be searched for detectable radio emission and a search strategy must form a compromise between sensitivity and confusion from background sources. The Australia Telescope Compact Array in Narrabri, New South Wales, Australia, consists of 6×22-m antennas on an east-west track, with baselines up to 6 km. At 6-cm wavelength the confusion level amounts to approximately 1 source per antenna beam (FWHM 9 arcmin) at flux densities of 0.4 mJy or greater and this noise level can be reached after a few tens of minutes' integration.

Simultaneous observations at 6- and 3.5-cm wavelength were made of 85 XRBs with the ATCA at the beginning of September 1995. In order to cover the large number of sources in the time available, 6 by 10-min scans were made of each field. The flux density scale was calibrated by daily measurement of 1934−638 and phase calibration by means of observations of nearby unresolved sources every ~ 40 minutes. Observations of the 6 sources in the sample not covered in September were made later with the ATCA in December 1995, but these have yet to be analysed.

The data were reduced using AIPS and maps made using HORUS after calibration. Some fields were contaminated by low-level extended emission from the galactic plane. This can be clearly seen in the data for 1744−299 for example (figure 1). To avoid the loss in sensitivity, data on the shortest baseline (367 m) were therefore ignored in the mapping process. Typical r.m.s. noise levels of 0.2 mJy were achieved in the maps.

3. The Detections

Preliminary results discussed here concentrate on the central 128×128 (arcsec)^2 of each map. In the 85 fields examined 8 sources were found with a peak brightness of greater than 1 mJy/beam at 6 cm. The images were produced using natural weighting of the u,v data and had beam sizes of approximately 2.7×2.0 arcsec in position angle $-3°$, varying slightly as the actual u,v coverage varies. Evidence of confusion indicated by side-lobe patterns showed that there were ~ 9 sources with positions outside the inner area. These sources will be investigated in a later paper, though a candidate for 1722−363 at ~ 3.5 arcmin from the field centre has been included here.

The 8 (plus 1722−363) detections are shown in table 1, together with the alternative name and type. The flux densities at 6 cm were found by fitting single Gaussian components to the maps. In all cases the sources were also detected at 8.4 GHz.

Source counts at low flux densities at 6 cm (Fomalont et al. 1984) show that roughly 1 in 20 fields of 128×128 (arcsec)^2 have one or more sources stronger than 1 mJy. We would

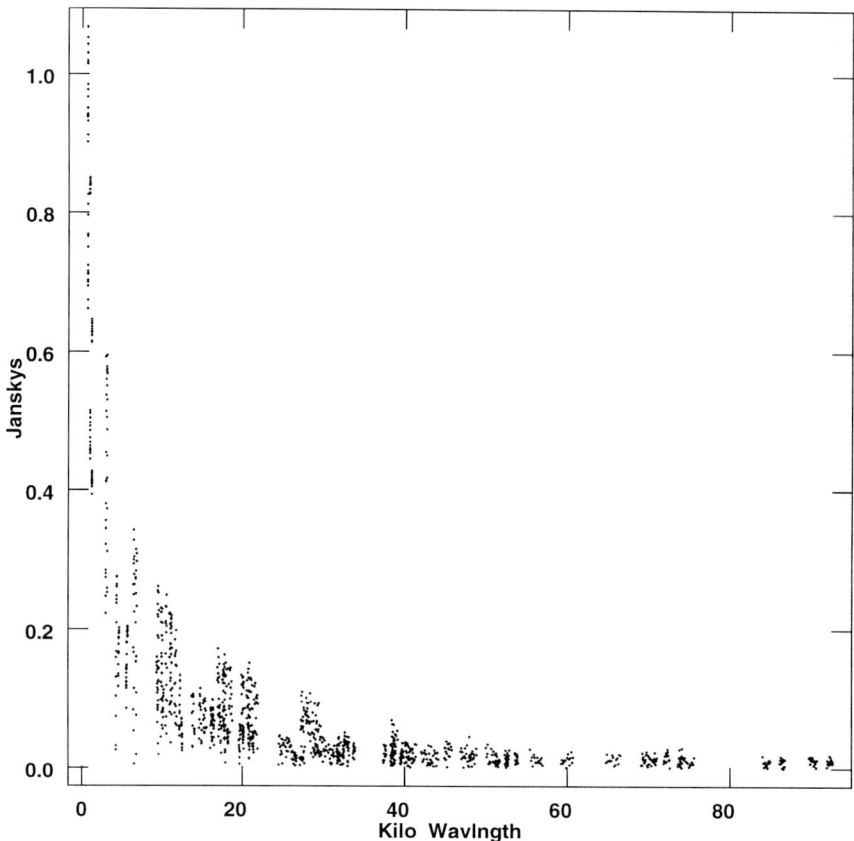

FIGURE 1. Plot of fringe amplitude vs u,v spacing for 1744−299 at 6 cm

therefore expect to find 2 or 3 background sources above this flux density level at high galactic latitudes and perhaps more at the low latitudes typical of the sources surveyed here. However several of the detections are quite strong and hence the confusing-source background should be much lower. The positions of 1516−569 and 1642−455 agree with the published positions of the X-ray sources to within less than 2 arcsec. The nominal X-ray position of 1722−363 is only given to 9 arcmin in the van Paradijs catalogue and so could be coincident with the radio source even though the source is 3.5 arcmin away from the field centre.

Approximately equal numbers of low-mass and high-mass systems were found at a total rate (9/85) not too dissimilar from the Nelson & Spencer survey. A wide range of types of object were found as for REXRB in the northern hemisphere. These results bring the number of radio-emitting high- and low-mass objects to 12 and 20 respectively but the proportion of X-ray binaries with radio emission is still small.

4. Further Work

Further analysis of this data set is proceeding. An investigation of the optical fields is needed particularly of the radio-emitting candidates in table 1. In addition deeper maps can be made by cleaning the whole 9 arcmin field of the ATCA at 6 cm. This also reduces the confusion problems caused by sources outside of the central 2 arc min. The radio spectra of the detections can also be found from the ATCA data. Follow-up observations using VLBI techniques to find the radio structure and further ATCA observations to search for variability of the the detected sources are also planned.

REFERENCES

DULDIG, M.L., GREENHILL, J.G., THOMAS, R.M., HAYNES, R.F., SIMONS, L.W.J., & MURDIN, P.G., 1979. *MNRAS* **187**, 567

FOMALONT, E.B. & GELDZAHLER, B., 1991. *ApJ* **383**, 289

FOMALONT, E.B., KELLERMANN, L.I., WALL J.V. & WEISTROP, D., 1984. *Science*, **225**, 23.

HJELLMING, R.M., 1988. In *'Galactic and Extragalactic Radio Astronomy'*, eds. G.L. Verschuur and K.I. Kellermann, p 381, Springer, Berlin

JOHNSTON, K.J., ET AL., 1986. *ApJ.* **309**, 707

MARGON, B., 1984. *ARA&A* **22**, 507

MIRABEL, I.F. & RODRIGUEZ, L.F., 1994. *Nature* **371**, 46

NELSON, R.F. & SPENCER, R.E., 1988. *MNRAS* **234**, 1105

STEWART, R.T., CASWELL, J.L., HAYNES, R.F. & NELSON, G.J., 1993. *MNRAS* **261**, 593

TINGAY, S.J., ET AL., 1995. *Nature* **374**, 141

VAN PARADIJS, J., 1995. In *'X-ray Binaries'*, eds W.H.G. Lewin and E.P.J. van den Heuvel, Cambridge, 1995

MERLIN astrometry of weak radio stars

By S. T. GARRINGTON[1], R. J. DAVIS[1],
L. V. MORRISON[2] AND R. W. ARGYLE[2]

[1] NRAL, The University of Manchester, Jodrell Bank, Macclesfield, UK

[2] Royal Greenwich Observatory, Cambridge, UK

We have used MERLIN at 5 GHz to determine the positions of about a dozen weak radio stars by phase referencing relative to IERS calibrators. By comparing the radio-star positions with the optical positions measured by HIPPARCOS, we have determined the offset between the fundamental extragalactic radio reference frame and the HIPPARCOS stellar reference frame. The post-fit residuals are a few mas, and can be accounted for by a combination of the errors in the HIPPARCOS proper motions and the MERLIN positions. The solution agrees well with that derived by Lestrade et al. from a largely independent set of stars using VLBI measurements.

1. Introduction

The measurement of the positions, parallaxes and annual proper motions of approximately 100,000 stars by the HIPPARCOS satellite with a precision of about 1.5 milliarcsec (mas) represents a milestone in the development of optical astrometry. At the same time, radio interferometry using VLBI is able to provide absolute positions of several hundred extragalactic objects (mainly quasars) to better than 0.5 mas. Linking these two sets of celestial co-ordinates is vital, because the extragalactic objects define an inertial frame against which the motions of stars can be measured accurately. At a more practical level, this link is necessary for the accurate registration of high-resolution optical and radio images, such as MERLIN and HST images with a common resolution of between 50 and 100 mas. Currently such registration is often only possible to within 0.5 arcsec, making the identifications of small-scale features uncertain.

The difficulty in achieving this link between the two frames arises because all quasars are too faint to be observed optically by HIPPARCOS and only a tiny fraction of stars are detectable at radio wavelengths. Even those few stars which can be detected by VLBI are generally too weak for absolute astrometry. However, their positions can be measured by phase-referencing relative to nearby quasars which define the extragalactic reference frame. Over the last 13 years J.-F. Lestrade and colleagues at JPL have succeeded in obtaining sub-mas astrometric solutions for about 10 stars and have computed a provisional link solution with an error of about 0.5 mas (Lestrade et al. 1995).

2. Astrometry with MERLIN

In order to obtain an independent check on the link solution obtained from the VLBI observations, and with a view to extending the number of stars used, an astrometric programme using MERLIN has been established. Although the resolution of MERLIN is rather less than that of VLBI, phase-referencing is relatively straight-forward. Furthermore, the real-time operation of MERLIN allows observations of these weak and highly variable radio stars while they are sufficiently bright to obtain positions to within a few mas.

Phase referencing is now a standard technique for MERLIN. The target data is corrected off-line by interpolating the atmospheric phase variations derived from the calibrator, typically within 5° of the target and using a cycle time of 5 to 10 minutes. The radio-star positions (relative to the VLBI calibrators) can be measured directly from the radio maps to within a fraction of the nominal MERLIN resolution of 50 mas at 5 GHz.

Accurate telescope positions are a key requirement for this technique: to first order, a 2-cm error in the position of the 32-m MERLIN telescope at Cambridge corresponds to 1.5 mas error in the separation of two sources over 5°. A series of observations of pairs of calibrator sources were used to determine MERLIN telescope positions, relative to the MKII telescope at Jodrell Bank. The derived telescope positions show a scatter of between 0.5 and 2 cm depending on the distance from Jodrell Bank, which gives a good indication of the differential delay errors for MERLIN phase referencing. The MERLIN correlator model does not yet include any ionospheric delay, but the observations were made close to sunspot minimum and at 5 GHz the differential delay should be 1 cm or less on the longest baselines. A tropospheric model, which uses real-time measurements of the atmospheric pressure at each telescope but a fixed contribution due to water-vapour, may contribute similar errors. In general, we estimate that the systematic errors in MERLIN astrometry will be in the range 0.5 to 1 mas per degree of target-calibrator separation.

Initial observations of four radio stars also being studied with VLBI were made in December 1992. These stars are relatively bright (10 – 70 mJy) and are quite close to their respective calibrators (both key requirements for inclusion in the VLBI observations). Comparison of the MERLIN and VLBI positions (kindly provided by J.-F. Lestrade) shows an rms difference of 1.2 mas (excluding the complex Algol system) clearly demonstrating the astrometric potential of MERLIN.

In order to find an independent set of stars for the link solution, pilot observations were made in Spring 1995 of about 25 RS CVn systems with a known history of radio activity. These data were analysed rapidly and observations of the ten detected stars continued over the following two weeks. Most stars showed considerable variability over this time, some disappearing completely. In the end, useful data were obtained on nine stars with flux densities of 2.3 – 12 mJy. The radio images have signal:noise ratios of 8 – 70, so their centroids can be determined to within about 2 mas in most cases. Including a systematic error term proportional to the target–calibrator separation, the estimated positional errors range from 2 to 7 mas (median value 3 mas).

As a check on these error estimates, the data for individual stars were split into subsets and processed independently. In all cases the differences are within the error estimates. This analysis also highlights the proper motions of some the stars, which can be measured by MERLIN even over an interval as short as two weeks (see Figure 1.)

The distribution of the stars detected is shown in Figure 1.

3. The link solution

We can now use the differences between the MERLIN radio positions, which are relative to the VLBI calibrators, and the HIPPARCOS optical positions to derive the offset of the HIPPARCOS stellar reference frame relative to the fundamental extragalactic reference frame defined by the VLBI calibrators. The definition of the link rotation matrix is given by Lindegren and Kovalevsky (1995) and the rotation angles are of order 20 mas. Without radio proper motions, we can only solve for the origin of the HIPPARCOS frame relative to the VLBI frame and not the time-dependent rotation, and the HIPPARCOS proper motions and parallaxes were used to correct the MERLIN positions to the HIPPARCOS epoch of 1991.25.

Three stars were excluded from the final solution, because they are problematic for HIPPARCOS, including the triple star Algol and 29 Dra, a probable HIPPARCOS dou-

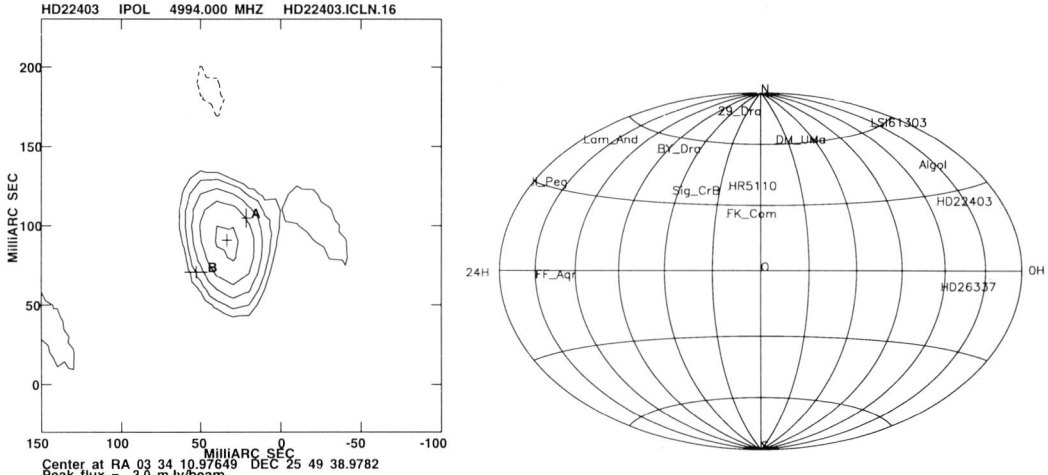

FIGURE 1. Left: Example of proper motion seen using MERLIN snapshots over a few weeks: the contours show the image of HD22403 at 10 May 1995, and the crosses A and B show the image centroid on 23 April and 27 May. Right: The stars used in deriving the link solution

ble. The post-fit residuals of the MERLIN positions have an rms value of 6 mas, which is attributed to the combined errors of the MERLIN positions (typically 3 mas) and the HIPPARCOS positions (typically 5 mas, including the proper motions). The solution has a standard error of just over 2 mas and is entirely consistent to within one sigma with the solution found by J.-F. Lestrade using VLBI observations of a different set of stars. The MERLIN results will be included in a global solution for the rotation of the HIPPARCOS frame to the extragalactic reference frame which will be published in 1996.

4. Conclusions

We have demonstrated the potential of MERLIN for differential astrometry over several degrees. In the best cases MERLIN may achieve sub-mas accuracy, but even for weak radio stars of only a few mJy and calibrator separations of several degrees, MERLIN can achieve accuracies of a few mas. By comparing MERLIN and HIPPARCOS positions for a sample of 13 weak radio stars, we have derived the offset between the HIPPARCOS and VLBI reference frames with an accuracy of about 2 mas. Future observations should enable us to obtain radio proper motions to within 1 mas/yr and thereby determine the time-derivative of the link matrix. Possible enhancements to MERLIN, such as upgrading the Lovell telescope and increasing the bandwidth, along with better corrections for atmospheric phase errors should allow MERLIN to achieve sub-mas positions for many more stars. In particular, it should be possible to use weaker calibrators much closer to the target sources for parallax and proper motion measurements. We hope, therefore, that MERLIN will play an important rôle in maintaining the HIPPARCOS-VLBI link.

REFERENCES

LESTRADE, J.-F. ET AL 1995. *A&A* **304**, 182

LINDEGREN, L. AND KOVALEVSKY, J. 1995. *A&A* **304**, 189

High-sensitivity radio observations of the 8.665-GHz ^{3}He$^+$ hyperfine line emission from planetary nebulae

By D. S. BALSER[1], T. M. BANIA[2], R. T. ROOD[3] AND T. L. WILSON[4]

[1]National Radio Astronomy Observatory, Green Bank, WV 24944, USA

[2]Astronomy Department, Boston University, Boston, MA 02215, USA

[3]Department of Astronomy, University of Virginia, Charlottesville, VA 22903, USA

[4]Max-Planck-Institut für Radioastronomie, 53121 Bonn, Germany

We report observations of the cosmic abundance of ^{3}He for a sample of six Galactic planetary nebulae. These abundances are derived from sensitive measurements of the 8.665-GHz hyperfine line of ^{3}He$^+$. We find abundances ranging from ^{3}He/H $= 10^{-4} - 10^{-3}$. These abundances are consistent with standard models of stellar evolution of low-mass stars and are a factor of ten larger than those found in Galactic H II regions.

1. Introduction

In cosmology ^{3}He plays a crucial role in testing Big Bang Nucleosynthesis (BBN) and may provide a lower limit to the baryon-to-photon ratio η (Wilson & Rood 1994). Recently ^{3}He has taken on new importance with the claimed detection of a high deuterium (D) abundance in a large-redshift galaxy (Songaila et al. 1994). To reconcile this D abundance with that observed in the local interstellar medium (Linsky et al. 1992), there must be substantial Galactic processing of D. Under "normal" circumstances D is converted into ^{3}He. To determine primordial abundances, however, we must first understand the production of ^{3}He in stars, that is, Galactic nucleosynthesis. In addition to BBN (including converted D), ^{3}He should be produced in significant quantities by stars of 1–2 $M_\odot$ (Rood et al. 1976). This hypothesis has been confirmed in at least one source by measurements of ^{3}He$^+$ in the planetary nebula (PN) NGC 3242 using the Max-Planck-Institut für Radioastronomie (MPIfR) 100-m telescope (Rood et al. 1992). Despite the simplicity of the case for stellar production of ^{3}He, only recently has this been included in chemical evolution calculations (e.g., Steigman & Tosi 1995). The results of these calculations are *in strong disagreement* with our observations of H II regions (Balser et al. 1994; Bania et al. this volume). Indeed, our H II region abundances indicate little if any production of ^{3}He since BBN. Several papers have recently appeared explaining how low-mass stars might not produce ^{3}He. Galli et al. (1994) hypothesize a low-energy resonance in the ^{3}He(^{3}He,2p)^{4}He reaction; Hogan (1994) suggests non-convective mixing processes in red giants. Our NGC 3242 result is a direct contradiction to the proposal of Galli et al.

The primary goal now is to establish an abundance pattern in the Galaxy. In the last few years we have made an attempt to measure ^{3}He$^+$ in a small sample of planetary nebulae (PNe) using the 100-m telescope. These observations are discussed in §2. Simple models are used to determine the ^{3}He/H abundance ratio in §3.

Source	Θ_s [arcsec]	^{4}He$^{++}/^4$He$^+$	T_L [mK]	^{3}He/H [10^{-3}]
IC 289	26	0.81	<2.2	<1.1
NGC 3242	40	0.23	6.0±0.3	1.1±0.3
NGC 6543	13	0.00	<2.6	<1.2
NGC 6720	60	0.30	<1.8	<0.15
NGC 7009	27	0.13	<2.4	<0.59
NGC 7662	14	0.40	<2.3	<1.1

TABLE 1. Planetary Nebulae Properties

2. Observations

Observations of the ^{3}He$^+$ line were made with the MPIfR 100-m telescope during six epochs between 1991 and 1995. Table 1 lists the PNe sample which was determined using the following criteria: (1) large angular size; (2) low-mass progenitor star; (3) low nitrogen abundance; and (4) little He^{++}. The PNe were ranked based on their expected 100-m main-beam brightness temperature assuming a constant ^{3}He$^+$ abundance. The effect of beam dilution was considered in this calculation which includes criterion (1). The PNe were then chosen from this list based on criteria (2–4). We eliminated PNe which had progenitor masses greater than $\sim 2\,M_\odot$ and PNe which had N/H abundances greater than about twice the solar value. High-mass stars are expected to burn ^{3}He into heavier elements and high N abundances indicate mixing which may destroy ^{3}He. Also, if the PN has a significant fraction of its helium in the form of He^{++} then measuring the ^{3}He$^+$ line will be even more difficult. In fact, IC 289 probably has too much He^{++} and we have stopped observing this source.

Table 1 lists the PN angular size, the fraction of doubly-ionized helium relative to singly-ionized helium taken from Cahn et al. (1992), the main-beam brightness temperature of the ^{3}He$^+$ line in mK, and the calculated ^{3}He/H abundance ratio in units of 10^{-3} (see §3). The 100-m telescope has an angular resolution of 84 arcsec at the ^{3}He$^+$ line rest frequency; thus even the largest PNe which meet criteria (1–4) are smaller than the 100-m beam.

We detect ^{3}He$^+$ emission only in NGC 3242. Limits for the other PNe are based on the 2σ values determined from the r.m.s. variations in the spectral baseline. Figure 1 shows the NGC 3242 spectrum where the vertical scale is the main-beam brightness temperature in units of mK. The vertical lines mark the expected positions of the He171η, H171η, and ^{3}He$^+$ lines. Figure 1 is composed of observations made during all six epochs. The probable detection quoted in Rood et al. (1992) was based on data from only one epoch in 1991. We now feel that this detection is solid.

3. Model Abundances

In order to convert these measurements into a ^{3}He/H abundance a model must be assumed for the nebulae. We have assumed a homogeneous, isothermal nebula where the ionized hydrogen and singly ionized helium zones are coincident. That is, we assume that all the helium is singly ionized inside the nebula. It is important to note that deviations from these assumptions in the form of density or ionization structure will *only increase* the determined ^{3}He/H abundance (Balser et al. 1994). Clearly, ionization structure exists at some level in most of these sources (cf. column (3) of Table 1).

The abundance for NGC 3242 of 1.1×10^{-3}, as well as the limits produced by the other PNe, are consistent with standard models of stellar evolution (see Rood et al.

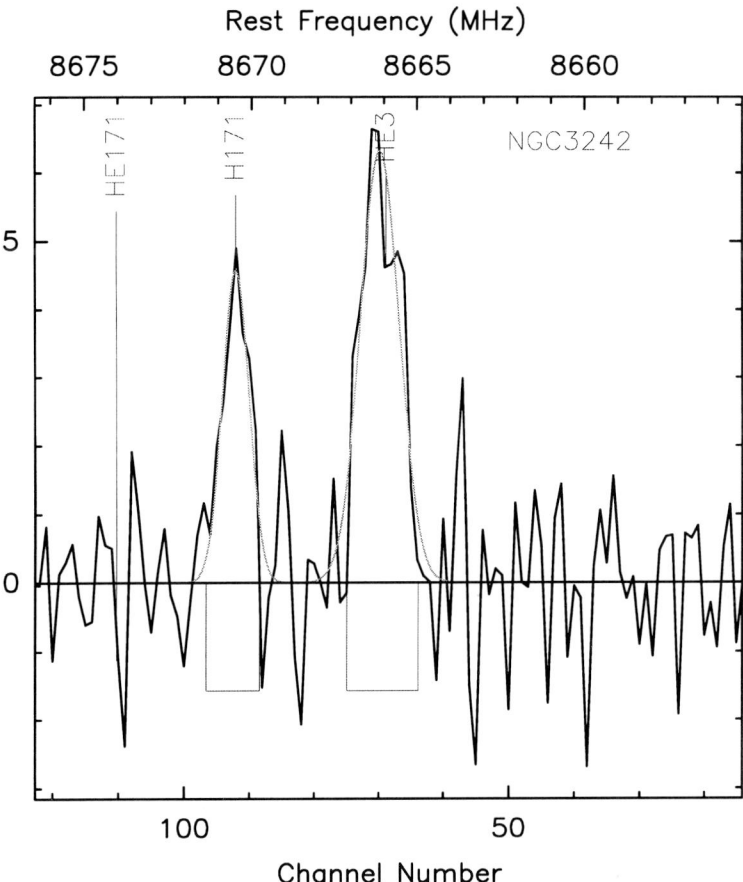

FIGURE 1. ^{3}He$^+$ spectrum for planetary nebula NGC 3242. The intensity scale is the main-beam brightness temperature in units of millikelvin. The spectrum has been smoothed to a velocity resolution of 6.8 km sec^{-1}. The vertical lines mark the expected positions of the He171η, H171η, and ^{3}He$^+$ lines. The dotted lines are Gaussian fits to the data.

1976). However, Galactic chemical evolution models which include stellar production of ^{3}He (e.g. Steigman & Tosi 1995) predict present day ^{3}He abundances which are significantly larger than abundances detected in Galactic H II regions (Bania et al. this volume). Although the data discussed here are consistent with standard stellar models, ^{3}He$^+$ must be detected in more PNe, or the limits in Table 1 must be substantially reduced, in order to make a case for or against the net production of ^{3}He by stars. Clearly we need a larger sample of PNe with measured ^{3}He abundances.

4. Future Measurements of ^{3}He$^+$ in Planetary Nebulae

We intend to continue observations of ^{3}He$^+$ in PNe with the 100-m telescope by expanding our source sample and by exploiting future improvements in the instrumentation. We have avoided PNe which are expected to have high-mass progenitors and have high nitrogen abundances because standard models predict that such objects should be depleted in ^{3}He. However, the details of stellar evolution in high-mass stars are complicated and mixing theories are not very well understood. Therefore, we plan to relax these criteria for a small sample of PNe which would produce "strong" ^{3}He$^+$ lines at the 100-m telescope for ^{3}He/H abundances on the order of 10^{-3}. The data discussed here are limited due to the severe problems with the instrumental spectral baselines of the 100-m telescope (Bania et al. 1993). The new autocorrelator, AK90, will provide 8,000 channels and should help to better define the spectral baseline. Moreover, metal spoilers will be installed on all four feed legs of the 100-m telescope to help remove ripples in the spectrum which are believed to be produced by reflections from the feed legs.

Because of the small angular size of many PNe we are also pursuing observations of ^{3}He$^+$ at the VLA which has a synthesised beam of 8 arcsec at the ^{3}He$^+$ line rest frequency in the D-configuration.

This research was supported by grant AST 91-21169 from the U.S. National Science Foundation. DSB was partially supported by a grant from the Deutscher Akademischer Austausch Dienst and the Max Planck Forschungpreis of the A. v. Humboldt-Stiftung.

REFERENCES

BALSER, D. S., BANIA, T. M., BROCKWAY, C. J., ROOD, R. T. & WILSON, T. L. 1994. *ApJ* **430**, 667

BANIA, T. M., ROOD, R. T. & WILSON, T. L. 1993. *Max-Planck-Institut für Radioastronomie Tech. Bericht* 75

CAHN, J. H., KALER, J. B. & STANGHELLINI, L. 1992. *Astron. Astrophys. Suppl.* **94**, 399

GALLI, D., PALLA, F., STRANIERO, O. & FERRINI, F. 1994. *ApJ* **432**, L101

HOGAN, C. J. 1994. *ApJ* **441**, L17

LINSKY, J. L., BROWN, A., GAYLEY, K., ANTHANASSIOS, D., SAVAGE, B. D., AYRES, T. R., LANDSMAN, W., SHORE, S. N. & HEAP, S. R. 1992. *ApJ* **402**, 694

ROOD, R. T., BANIA, T. M., & WILSON, T. L. 1992. *Nature* **355**, 618

ROOD, R. T., STEIGMAN, G. & TINSLEY, B. M. 1976. *ApJ* **207**, L57

SONGAILA, A., COWIE, L. L., HOGAN, C. J. & RUGERS, M. 1994. *Nature* **368**, 599

STEIGMAN, G. & TOSI, M. 1995. *ApJ* **453**, 173

WILSON, T. L. & ROOD, R. T. 1992. *ARA&A.* **32**, 191

High-resolution observations†‡ of young planetary nebulae

By M. BRYCE[1], A. PEDLAR[2],
T. MUXLOW[2], P. THOMASSON[2], G. MELLEMA[3],
J. MEABURN[1] AND I. BAINS[1]

[1]Department of Physics and Astronomy, The University of Manchester, Oxford Road, Manchester M13 9PL, UK

[2]University of Manchester, Nuffield Radio Astronomy Laboratories, Jodrell Bank, Cheshire SK11 9DL, UK

[3]Stockholm Observatory, S-13336 Saltsjöbaden, Sweden

High spatial resolution 6-cm radio continuum emission maps of two well-known young planetary nebulae (NGC 7027 and BD+30 3639) have been obtained by combining MERLIN and VLA observations. These maps, the first MERLIN images of planetary nebulae, are compared with WFPC2 images of the two nebulae obtained from the HST archive¶.

Planetary nebulae (PNe) are thought to evolve from low mass ($\sim 0.8 - 8 M_\odot$) stars, which shed their outer envelopes during the red giant and asymptotic giant branch phases. This cool, slow-moving, dense circumstellar material is ideal for molecular and dust formation. The remaining core star, destined to become a white dwarf, then emits a fast, tenuous, hot wind which sweeps up the red giant envelope. The stellar uv radiation field photoionises the expanding shells of gas and also, along with the nebular Lα and local shock regions, will tend to dissociate the dust. A huge variety of shapes and sizes of PNe are observed; many of these can be successfully reproduced within the generalised Interacting Stellar Winds models (eg Frank et al, 1993).

The advent of HST imagery means that for the first time, well-resolved optical images of young PNe can now be obtained, although the dust present in many of these objects can obscure the true extent of the ionised gas. High-resolution radio maps show directly this distribution and therefore can be used in conjunction with optical images to reveal the extent of dust in a given nebula.

BD+30 3639 was known from previous ground-based optical (eg Balick, 1987) and VLA radio (eg Basart & Daub, 1987) observations to be an elliptical shell ($\sim 6'' \times 4''$). This PN was observed at 6 cm using MERLIN in April 1995. The resulting map, (beam size 57×47 mas) showed that this shell is very rectangular. In order to improve and stabilise the map, VLA A-array data obtained in June 1995 were added to fill in the shorter baselines and improve the signal to noise ratio. The final map (Fig. 1a) has a synthesised beam size of 82×78 mas and a peak flux density of 7×10^{-4} Jy/beam ($T_B = 5450$ K). As expected from the original MERLIN map, this map shows a remarkably rectangular shape, with well resolved limbs containing small, resolved, bright features. Basart & Daub (1987) derived a maximum optical depth of $\tau = 0.4$ from their VLA observations of this nebula, measured in the same region of the northern limb as the peak flux density

---

† Based on observations made with MERLIN, a national facility operated by the University of Manchester on behalf of PPARC

‡ The National Radio Astronomy Observatory is operated by Associated Universities, Inc., under cooperative agreement with the National Science Foundation.

¶ Based on observations made with the NASA/ESA Hubble Space Telescope, obtained from the data archive at the Space Telescope Science Institute. STSCI is operated by the Association of Universities for Research in Astronomy, Inc. under NASA contract NAS 5-26555.

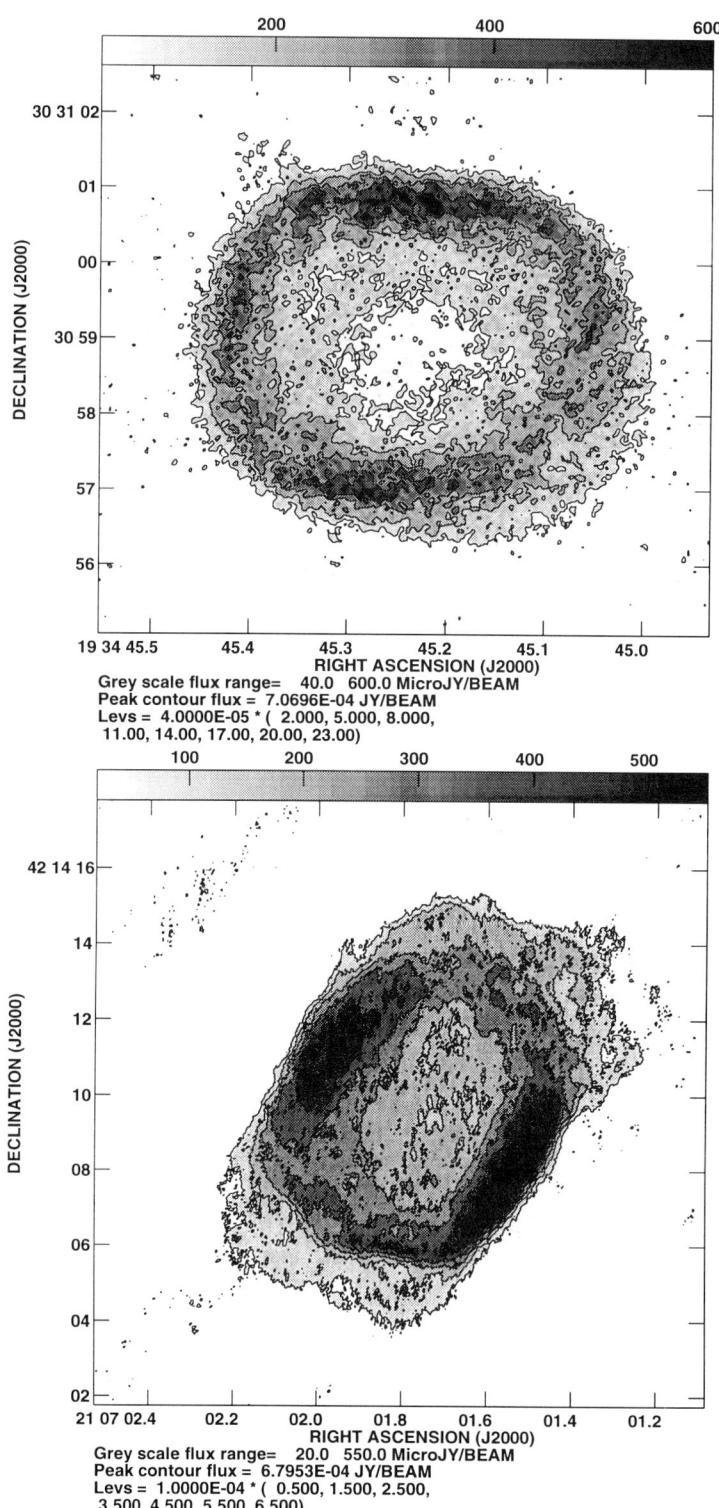

FIGURE 1. Radio maps of a) BD+30 3639 and b) NGC 7027, produced by combining 6-cm data from MERLIN and the VLA.

in the new map. Using their derived value of $T_e = 5690$ K at this point and the measured peak flux density per beam from the new, higher-resolution map, we find a value of $\tau = 3.7$ which shows that there are very compact, optically thick clumps within the basic elliptical structure of this nebula. The distance to BD+30 3639 has been measured from proper motions to be 1500 pc (Kawamura & Mason, 1996) and at this distance, the clumps have dimensions of $\leq 2 \times 10^{15}$ cm ($N_e \sim 4 \times 10^5$ cm^{-3})

Previous VLA observations of NGC 7027 again revealed an elliptical shell of radio emission (eg Basart & Daub, 1987) but ground-based optical observations show a very bright spot, offset NW of the central star (eg Balick, 1987), surrounded by faint, irregular, highly polarised emission (Walsh & Clegg, 1994). This PN, with a major dimension of $\sim 10''$ is rather large for MERLIN, even in wide field mode, and therefore we have combined the MERLIN observations with A- and B-array data from the VLA. The resulting map (Fig. 1b) has a beam size of 62×48 mas and reveals the extremely sharp edges of the two long limbs as well as various other smaller-scale features. The highest flux density measured in the new map is 6.8×10^{-4} Jy/beam ($T_B = 11\,500$ K). Basart & Daub (1987) derived an electron temperature of 13 200 K and an optical depth of 1.0 in this region of the SW limb. Taking their value of T_e and the peak flux density per beam from the new map, we derive an optical depth of 2.05, showing that as in BD+30 3639, the elliptical shell contains substantial sub-structure not resolved by previous observations.

HST images of BD+30 3639 appear very similar to the new radio map. To aid comparison, the radio map and Hβ image were convolved to a common beam size of 83 mas in diameter and then the Hβ image was scaled in brightness (somewhat arbitrarily) to the radio map using the small diagonal feature common to both maps in the western limb. Finally the radio map was divided by the scaled optical image; the resultant difference map is shown in Fig. 2a. The difference map reveals several compact, dusty regions (dark tones) which are particularly prevalent in the northern limb. These results confirm the findings of Lame et al (1995) who compared various HST images with VLA radio data. The radio and optical images of NGC 7027 are so different that it was not thought worthwhile to construct a difference map. However, a direct comparison (Fig. 2b) shows that despite the apparent differences, the radio contours follow closely the basic shape of the optical nebula and indeed several small-scale features such as the hook of emission protruding west of the bright optical spot appear in both images. The bipolar optical shell is largely obscured by patchy dust clouds yet must be closely associated with the (apparently morphologically different) ionised radio emission region.

We thank Nial Tanvir, (UK HST coordinating facility) for help in processing these images, staff at the VLA centre in Socorro and at Jodrell Bank for their help in reducing the radio data, Y. Terzian who provided the B-array data from NGC 7027 and J.P. Harrington who made available results from his HST images of BD+30 3639 (Lame et al, 1995) in advance of publication. This work relied largely on computing facilities provided by the STARLINK project.

REFERENCES

BALICK, B. 1987. *AJ* **94**, 671.
BASART, J.P. & DAUB, C.T. 1987. *ApJ* **317**, 412.
FRANK, A., BALICK, B., ICKE, V. & MELLEMA, G. 1993. *ApJ* **404**, L25.
KAWAMURA, J. & MASSON, C. 1996. *ApJ* **461**, 282.
LAME, N.J., HARRINGTON, P., BORKOWSKI, K. & WHITE, S. 1995. *BAAS* **187**, 8106L.
WALSH, J.R. & CLEGG, R.E.S. 1974. *MNRAS* **268**, L41

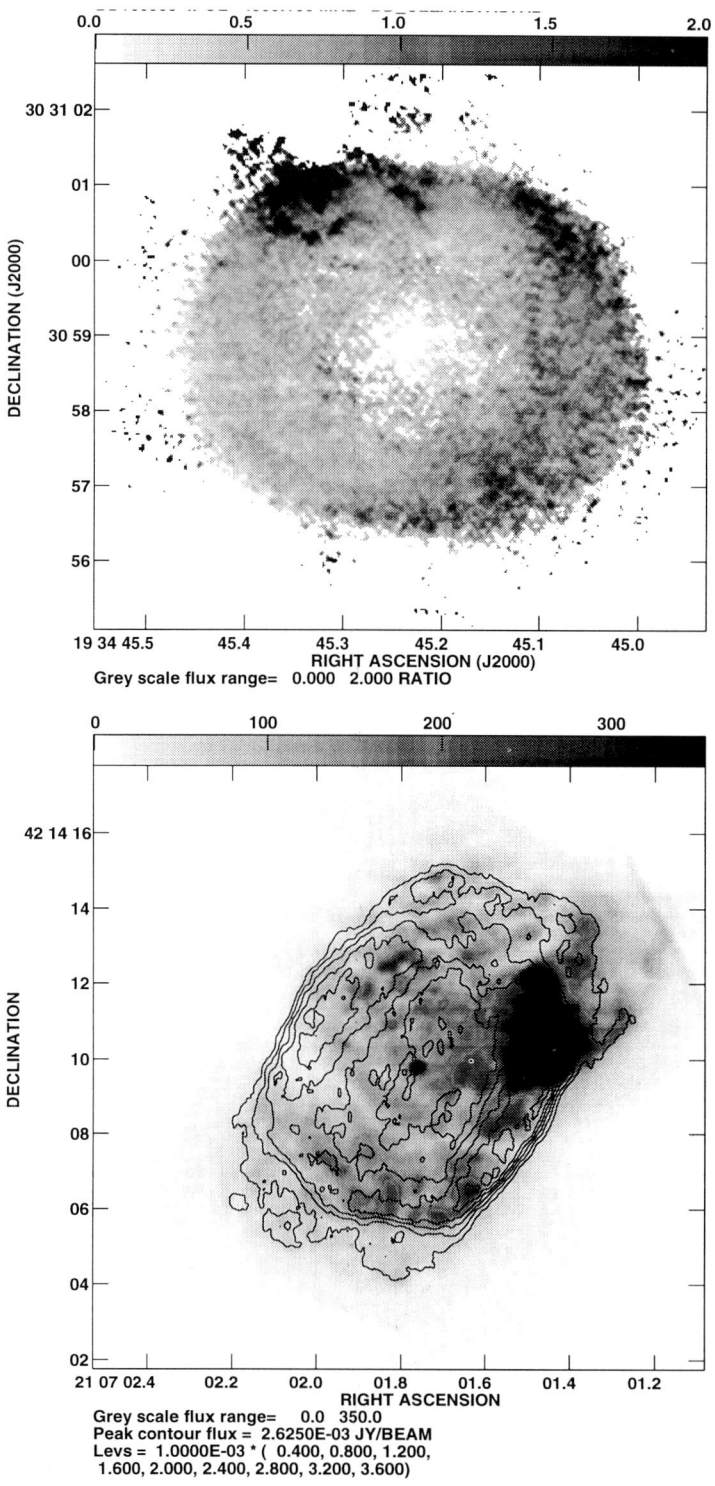

FIGURE 2. a) A difference map of BD+30 3639 made by dividing the radio map by an appropriately scaled HST Hβ image. Dark tones show an excess of radio emission indicating optical obscuration by dust. b) A greyscale HST image of NGC 7027 obtained through the F547M filter superimposed with contours from the radio map showing that although the two images at first seem very different, in fact the correspondence is quite close in many respects.

Continuum and polarization mapping of diffuse Galactic radio sources

By JOHN R. DICKEL[1]

[1]Astronomy Department, University of Illinois at Urbana-Champaign, Urbana IL 61801, USA

Many Galactic radio sources are very extended but show a wealth of fine filamentary, jet, and other structures. Often the full resolution of current aperture-synthesis instruments cannot be used to investigate the details of these features because of limited sensitivity. Many filaments are observed optically to break up into sub-filaments but their surface brightnesses are too low for detection at the highest radio resolution available. This is particularly true for the polarized emission from supernova remnants. A very large collecting area with many individual elements well spread over tens of km is necessary to obtain sufficient sensitivity to all the scale sizes present in these sources.

1. Introduction

There are three types of extended Galactic radio sources which present significant challenges to high-sensitivity and high-resolution radio observing. They are a) supernova remnants (SNRs) which generally have shell morphologies. These objects produce linearly polarized non-thermal synchrotron radiation with typical spectral indices near -0.5. b) HII regions with complex morphologies. Their continuum emission is thermal bremsstrahlung which is generally optically thin with an almost flat radio spectrum. c) Molecular clouds which produce emission from dust at millimetre wavelengths. The dust emission can have different spectral indices depending on the optical depth.

SNRs with their complex structure, possible spatially variable non-thermal spectra, and significant polarization offer the widest range of instrumental challenges so we shall emphasize the observations of those objects. The need for high resolution and small distortion of the polarized emission by Faraday rotation argue for observations at short wavelengths but the non-thermal spectra argue for long wavelengths. Thus most SNRs are observed primarily in a compromise region of the spectrum at cm wavelengths.

In the section 2 we shall discuss the scientific objectives of the study of SNRs and then in section 3 we will describe some of the requirements for and results of the observations. Section 4 will contain concluding remarks.

2. Scientific Objectives

The explosion of a supernova releases about 10^{51} ergs of energy into the interstellar medium (ISM) primarily in kinetic motion of expansion. As surrounding material is swept up, the expansion, which began at a speed of about 10,000 km/sec, will slow but remain highly supersonic so that a strong shock leads the ejecta. In addition, a reverse shock propagating back into the ejecta is created at the interface between the ejecta and the swept-up material. These shocks accelerate relativistic particles and will also compress the ambient magnetic field somewhat. In addition, the interface between the ejecta and the surrounding medium is Rayleigh-Taylor, and later Kelvin-Helmholtz unstable. These instabilities can stretch and greatly increase the magnetic field strength. Other shocks and instabilities can be created in clumps overrun by the expanding piston of the SNR. These conditions are responsible for the strong synchrotron emission observed from supernova remnants. As a remnant ages, compression of cool clumps in the hot, shocked medium of the SNR can also increase the synchrotron emissivity.

There are still many questions about the acceleration and amplification processes that remain unanswered. For example, the spectral index of the emission from an ensemble of particles undergoing shock acceleration should be -0.5 but most young SNRs which should have the most significant acceleration have steeper spectral indices and only the older ones are near -0.5 (Dickel 1991). Cowsik and Sarkar (1984) explained the spectral flattening by stochastic acceleration but the original injection spectrum and the location of the specific regions of acceleration remain uncertain. We clearly need to probe the detailed sites of particle acceleration and search for spectral index changes across individual remnants. Consistent observations repeated over several years are also proving useful to study the evolution of the emission from young remnants (Anderson & Rudnick 1995).

Because of the stretching processes, the magnetic fields in young remnants are generally radial but in the older ones they can be mixed or occasionally tangential to the shell. Other interesting phenomena associated with the magnetic field include the fact that the fractional polarization is generally low everywhere, typically less than 10% (Milne & Dickel 1975), and it does not change significantly with telescope resolution. There must be a lot of very fine scale structure in the magnetic field as well as cells on all scale sizes that can be resolved. Also, the strength and the organization of the magnetic fields do not track one another. Thus we often see the greatest polarized intensity where the total intensity is low.

Consideration of the large energy release leads to the question: Do supernovae control the structure of the ISM or does the ISM control the expansion of SNRs? The answer, of course, is some of each. In some cases, nearby molecular clouds will be dense enough to quickly stop the expansion (Dickel, Dickel, & Crutcher 1976) but the cloud is also heated with resultant dissociation and changes in chemistry. Much of the original energy ends up being deposited in heat and radiation. In other cases, SNRs can break into holes or tunnels in the interstellar medium and carry the energy off to vast distances from the original site of the supernova (e.g. Milne, Caswell, & Haynes 1993). Whether the SNR has blown the tunnel or merely connected to it remains uncertain. More polarimetry to see the details of the magnetic field structure of these extended regions and pressure measurements with the aid of X-ray observations should help to better establish the physical cause of these apparent breakouts. In either case this process should tend to enhance the differences between the cold, warm, and hot phases of the ISM.

3. Observing considerations

3.1. *Structure and morphology*

3.1.1. *Characteristics*

SNRs and other galactic sources are often very extended (tens of arcmin) and some of the breakout features extend even further. The same sources have fine structure on a scale of < 1 arcsec. Typically over 9/10 of the integrated flux density is from smoothly distributed emission (e.g. Mufson et al. 1986). This smooth emission is faint and requires high sensitivity to detect. Spectral index variations across individual SNRs are theoretically expected and observationally suggested, but remain to be firmly established.

3.1.2. Requirements and implications

The field of view must cover about 1° but the spatial frequency response must extend over a range of about 5000/1. These specifications are important to ensure that both the smooth and fine-scale components are included for the accurate surface brightness measurements needed in spectral index studies. Good scaled arrays at several wavelengths are also important for good spectra.

Many small antenna elements, very efficient procedures for large mosaics, and a large single dish associated with an array can be used to help satisfy the above needs. Bandwidth smearing becomes important for the large fields of view needed and so we also need many narrow frequency channels with $\Delta\nu/\nu \approx 1/5000$. The total bandwidth is ultimately limited by the ability to do multifrequency synthesis in the presence of possible spectral index variations and changes in instrumental parameters with frequency. This limits the total $\Delta\nu$ to about 0.1 or 0.2 ν.

3.2. Brightness measurements

3.2.1. Characteristics

The filaments observed in SNRs often appear to break up into subfilaments and so individual features can become very faint. The surface brightness is similar on all spatial scales. The associated breakout or tunnel features so far discovered are brighter than 1/100 of the associated SNR shells. There is seldom a strong enough small-diameter source in the field for self calibration but confusion noise can limit the system sensitivity.

3.2.2. Requirements and Implications

Comparable high sensitivity is needed for all spatial scales so a well spread out array with similar numbers of long and short baselines is appropriate. Dynamic range is not a major concern and values of about 200/1 should be sufficient. Accurate flux-density calibration is essential for spectral measurements, particularly at high frequencies where the earth's atmosphere can rapidly change the amplitude and phase of signals. Atmospheric monitoring and reliable information on good, nearby calibration sources must be maintained and the calibrators must be observed regularly throughout an observing session. Finally, we anticipate that an rms sensitivity of about 3 μJy is needed for the next significant advance in detecting fine-scale features in SNRs.

3.3. Polarization measurements

3.3.1. Characteristics

The polarized emission from SNRs often appears quite cellular and there is no good match of the polarized and total surface brightnesses. They often peak in different places and appear to have different spatial frequency structures. Current resolution of about 1 arcsec has not yet reached the lower limit to the size of the polarization cells. Greater sensitivity will be needed to detect the polarization on those small scales.

The magnetoionic medium both within SNRs and along the line of sight through the Galaxy causes Faraday rotation which changes the observed position angles of the electric vectors of the polarized emission. The rotation depends on the square of the wavelength and values can reach about 500 rad/m^2. This Faraday rotation can also cause apparent depolarization at long wavelengths as the emission from the back of the SNR can be rotated with respect to that from the front and the summation of the vectors will be reduced below the intrinsic value.

3.3.2. Requirements and implications

Because the polarized intensity is generally less than 10% of the total intensity, an rms noise level of about 0.1 μJy is needed for complete polarimetry. It is generally possible to reach the theoretical noise level in the polarization measurements because the confusing polarized signals are also weak and thus below the noise.

Multiple wavelengths are needed to measure the Faraday rotation and determine any $n\pi$ rotations between individual pairs of wavelengths. Depolarization is dependent on the Faraday rotation and thus also a function of wavelength^2. Therefore evaluation of the depolarization also takes both long and short wavelengths. At the longer wavelengths bandwidth depolarization can be significant because the vectors at each end of the observing band are rotated by different amounts. This limits the available bandwidth at 20 cm wavelength to about 50 MHz if we want the polarized signal to be above 0.9 of the true value for a rotation measure of 500 rad/m^2. Shorter wavelengths, of course, can have larger bandwidths.

4. Concluding remarks

The high sensitivity (rms noise $<$ 1μJy) needed to make a significant advance in our understanding of the radio emission from supernova remnants will require very sensitive receivers and large apertures. It is not possible to use large bandwidths to gain sensitivity because the presence of possible spectral index variations within individual SNRs limits the ability to do multi-frequency synthesis over an extended frequency interval. In addition, depolarization across large bandwidths will hurt polarimetry.

While a large total collecting area is necessary, the individual elements of the telescope must cover a large field of view. Good sensitivity is needed for smooth structure as well as fine-scale features so the array can be sparsely populated to allow similar coverage of all spatial scales.

Many observed properties of SNRs cannot yet be explained theoretically but as advances in computer technology quickly increase the speed and memory of the theorists machines, they will soon be able to make accurate 3-dimensional models with better resolution and sensitivity than available to the observers. They will then be demanding answers which can be supplied only by the next generation of radio telescopes.

I thank D. K. Milne for stimulus and exchange of ideas over many years. The research has been supported in part by the U. S. National Science Foundation, NASA, and the Campus Honors Program of the University of Illinois at Urbana-Champaign.

REFERENCES

ANDERSON, M. & RUDNICK, L. 1995. *ApJ* **441**, 307
COWSIK, R. & SARKAR, S. 1984. *MNRAS* **207**, 745
DICKEL, J. R. 1991. In: *Supernovae* (ed. Woosley, S. E.). pp. 675–678. Springer.
DICKEL, J. R., DICKEL, H. R., & CRUTCHER, R. M. 1976. *PASP* **88**, 840
MILNE, D. K., CASWELL, J. L., & HAYNES, R. F. 1993. *MNRAS* **264**, 853
MILNE, D. K. & DICKEL, J. R. 1975. *Aust. J. Physics* **28**, 209
MUFSON, S. L., MCCOLLOUGH, M. L., DICKEL, J. R., PETRE, R., WHITE, R., & CHEVALIER, R. A. 1986. *AJ* **92**, 1349

High-sensitivity observations of circumstellar masers

By R. J. COHEN

University of Manchester, Nuffield Radio Astronomy Laboratories, Jodrell Bank, Macclesfield, Cheshire SK11 9DL, U.K.

Recent progress in the study of circumstellar masers is reviewed. With improvements in receiver technology many new maser transitions have been identified in recent years. Some are particularly valuable as they provide unique information on the hot dense regions of the circumstellar envelope where mass-loss originates. High-sensitivity imaging by MERLIN and VLBI is now providing our first pictures of mass-loss in real time, through proper-motion measurements of OH, H_2O and SiO masers. There are several indications of bipolar symmetry in the mass-loss. Possible reasons for the bipolarity include the influence of magnetic fields. The first measurements of circumstellar magnetic field structure using OH maser polarization fall tantalizingly short of providing direct evidence for this connection. Prospects for high-sensitivity observations are also discussed. Deep galactic surveys for circumstellar masers are likely to yield over 10^4 sources, and lead to fundamental contributions to our understanding of stellar evolution and galactic dynamics.

1. Introduction

Circumstellar masers are important signposts for the late stages of stellar evolution associated with mass-loss. The maser stars are usually Miras or semi-regular variables on their way to becoming planetary nebulae, or supergiants on their way to becoming Wolf-Rayet stars or supernovae. They have periods of typically 1–5 yr, and mass-loss rates of typically 10^{-8} to 10^{-4} $M_\odot$ yr^{-1}. Masers offer a high–precision tool with which to study the mass-loss process at high angular resolution and to investigate the kinematics and dynamics of the circumstellar envelope. They also trace stages in the chemical evolution of the envelope, and in the case of SiO masers they may trace dust formation.

Masers are an interesting and challenging phenomenon in their own right. The circumstellar masers are in many ways the best understood, and so provide an excellent opportunity to test maser models. Auxiliary data from thermally excited molecular lines, infrared and in some cases optical data, together help to define the physical conditions in which the masers operate. This is in stark contrast to the situation with masers in star-forming regions, where many of the basic parameters cannot be measured directly.

Above all masers are bright, which makes them valuable astrophysical tools even when the detailed physics is uncertain. Circumstellar masers are already used for galactic distance measurements, and in galactic dynamics. They give a unique probe of circumstellar magnetic fields. Finally as bright sources they are generally useful as phase reference sources and calibrator sources for astronomers whose main interest lies elsewhere.

I will review recent high–sensitivity observations of circumstellar masers. The field is a rapidly expanding one, so I will concentrate on developments in the last few years. Earlier work is covered in reviews by Bowers (1993) and Cohen (1992). I will also try to indicate some directions which future research might take with improving sensitivity, in particular what we might expect from deep surveys, searches for new maser lines, and from high–sensitivity imaging.

2. OH–IR sources

I would like to begin with some historical milestones. The first circumstellar masers to be discovered were the OH-IR sources, which were found by Wilson & Barrett (1968) in a targetted search of infrared sources. They radiate in the OH 1612-MHz line, and have a twin-peaked spectrum characteristic of a uniformly expanding shell of OH. The midpoint of the line profile indicates the stellar velocity, while the separation of the peaks indicates twice the expansion velocity. The shell geometry arises from photodissociation of water molecules in the expanding envelope by external ultraviolet radiation (Goldreich & Scoville 1976). For a shell expanding at a nearly constant velocity the masers are radially beamed, which accounts for the twin-peaked structure of the OH 1612-MHz profile (Olnon 1977; Reid et al. 1977). The stellar variability led to the identification of a far-infrared pump for the maser (Elitzur, Goldreich & Scoville 1976). The masers vary in phase with the infrared radiation, but because of the light travel time across the shell we see a phase-lag between the near side (blue-shifted OH peak) and the far side (red-shifted OH peak), which was first measured by Schultz, Sherwood & Winnberg (1978).

A key stage in the study of OH–IR sources was the first imaging of an OH shell at 1612 MHz (Booth et al. 1980). This was the first MERLIN spectral line data cube, made with only three telescopes and two baselines. It showed the shell structure beautifully, in fact it seemed almost too good to be true. Some of the authors are here today, and they will remember the long earnest discussions they had with Tim Cornwell concerning possible mapping artefacts. After six months of deliberation they were convinced and published, only just in time. Two months later came Bowers, Johnston & Spencer (1980), with the first VLA images of OH 1612-MHz sources.

Comparing the angular diameter and phase-lag linear diameter give us a way to estimate the distance to the OH–IR source (Baud 1981). There are large programmes currently underway to exploit this technique. The best measurements yield distances accurate to 5% (West et al. 1992). There are hopes that the technique may one day give the Galactic centre distance to this accuracy or better. First however there are several issues to be resolved. The results obtained from the phase-lag monitoring can depend on details of the analysis procedure (van Langevelde, van der Heiden & van Schooneveld 1990; David, Etoka & Le Squeren 1996). For the most distant OH–IR sources interstellar scattering can be a problem, as it smears the images of the shells (van Langevelde et al. 1992). Finally there are fundamental questions about the assumption of spherical symmetry, and the effects of maser beaming (Moran 1993), since the method compares the front-to-back emissivity of the shell determined from phase-lag monitoring with the side-to-side emissivity determined from imaging. High-sensitivity imaging of the weak emission near the stellar velocity has a crucial role to play in this respect.

Finally a personal note. I began my own work on circumstellar masers and OH–IR sources in 1981. The following year the first GLONASS satellite was launched. Interference from GLONASS satellites has hampered work on OH–IR sources ever since. Thus I became interested in masers and in radio interference at much the same time.

3. Maser surveys

Up until now surveys for circumstellar masers have mainly been targetted searches, for example searches of visible stars or infrared sources. In particular the OH–IR sources are readily identifiable by their infrared colours (Olnon et al. 1984). Following the publication of the IRAS database there were many very successful searches for 1612-

MHz emission from suitable IRAS sources. Hundreds of new sources were discovered (e.g. Engels et al. 1984; Eder, Lewis & Terzian 1988; Sivagnanam et al. 1990; te Lintel Hekkert et al. 1991). This seems impressive until we realize the scale of the whole job that is to be done. The IRAS Point Source Catalogue contains tens of thousands of candidate sources. It will take many years to search them all to the appropriate sensitivity, and the final yield will be not hundreds but thousands and perhaps tens of thousands of circumstellar masers.

We can get a preview of what these new surveys might bring by looking at the results of the Arcetri survey of known 22-GHz H_2O masers, which have been analysed by Palagi et al. (1993). They divided their detections into star-forming regions and stars. The star-forming regions span more than six orders of magnitude in luminosity. The sensitivity of the survey is sufficient to detect a source like Orion-KL anywhere in the Galaxy, but for the lower luminosity sources our view is very much biassed to the solar neighbourhood (their Figure 6). For the stars this bias is even stronger. The great majority of circumstellar H_2O masers lie within 1 kpc or so of the Sun. At larger distances we see only the few sources with exceptionally high luminosities. It is clear then that deeper surveys will yield enormous numbers of new circumstellar maser sources at ever larger galactic distances. To find them it may be best to simply carry out systematic searches along the galactic plane. The next generation of survey could go 100 times deeper than the Arcetri survey right now. Given problems of confusion and identification it may be best to survey using an interferometer.

4. New masers

One of the major growth areas in recent years has been the detection of new maser lines. By and large these have been new maser transitions of previously known maser molecules such as SiO and H_2O. For example Cernicharo, Bujarrabal & Santaren (1993) made first detections of four SiO maser lines and presented near–simultaneous SiO maser spectra of the supergiant VY CMa in rotational transitions from J=1–0 up to J=6–5 in the vibrational states v=1, v=2, v=3 and v=4. Theoretical work on SiO excitation suggests that masers should also occur in higher rotational states, and some of these are now being detected up to J=7–6 and even higher (Gray et al. 1995, and private communication). Such is the rate of progress that systematic studies are also being made of SiO in many sources and many transitions. Cho et al. (1996) observed over 100 sources in 6 maser transitions of SiO, including the rare isotopic species ^{29}SiO. Discoveries of new H_2O masers have been dominated by the CfA group, whose latest work includes the discovery of strong maser emission at 658 GHz from vibrationally excited H_2O (Menten & Young 1995). Like several of the other recently discovered lines of H_2O, the 658-GHz line is as luminous and as widespread as the well–known 22-GHz maser.

It is a sign of maturity in the subject that many of the new maser lines were predicted as masers on the basis of numerical modelling of maser excitation. Nonetheless we are entitled to ask, does all the extra information help? One of the gains has been that many of the new masers trace hotter regions of the circumstellar envelope than have previously been available for study. The 658-GHz line of H_2O for example has an excitation temperature of over 2300 K, and the v=4 lines of SiO have excitation temperatures of around 7000 K. Thus they probe warm inner regions of the envelope where mass-loss originates. The modellers sometimes hold out the prospect that maser line–ratios may be useful in bracketting the physical conditions in these astronomically very interesting regions. I would just like to add a word of caution on this, and share my doubts with you. We recently conducted a study of three H_2O maser lines, at 22, 321 and 325 GHz (Yates & Cohen 1996). The lines were observed at 4 epochs spaced over a year. At each epoch the observations were as close to simultaneous as we could arrange, usually within a day of each other, so as to obtain reliable snapshots of the maser action. All 9 sources we looked at showed strong variations, but unfortunately the variations did not correlate at all well, and in some cases they were completely uncorrelated. There is no simple model to explain the different variability patterns. Some of the differences may be accounted for by anisotropy of the maser radiation and by saturation effects. Maser beams moving in different directions through the envelope will be competing for the finite pumping resources. Thus for example a powerful maser beam moving across our line of sight could control the flux of the beam that we happen to be observing. How reliably could we deduce physical conditions from line ratios in circumstances like this? Until the maser models include effects like these we will not be in a position to answer such a question.

5. Developments in imaging

It is ten years since Chapman & Cohen (1986) published their paper on the imaging of circumstellar masers, which showed clearly for the first time the hierarchy of circumstellar masers at different distances from the star. Progress since then has been in three main directions. Firstly the extension of MERLIN and VLBI techniques has led to full synthesis imaging of the 22-GHz H_2O and 43-GHz SiO masers (shown as unresolved spots by Chapman & Cohen). Secondly the improved precision of the measurements has enabled proper motions to be detected (mass-loss observed in real time). Thirdly the implementation of polarization mapping has led to the first attempts at mapping the circumstellar magnetic field.

The first images of SiO masers have come from the VLBA. Diamond et al. (1994) published maps of the SiO masers around TXCam and UHer. In each case the masers lie in a ring at 2–4 stellar radii, in the extended atmosphere of the star, but inside the dust-formation point. The ring structure suggests an ordered outflow, with tangential beaming of the masers (as expected in a region of strong radial acceleration). Individual maser spots were resolved. VLBA observations of the supergiant VX Sgr by Greenhill et al. (1995) show similar properties, with a ring-like distribution at 1.3 stellar radii. Concurrent infrared data confirm that the masers lie well inside the dust-formation radius, which was measured to be 6 stellar radii. The arrangement of maser emission suggests dense velocity–coherent structures with characteristic sizes of 0.4 AU. It is worth noting that these maps contain only one quarter of the total flux, which occurs in high brightness spots of milliarcsecond extent. The rest presumably comes from a more extended halo around the bright cores.

Recent images of H_2O masers by the VLA and MERLIN have achieved the highest sensitivity ever, with rms noise levels of 10 mJy/beam or less (Bowers & Johnston 1994; Richards, Cohen & Yates 1995). The MERLIN data resolve the structure of individual maser spots and allow brightness temperature measurements. Data for RTVir are shown in Figure 1, taken from Bains (1995). The plot shows brightness temperatures calculated for each spectral channel across each maser feature. The values range from 10^{10} K down to 10^5 K (the detection level), in good agreement with the predictions by Cooke & Elitzur (1985). The distribution of masers in this source is shown schematically in Figure 2, where each symbol indicates the position of the emission centroid in a particular spectral channel. The distribution is surprising in two respects. Firstly the individual maser features appear to be elongated filaments, barely resolved by the MERLIN beam. Their position shifts systematically with velocity, and they appear in Figure 2 to trace short

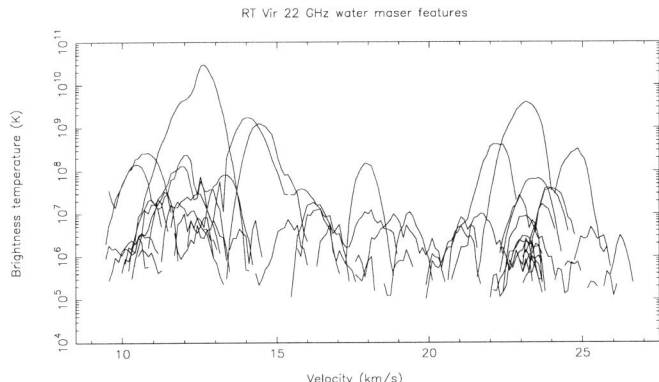

FIGURE 1. Brightness temperature profiles for 22-GHz H_2O maser features in the circumstellar envelope of RTVir (Bains 1995).

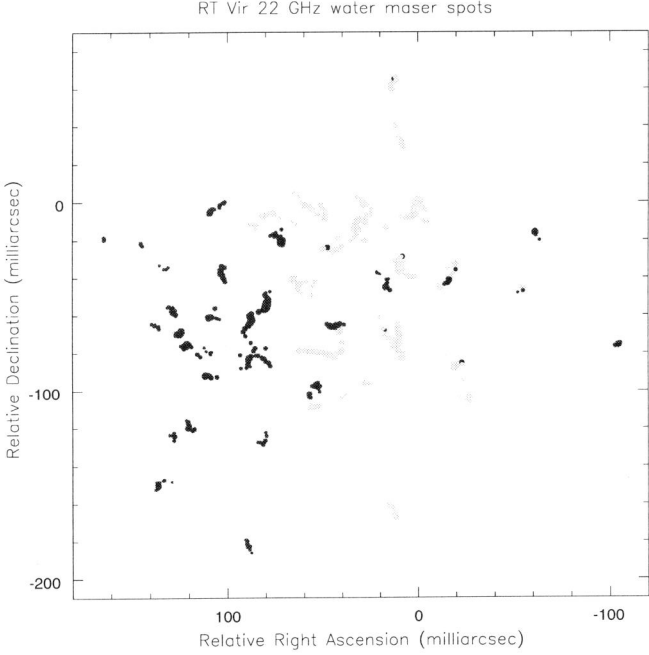

FIGURE 2. Distribution of 22-GHz H_2O maser spots in the circumstellar envelope of RTVir. Black dots indicate red–shifted masers and grey dots indicate blue–shifted masers (Adapted from Bains 1995).

tracks 10–20 mas in length. Secondly there is a systematic displacement of the red–shifted masers (black) to the East and the blue–shifted masers (grey) to the West. This suggests that the source may have an intrinsic bipolarity, appearing roughly spherical only because it is seen nearly end–on.

The first proper motions of circumstellar OH and H_2O masers have been measured using MERLIN. Chapman, Cohen & Saikia (1991) measured a ring-like distribution of OH 1665-MHz masers around UOri, which had expansion proper motions of 5.4 mas yr^{-1} over four years. The expansion rate matches the expansion deduced from the radial velocities, for the assumed distance of 260 pc. Yates & Cohen (1994) measured a general expansion of H$_2$O maser features around IKTau, at a rate of 6 mas yr^{-1}, while Richards, Yates & Cohen (poster at this conference, and MNRAS in press), found proper motions of 1.8 mas yr^{-1} of widely spaced H$_2$O maser features on opposite sides of NMLCyg, indicating a bipolar outflow. Further H$_2$O and SiO proper motion studies are underway using the VLBA (Marvel & Diamond, in press; Diamond & Kemball, in press).

Proper motions have also been measured for OH 1612-MHz masers around the supergiant IRC+10420 by Kemball (1992) using the European VLBI Network (EVN). Data taken in 1982 and 1987 indicate an elliptical distribution of the blue-shifted masers, with an ellipsoidal outflow. By comparing the proper motion data with an ellipsoidal model by Nedoluha & Bowers (1992), Kemball finds a distance of 3.3 kpc to the source. The general picture which emerges from all these proper motion results is of clumpy mass-loss, with discrete maser clouds moving through a (stationary?) region where maser action is excited.

The EVN measurements of IRC+10420 recorded signals of both circular polarizations, and Kemball was able to carry out the first VLBI spectropolarimetry. Preliminary results of this analysis have been reported by Kemball & Diamond (1993). The more complete analysis in Kemball (1992) indicates two likely Zeeman pairs with magnetic field strengths of -1.1 and +2.7 mG, and numerous features with significant elliptical polarization. There is no clear relation between the magnetic field direction and the axis of the ellipsoidal outflow. Nevertheless these results are a pointer to the more detailed information which should become available in the years ahead using modern correlators which can provide full velocity resolution to 0.1 km s^{-1} or better.

6. Future prospects

Some of the goals for the coming years have already been covered earlier in this review. High-resolution imaging at multiple epochs is expected to address many unanswered questions about mass-loss (Section 5). We can look forward before too long to movies showing mass-loss in different zones, traced by the different circumstellar masers. These data will also contribute to our understanding of the physics of circumstellar masers, and the nature of maser variability.

Deep galactic surveys for circumstellar masers promise to increase the numbers of known sources by one or even two orders of magnitude, as the full galactic distribution of the long-period variables is revealed. The work by Lindqvist et al. (1992) on OH–IR sources in the galactic centre has demonstrated the power of interferometric techniques for this kind of survey. Distance measurements of the distant OH–IR sources using the phase-lag technique should make a major contribution to the determination of an accurate galactic distance scale (cf. van Langevelde et al. 1993). In addition it may be possible to measure orbital proper motions for some of the sources. This would allow new kinds of investigation into galactic dynamics, extending the kind of work which has been possible up to now using only radial velocities, for example in the galactic centre region (Lindqvist, Habing & Winnberg 1992). Finally it is not too optimistic to consider the extension of these techniques beyond our Galaxy. The detection of OH and SiO maser emission from OH–IR sources in the Magellanic Clouds is an indication that this could happen before too long (Wood et al. 1992; van Loon et al. 1996).

It is a pleasure to thank my students, with whom I have enjoyed learning about masers.

REFERENCES

BAINS, I. 1995. The circumstellar envelope of RT Virginis. MSc thesis, The University of Manchester, Manchester, U.K.

BAUD, B. 1981. *ApJ* **250**, L79

BOOTH, R. S., KUS, A. J., NORRIS, R. P. & PORTER, N. D. 1980. *Nature* **290**, 382-384.

BOWERS, P. F. 1993. Recent studies of circumstellar masers. In *Astrophysical Masers* (ed. A. W. Clegg & G. E. Nedoluha), pp. 321-329. Springer-Verlag, Berlin.

BOWERS, P. F. & JOHNSTON, K. J. 1994. *ApJS* **92**, 189

BOWERS, P. F., JOHNSTON, K. J. & SPENCER, J. H. 1980. *Nature* **291**, 382-385.

CERNICHARO, J., BUJARRABAL, V. & SANTAREN, J. L. 1993. *ApJ* **407**, L33

CHAPMAN, J. M. & COHEN, R. J. 1986. *MNRAS* **220**, 513

CHAPMAN, J. M., COHEN, R. J. & SAIKIA, D. J. 1991. *MNRAS* **249**, 227

CHO, S. -H., KAIFU, N. & UKITA, N. 1996. *A&AS* **115**, 117

COHEN, R. J. 1992. Compact Maser Sources. In *Observational Astrophysics* (ed. R. E. White), pp. 65-131. Institute of Physics Publishing, Bristol.

COOKE, B. & ELITZUR, M. 1985. *ApJ* **295**, 175

DAVID, P., ETOKA, S. & LE SQUEREN, A. M. 1996. *A&AS* **115**, 387

DIAMOND, P. J. & KEMBALL, A. J. 1996. Structural Changes of SiO Maser Emission from Stars. In *Radio Emission from the Stars and the Sun* (ed. A. R. Taylor & J. M. Paredes), in press. Publ. Astron. Soc. Pacific, Conference Series.

DIAMOND, P. J., KEMBALL, A. J., JUNOR, W., ZENSUS, A., BENSON, J. & DHAWAN, V. 1994. *ApJ* **430**, L61

EDER, J., LEWIS, B. M. & TERZIAN, Y. 1988. *ApJS* **66**, 183

ELITZUR, M., GOLDREICH, P. & SCOVILLE, N. 1976. *ApJ* **205**, 384

ENGELS, D., HABING, H. J., OLNON, F. M., SCHMID-BURGK, J. & WALMSLEY, C. M. 1976. *A&A* **140**, L9

GRAY, M. D., IVISON, R. J., YATES, J. A., HUMPHREYS, E. M. L., HALL, P. J. & FIELD, D. 1995. *MNRAS* **277**, L67

GREENHILL, L. J., COLOMER, F., MORAN, J. M., BACKER, D. C., DANCHI, W. C. & BESTER, M. 1995. *ApJ* **449**, 365

KEMBALL, A. J. 1992. Data reduction techniques for very long baseline interferometric spectropolarimetry. PhD thesis, Rhodes University, Grahamstown.

KEMBALL, A. J. & DIAMOND, P. J. 1993. VLBI Spectral Line Polarization Observations of the OH Masers in the Supergiant IRC+10420. In *Astrophysical Masers* (ed. A. W. Clegg & G. E. Nedoluha), pp. 369-372. Springer-Verlag, Berlin.

LINDQVIST, M., HABING, H. J. & WINNBERG, A. 1992. *A&A* **259**, 118

LINDQVIST, M., WINNBERG, A., HABING, H. J. & MATTHEWS, H. E. 1992. *A&AS* **92**, 62

MARVEL, K. & DIAMOND, P. J. 1996 First Proper Motion Measurements of Circumstellar Water Masers. In *Radio Emission from the Stars and the Sun* (ed. A. R. Taylor & J. M. Paredes), in press. Publ. Astron. Soc. Pacific, Conference Series.

MENTEN, K. M. & YOUNG, K. 1995. *ApJ* **450**, L67

MORAN, J. M. 1993. Distance measurements through observations of masers. In *Sub-arcsecond Radio Astronomy* (ed. R. J. Davis & R. S. Booth), pp. 62-67. Cambridge University Press, Cambridge U.K.

NEDOLUHA, G. E. & BOWERS, P. F. 1992. *ApJ* **392**, 249

OLNON, F. M. 1977. Shells around stars. PhD thesis, Rijksuniversiteit te Leiden, Leiden, The Netherlands.

Palagi, F., Cesaroni, R., Comoretto, G., Felli, M., & Natale, V. 1993. *A&AS* **101**, 153

Reid, M. J., Muhleman, D., Moran, J. M., Johnston, K. J. & Schwartz, P. R. 1977. *Astrphys. J.* **214**, 60

Richards, A. M. S., Cohen, R. J. & Yates, J. A. 1995. The Circumstellar Envelope of S Persei. In *Circumstellar Matter 1994* (ed. G. D. Watt & P. M. Williams), pp. 545 Kluwer Academic Publishers, Dordrecht, The Netherlands.

Schultz, G. V., Sherwood, W. A. & Winnberg, A. 1978. *A&A* **63**, L5

Sivagnanam, P., Braz, M. A., Le Aqueren, A. M. & Tran Minh, F. 1988. *A&A* **233**, 112

te Lintel Hekkert, P., Caswell, J. L., Habing, H. J., Haynes, R.F., & Norris, R. P. 1991. *A&AS* **90**, 327

Van Langevelde, H. J., Frail, D. A., Cordes, J. M. & Diamond, P. J. 1992. *ApJ* **396**, 686

Van Langevelde, H. J., Janssens, A. M., Goss, W. M., Habing, H. J. & Winnberg, A. 1993. *A&AS* **101**, 109

van Langevelde, H. J., van der Heiden, R. & van Schooneveld, C. 1990. *A&A* **239**, 193

van Loon, J. Th., Zijlstra, A. A., Bujarrabal, V. & Nyman, L.-A. 1996. *A&A* **306**, L29

West, M. E., Gaylard, M. J., Combrinck, W. L., Cohen, R. J. & Shepherd, M. C. 1992. Mira and OH/IR maser characteristics and distance determination. In *Variable Stars and Galaxies* (ed. B. Warner), ASP Conference Series, Vol. **30** pp. 277-283. Astronomical Society of the Pacific, San Francisco.

Wilson, W. J., & Barrett, A. H. 1968. *Science* **161**, 778

Wood, P. R., Whiteoak, J. B., Hughes, S. M. G., Bessell, M. S., Gardner, F. F. & Hyland, A. R. 1992. *ApJ* **397**, 552

Yates, J. A. & Cohen, R. J. 1994. *MNRAS* **270**, 958

Yates, J. A. & Cohen, R. J. 1996. *MNRAS* **278**, 655

Megamasers

By WILLEM A. BAAN[1]

[1] Arecibo Observatory, Cornell University, Arecibo, PR 00613, USA

The characteristics of the prominent classes of OH and H_2O megamasers reveal two parent populations: FIR-luminous and dusty merger galaxies for OH and relatively dust-free AGN/Seyfert galaxies for H_2O. The 60-μm FIR luminosity functions and the effective opening angles of the (amplifying) molecular structures are very different for the two classes. For the H_2O megamasers the population inversion results from X-ray heating and collisional pumping of the torus molecular gas. The H_2O megamaser luminosity varies with both the nuclear radio and X-ray luminosities and varies linearly with their product. Partial amplification appears to play a role in both types of sources.

1. Introduction

The history of megamaser activity is relatively short but within this short timespan the megamasers have provided much information about nuclear activity and the physics of galactic nuclei. Extragalactic watermaser sources were found in 1979 with the discovery of the masers in the LMC. The first true megamaser was found in NGC 3079 in 1984. Extragalactic CH megamasers have been known since 1980 and currently about 10 sources are known. The first OH megamaser was found in 1982; presently, close to 60 megamasers are known. Formaldehyde megamasers were found in 1986; currently, there are 12 sources.

The emission of the megamasers is found to be non-isotropic because the molecular emissions result from some amplification scheme close to the plane of the molecular disk in the nuclear region. The opening angle of the axi-symmetric (disk-like) emission pattern may vary with each molecule. Hydroxyl and formaldehyde megamasers show a quadratic relation between the molecular-line luminosity and the 60 micron FIR luminosity. This quadratic relation may be explained by amplification of background continuum by foreground inverted molecular gas. The pumping agent for the OH constituent of the molecular gas is the FIR radiation field. The formaldehyde inversion is not well understood; it is likely to be related to the radio and FIR radiation fields. On the other hand, the watermaser activity is due to collisional processes in the dense medium of a molecular torus resulting from exposure to shocks or, more likely, to X-rays from the nucleus.

In the following we will compare the characteristics of the OH and H_2O megamasers. These two types of sources have quite different characteristics and they emphasize different types of nuclear activity. An earlier review of megamaser properties has been presented by Henkel, Baan & Mauersberger (1991).

2. OH Megamasers

The characteristics of OH megamasers have been discussed extensively in the past (Baan 1989, 1991; Henkel et al. 1991). The molecular emissions in OH galaxies provide new means for looking inside the nuclear regions and reveal the dynamics of the system. The prototype galaxy Arp 220 (IC 4553) has very prominent OH and H_2CO megamaser emission centred on the two merging (radio) nuclei (Baan & Haschick 1995). VLBI studies with EVN reveal three high-brightness OH components: two components coincide with the eastern nucleus and one with the western nucleus (Diamond et al. 1989). However, these three OH components have a position angle that is tilted relative to that

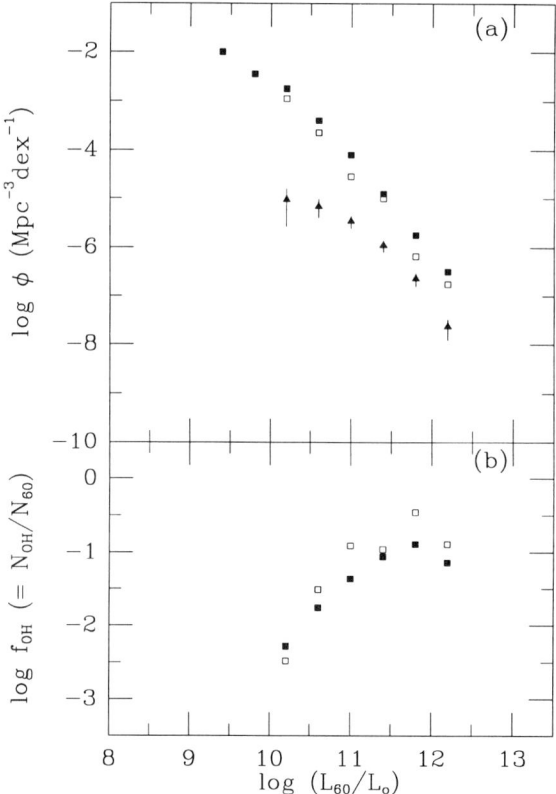

FIGURE 1. The 60-μm FIR luminosity function for OH megamasers compared with that of all FIR galaxies. In frame (a) the open and filled squares represent two different luminosity functions of FIR galaxies (Lawrence et al. 1987; Soifer et al. 1987). The filled triangles represent the 60-μm FIR luminosity function for OH megamaser galaxies. In frame (b) f_{OH} represents the fraction of OH megamasers among the FIR population.

of the continuum sources. Similarly, the VLA-A H$_2$CO emission structure with peak emission of 4 mJy is tilted in the opposite direction from that of the continuum sources. A simple model using the velocity field suggests a consistent picture for the motion of the two orbiting nuclei, with the OH and H$_2$CO emissions occurring respectively slightly in front of and behind the orbiting nuclei (Baan & Haschick 1995). The OH VLBI components appear not to coincide with the nuclei themselves as predicted for an AGN/torus configuration (Lonsdale et al. 1994).

2.1. A Luminosity Function of OH Megamasers

The FIR luminosity function of the OH megamasers can be effectively compared with the luminosity function of the luminous IRAS galaxies (Figure 1; see Baan 1991 for an earlier version). The luminosity function of the current OH megamaser catalogue based on a 60-μm flux-limited sample is presented in Figure 1a together with two representations of the 60-μm FIR luminosity function of bright IRAS galaxies (Lawrence et al. 1987; Soifer et al. 1987). Considering that any incompleteness of the OH megamaser sample mainly affects the lower-luminosity range of the luminosity function, the fraction of FIR galaxies with OH-megamaser activity, f_{OH}, significantly increases with 60-μm FIR luminosity and varies almost linearly from 0.003 to 0.2 (Figure 1b).

2.2. OH Gigamasers and Cosmology

Megamaser searches of galaxies at high radial velocities have in the past been severely limited by receiver capabilities and spectral coverage. Powerful megamasers have now been found to about the same redshifts as luminous (thermal) CO sources; IR 14070+0525 with isotropic luminosity $L_{OH} = 1.4 \times 10^4$ $L_\odot$ has a redshift of 0.265 (V = 79600 km s^{-1}) (Baan et al. 1992), IR 20100-4156 has $L_{OH} = 10^4$ $L_\odot$ at z = 0.127 (V = 38000 km s^{-1}) (Stavely-Smith et al. 1989), and IR13218+0552 has $L_{OH} = 1.9 \times 10^3$ $L_\odot$ at z = 0.205 (V = 61488 km s^{-1}) (Bragg, Baan & Haschick 1996). The limited extent of these surveys shows that OH megamaser activity is indeed rather common at high redshifts and sources can be detected quite easily with existing high sensitivity telescopes. The velocity structure of the highest-redshift OH masers suggests that the dynamic structure of these sources is also quite different to that observed at smaller redshifts. For instance, the source IR 14070+0525 has a total linewidth of about 2200 km s^{-1}.

The quadratic luminosity relation observed for the OH sources renders OH megamasers observable to very high redshifts. Using the standard criteria for past searches, a megamaser is required to have only $L_{OH} = 1400$ $L_\odot$ to be detectable beyond the z = 3 barrier. This is easily surpassed in the present sample of masers and the observed range of FIR and OH luminosities. OH megamasers will thus prove very useful in the study of the molecular evolution of galactic nuclei.

2.3. Classification of OH Megamasers

OH megamasers are a dominant sub-group of the (super-) luminous FIR galaxies, which at the high luminosity end show mostly optical AGN characteristics and are in part interacting systems. At lower FIR luminosities the sample is predominantly composed of starburst (SBN) nuclei. Optical classification studies of the OH-megamaser sample support this general trend with 77 percent showing AGN or composite AGN/SBN characteristics and the rest showing pure starburst characteristics (Baan, Salzer & LeWinter 1996). On the other hand, a classification on the basis of the radio properties suggests that the majority of 74 percent of the same sample are SBNs or look-alikes, which is consistent with other radio surveys (Baan, Haschick & Besenfelder 1996). These results suggest that a megamaser galaxy may look like an AGN in the optical and like a SBN in the radio. Among the luminous FIR galaxies, more than half of the sources have weak VLBI components (Lonsdale et al. 1993), while nearly half the (radio loud) Seyfert 2's and a quarter of the Seyfert 1's have compact cores (Norris et al. 1990).

The discrepancy of the optical and radio classifications of the megamaser sample confirms the uniqueness of these sources. Large concentrations of molecular gas and dust and galactic interactions provide unusual conditions in the nuclear region, which may confuse the standard nuclear diagnostic schemes. An initial mass function (IMF) for the nuclear starburst shifted to higher masses would harden the ionizing radiation field. A higher metallicity for the nuclear gas may also raise the standard metal/hydrogen line ratios, making the SBN look more like an AGN. It is uncertain if these effects are strong enough to explain the optical-radio classification discrepancies. Alternatively a weak AGN could be embedded within the nuclear starburst such that the nuclear (and circumnuclear) starburst dominates the radio properties and the nuclear ionizing radiation field determines the optical characteristics.

3. H$_2$O Megamasers

A total of sixteen H$_2$O megamaser emission lines have been found in the nuclear regions of relatively nearby galaxies with ten being Seyfert 2 nuclei and six LINERs (see Braatz,

Wilson & Henkel 1996 for summary). In analogy with the OH megamasers, the H_2O emission is confined to a molecular disk with a relatively small opening angle (Baan 1985; Haschick et al. 1990; Braatz et al. 1994). In AGN classification schemes such a scenario would naturally be associated with Seyfert 2/LINER sources. Two types of H_2O emission profiles can be distinguished: most sources show spectra made up of narrow features as exemplified in NGC 4258 (see Haschick, Baan & Peng 1994); two sources show a broad continuous spectrum without distinct features as exemplified in the early-type galaxies NGC 1052 and TXFS 2226-184 (Braatz et al. 1994; Koekemoer et al. 1995).

The torus model for H_2O megamasers has been proposed on the basis of single-dish monitoring and VLBI observations (Watson & Wallin 1994; Haschick et al. 1994; Greenhill et al. 1995b; Miyoshi et al. 1995). The maser emission in the H_2O sources originates in a thin edge-on molecular disk with the nuclear radio continuum possibly providing seed radiation. If the gain varies with position and direction in the molecular disk, the maser emission may display distinct narrow features. If the gain is the same in all directions a broad spectrum may result. The Keplerian motion of the masering clouds in the disk may result in a steady drift of (long-lasting) features in velocity space (gravitational acceleration proportional to V_{orb}^2/R_{disk}). Such drifts have been seen in NGC 4258 (Haschick et al. 1994; Greenhill et al. 1995a), NGC 2639 (Wilson, Braatz & Henkel 1995), and NGC 1068 and NGC 3079 (Baan & Haschick 1996). Weak emission features at $V_{sys} \pm V_{orb}$ may identify the edge sections of the molecular disk. These have been observed in NGC 4258 (Miyoshi et al. 1995) but have not been found yet in other sources. High-resolution interferometric data may reveal a spatial (linear) velocity gradient across the source in addition to the velocity drift of individual features (spatial velocity gradient proportional to V_{orb}/R_{disk}).

The combination of the velocity drift of spectral features over time and the spatial velocity gradient of VLBI features at V_{sys} in the prominent megamaser NGC 4258 have allowed a first order determination of the dynamics of the molecular disk (Wallin and Watson 1994; Haschick et al. 1994). VLBI data on the spatial distribution of the velocity features in NGC 4258 at V_{sys} and $V_{sys} \pm V_{orb}$ allowed for a more detailed determination of the Keplerian molecular disk and the mass distribution of the (black hole) nucleus (Miyoshi et al. 1995; Moran et al. 1995). A similar attempt has been made to determine the disk structure for NGC 1068 using VLA-A data at two epochs (Gallimore et al. 1996). However, a spatial (Keplerian) velocity gradient in this VLA-A data at the nucleus is too small to be consistent with the velocity drifts observed in the single dish monitoring data (Baan & Haschick 1996). Taking the steepest spatial velocity gradient possible within the existing VLA-A data of Gallimore et al. and the observed velocity drift of about $1 \,\mathrm{km\, s^{-1}\, yr^{-1}}$ gives an orbital velocity of $1440 \,\mathrm{km\, s^{-1}}$ and an inner radius of the molecular torus of $2.17 \,\mathrm{pc}$ (Baan & Haschick 1996). No detailed modelling is yet possible for NGC 2639 and NGC 3079.

3.1. A Luminosity Function of H_2O Megamasers

The rather limited sample of H_2O megamasers allows the construction of a preliminary luminosity function. Although the FIR activity may have no relation with the H_2O megamaser activity, the 60-μm FIR luminosity function for H_2O megamasers allows comparison with the luminosity function for OH sources. Figure 2a shows the two luminosity functions of bright FIR galaxies, a luminosity curve of Seyferts based on estimates of Miley, Neugebauer & Soifer (1985), and the luminosity function of H_2O sources. The AGNs/Seyferts likely represent the parent population of H_2O megamasers. Although the redshift-limited H_2O sample may still be incomplete because of maser variability, the fraction f_{H_2O} of megamasers among the Seyferts decreases with increasing FIR luminosity

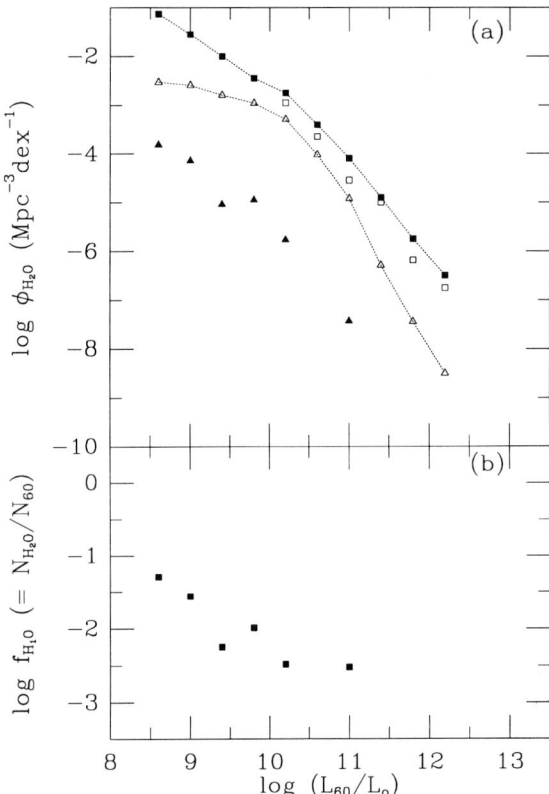

FIGURE 2. The 60-μm FIR luminosity function for H_2O megamasers compared with that of all FIR galaxies. In frame (a) the open and filled squares represent two different luminosity functions of FIR galaxies (Lawrence et al. 1987; Soifer et al. 1987). The connected open triangles represent the 60-μm FIR luminosity function for all AGN/Seyfert galaxies based on the estimates of Miley et al. (1985). The unconnected filled triangles represent the luminosity function of the H_2O megamasers. In frame (b) the fraction f_{H_2O} of H_2O megamasers is given relative to the Seyfert population.

from roughly 0.1 to 0.003 (Figure 2b). Many of these H_2O galaxies are optically-selected Seyferts having a relatively low FIR luminosity. They differ from FIR-selected Seyferts in that they are less enshrouded in dust.

4. Comparisons of OH and H_2O megamasers

4.1. *The opening angle of the molecular disks*

The occurrence of OH-megamaser activity depends on the efficiency of the FIR pumping scheme, the FIR spectral temperature and the spatial distribution of the molecular gas. Therefore, the factor f_{OH} from Figure 1b expresses the combined effects of the pumping efficiency and the spatial structure of the emission pattern. Simply converting this ratio into a disk opening angle gives the data points in Figure 3 (see Baan 1991). The OH opening angle increases with L_{FIR} from a few degrees to about 16 degrees as the observable disk gets thicker or more of the disk gets exposed.

In analogy with the OH megamasers, the fraction f_{H_2O} from Figure 2b reflects the opening angle of the molecular torus in H_2O sources. This angle is very small and varies

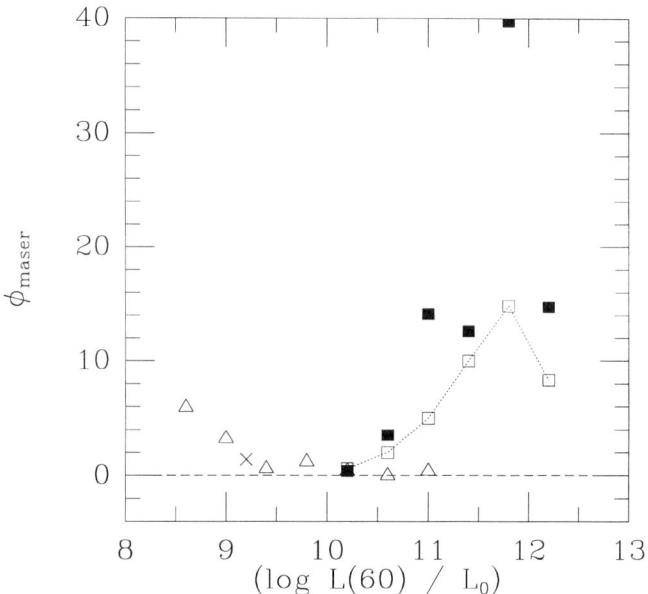

FIGURE 3. Effective opening angle of amplifying disk in OH and H_2O megamaser galaxies. The OH megamasers are indicated with open squares using the results of Figure 1b, while the H_2O sources are indicated with open triangles using the results of Figure 2b. The open (connected) and filled (unconnected) squares are based on OH comparisons with two different luminosity functions for IRAS galaxies of Figures 1 and 2. The cross represents the deduced opening angle from the observational data of NGC 4258 (Miyoshi et al. 1995).

from 6 degrees to almost zero at higher FIR luminosity (Figure 3). Recent directed H_2O surveys suggest that at low redshifts (V $\leq$ 2000 km s^{-1}) the fraction of megamasers among searched active galaxies is 0.11, while for higher velocity (V $\leq$ 7000 km s^{-1}) the fraction is 0.054. The factors f_{H_2O} found in Figure 2b relative to all Seyferts are lower by a factor of four than the survey values. The H_2O time variability and the constraints of the search sample are likely to have limited the completeness of the sample. The fraction of 0.054 found for the whole H_2O sample corresponds to an opening angle of about 6 degrees. For comparison, the only known opening angle of 1.4 degrees for NGC 4258 (Miyoshi et al. 1995) is found to be consistent with the opening angles predicted from the luminosity functions (Figure 3). The present results may suggest that H_2O megamaser activity disappears at higher FIR and X-ray luminosities.

4.2. Radio and X-ray Characteristics of OH and H_2O Megamasers

A direct comparison of the characteristics of OH and H_2O megamaser galaxies is possible using their 4.8-GHz radio and FIR luminosities. Figure 4 shows the luminosity of the nuclear radio luminosities versus the (integrated) FIR luminosity of OH and H_2O megamaser galaxies. The nuclear radio data for the galaxies are based on VLA-A data and the FIR luminosity from the IRAS data base (Baan, Haschick & Besenfelder 1996). There is a distinct separation between the OH and H_2O sources samples. Only two OH megamaser sources are radio loud Seyferts, while all H_2O sources are relatively radio loud. The "star formation efficiency" (the production of FIR radiation relative to the radio) of the OH sources is significantly larger than for H_2O sources. The H_2O sources have lower

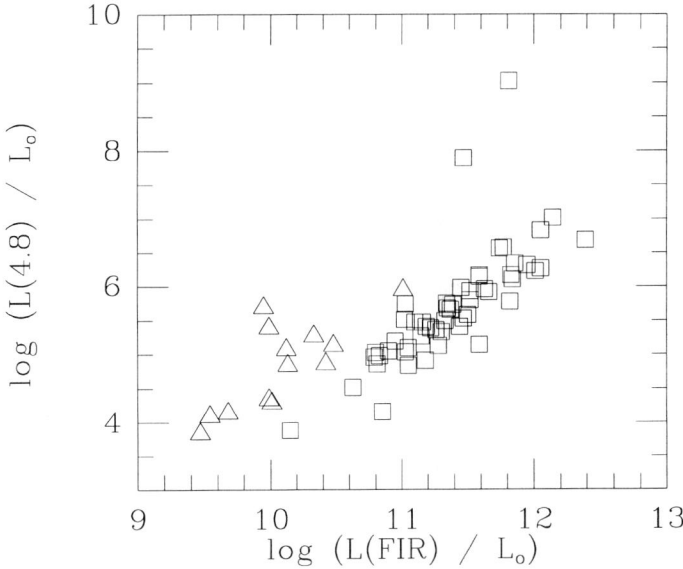

FIGURE 4. Radio luminosity versus FIR luminosity for OH and H_2O megamasers. The OH megamasers are indicated with open squares (Baan et al. 1996a), while the H_2O megamasers have been indicated with filled triangles. The locus of the H_2O sources lies about a factor of ten about that of the OH sources. The two radio loud OH sources are prominent Seyfert 2 galaxies.

FIR and radio luminosities and constitute a different population of "naked" Seyferts lacking the significant amounts of dust characteristic of the OH megamaser galaxies.

4.3. Pumping Considerations

The true nature of the masering process in H_2O megamasers is as yet unknown. The observed quadratic OH - FIR and H_2CO - FIR luminosity relations have a simple explanation in terms of exponential amplification of part of the radio continuum (Baan 1989; Baan, Haschick & Uglesich 1993). In order to find similar correlations, the H_2O megamaser luminosity has been plotted against the 4.8-GHz radio luminosity, the soft X-ray luminosity, the FIR luminosity, and the 60-μm FIR luminosity. The infrared data did not show any significant correlations and have not been presented here. However, both the radio and X-ray luminosity diagrams in Figure 5 show correlations: $L_{H_2O} \propto L_{4.8GHz}^{2.3}$ and $L_{H_2O} \propto L_{Xray}^{1.9}$. If the 4.8-GHz nuclear radio continuum is being used as seed radiation and the X-ray luminosity is responsible for the inversion of the molecular gas in the torus, then both correlations may be expected.

An (unsaturated) exponential amplification picture based on these concepts would predict $L_{H_2O} \propto (L_{4.8GHz} L_{Xray})$ for small optical depths. The diagram in Figure 6 shows a roughly linear correlation between the H_2O megamaser luminosity for fifteen sources (including upper limits) and the above luminosity product as predicted for amplification. The two sources with broad emission lines (NGC 1052 and TXFS 2226-184) do not stand out within the sample.

The pumping of the H_2O molecules in the molecular torus may be achieved with X-ray heating of the molecular gas at the inner edge of the torus, which results in collisional pumping of the molecules (Neufeld, Maloney & Conger 1994; Collison and Watson 1996). The observed narrow emission components are at V_{sys} (in front of the radio nucleus) and (possibly) at $V_{sys} \pm V_{orb}$, and there is little or no emission in between. These results

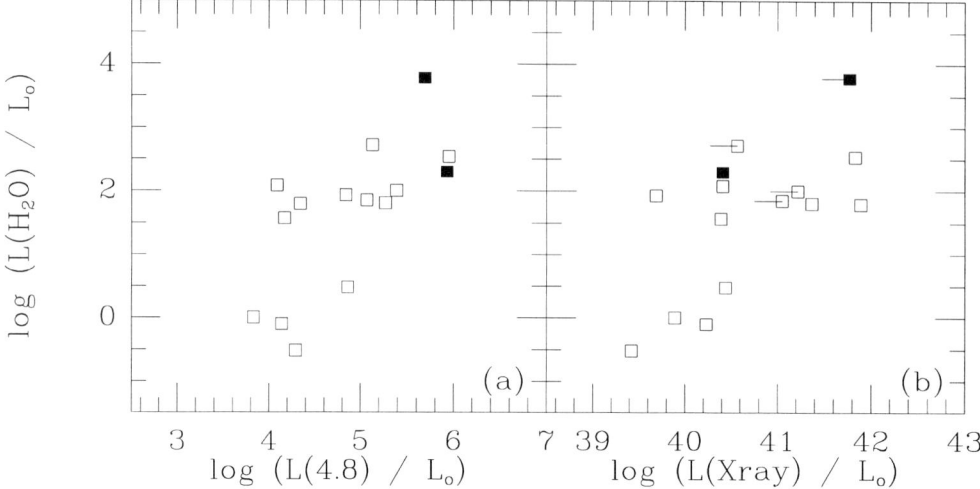

FIGURE 5. H_2O megamaser luminosity versus radio and X-ray luminosity: a) the H_2O luminosity versus the 4.8 GHz radio continuum luminosity, b) the H_2O luminosity versus the soft X-ray luminosity. The filled data points are the two broad profile megamasers. Upper limits in the X-ray data have been marked.

are consistent with X-ray-induced pumping schemes and suggest a spatial confinement of the emission regions by velocity gradients in the torus and/or the background continuum structure. The current galaxies with detected megamasers have relatively weak X-ray sources, and the luminosity of the H_2O increases with the X-ray luminosity. However, strong X-ray sources may destroy the molecular gas in the nuclear torus.

5. Conclusions

The two prominent classes of OH and H_2O megamasers have strikingly different characteristics. The OH megamasers occur among dusty FIR luminous galaxies showing a mixture of AGN and starburst characteristics. The H_2O megamasers occur among dust-arm (naked) low luminosity FIR galaxies with clear AGN/LINER characteristics. Except for two galaxies, the OH galaxies are radio-quiet, while most of the H_2O sources are relatively radio-loud.

The ratio of the number of megamasers relative to their parent population provides a rough measure of the opening angle of the molecular disk structure responsible for the masering. For H_2O this fraction/opening angle is relatively small and decreases with (FIR and X-ray) luminosity, while for OH sources this fraction increases roughly linearly with FIR luminosity. This is consistent with the notion that the H_2O masering torus/disk is close to the nucleus, as seen for several galaxies, while the OH masering (flaring) disk section is at larger radii, where FIR heating/pumping can be more effective.

The level of saturation of the OH and H_2O maser emissions is still under discussion. The observed quadratic OH - FIR and H_2CO - FIR luminosity relations suggest at least partial amplification of the nuclear radio continuum. Although unsaturated masering is still difficult to visualize for the H_2O megamasers, a similar H_2O - $(L_{4.8}.L_{Xray})$ relation also suggests partial amplification of the radio continuum. This idea is further supported

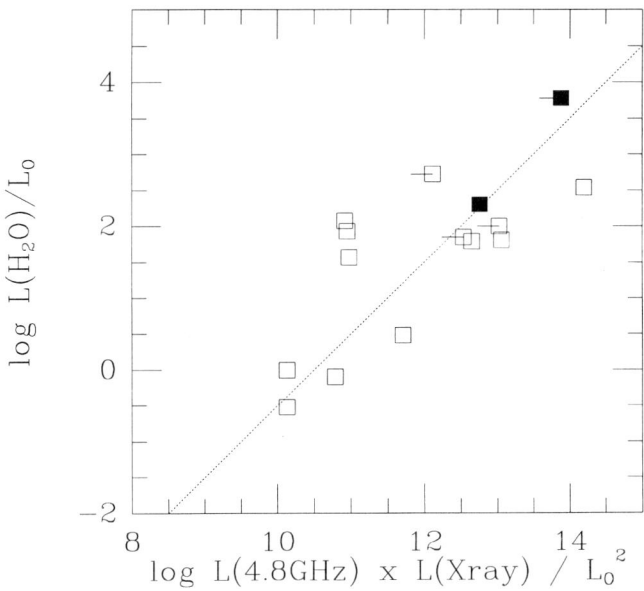

FIGURE 6. H_2O megamaser luminosity versus the product of the radio and X-ray luminosity. A linear relation (dotted line) is expected for a simple amplification picture involving the radio continuum background and pumping due to X-ray heating of the molecular gas in the foreground. The two filled data points represent the two broad profile megamasers.

by the spatial distribution of maser features in NGC 4258, with features occurring only at the nucleus and the edges of the torus.

OH megamasers have much potential as cosmological tools. The observed luminosities for OH megamasers will allow observation of "normal" megamasers beyond the $z = 3$ boundary. Although OH megamasers cannot yet be designated as standard candles, the occurrence of OH megamaser activity at high redshift is of great interest. On the other hand, the observed luminosities of H_2O megamasers are still insufficient for use as high-redshift cosmological tools. Searches for H_2O megamasers at high redshifts may reveal their actual luminosity range and show their usefulness for cosmological studies.

Arecibo Observatory is part of the National Astronomy and Ionosphere Center and is operated by Cornell University under a cooperative agreement with the National Science Foundation. This research has used the NASA/IPAC Extragalactic Database (NED), which is operated by the Jet Propulsion Laboratory, California Institute of Technology, under contract with the National Aeronautics and Space Administration.

REFERENCES

BAAN, W.A. 1985. *Nature,* **315**, 26

BAAN, W.A. 1989. *ApJ* **338**, 804

BAAN, W.A. 1991. *Skylines, Third Haystack Observatory Conf.,* eds. A.D. Haschick and P.T. Ho, *P.A.S.P. Conf. Series,* **16**, 45

BAAN, W.A & HASCHICK, A.D. 1995. *ApJ* **454**, 745

BAAN, W.A & HASCHICK, A.D. 1996. *ApJ* in the press

BAAN, W.A., HASCHICK, A.D. & UGLESICH, R. 1993. *ApJ* **415**, 140
BAAN, W., RHOADS, J., FISHER, K., ALTSCHULER, D. & HASCHICK, A. 1992. *ApJ* **396**, L99
BAAN, W., HASCHICK, A., & BESENFELDER, E. 1996. *ApJ* submitted
BAAN, W., SALZER, J. & LEWINTER, R. 1996. *ApJ* , submitted
BRAATZ, J., WILSON, A., & HENKEL, C. 1994. *ApJ* **437**, L99
BRAATZ, J., WILSON, A., & HENKEL, C. 1996. *ApJ* in the press
BRAGG, A., BAAN, W.A. & HASCHICK, A.D. 1996. in preparation,
COLLISON, A., & WATSON, W. 1996. *ApJ* **340**, L17
DIAMOND, P., NORRIS, R., BAAN, W. & BOOTH, R. 1990. *ApJ* **340**, L49
GALLIMORE, J., BAUM, S. & O'DEA, C. 1996. *ApJ* , in the press
GREENHILL, L., HENKEL, C., BECKER, R., WILSON, T. & WOUTERLOOT, J. 1995a. *A&A* **304**, 21
GREENHILL, L., JIANG, D., MORAN, J., REID, M., LO, K.Y., & CLAUSSEN, M. 1995b. *ApJ* **440**, 619
HASCHICK, A., BAAN, W., SCHNEPS, M., REID, M., MORAN, J., & GÜSTEN, R. 1990. *ApJ* **356**, 149
HASCHICK, A., BAAN, W., & PENG, E. 1994. *ApJ* **437**, L35
HENKEL, C., BAAN, W.A. & MAUERSBERGER, R. 1993. *Astron. Ap. Rev.,* **3**, 47
KOEKEMOER, A., HENKEL, C., GREENHILL, L., DEY, A., VAN BREUGEL, W., CODELLA, C. & ANTONUCCI, R. 1995. *Nature,* **378**, 697
LAWRENCE, A., WALKER, D., ROWAN-ROBINSON, M., LEECH, K. & PENSTON, M. 1987. *M.N.R.A.S.,* **219**, 687
LONSDALE, C.J, SMITH, H.E. & LONSDALE, C.J. 1993. *ApJ* **405**, L9
LONSDALE, C., DIAMOND, P., SMITH, H. & LONSDALE, C. 1994. *ApJ* **370**, 117
MILEY, G., NEUGEBAUER, G. & SOIFER, B. 1985. *ApJ* **293**, L11
MIYOSHI, M., MORAN, J., HERNSTEIN, J., GREENHILL, L., NAKAI, N., DIAMOND, P., & INOUE, M. 1995. *Nature,* **373**, 127
MORAN, J.M., GREENHILL, L., HERRNSTEIN, J., DIAMOND, P., MIYOSHI, M., NAKAI, N. & INOUE, M. 1995. *Proc. Nat. Ac. Sc.,* in the press
NEUFELD, D., MALONEY, P., & CONGER, S. 1995. *ApJ* **436**, L127
NORRIS, R., ALLEN, D., SRAMEK, R., KESTEVEN, M. & TROUP, E. 1990. *ApJ* **359**, 291
STAVELY-SMITH, L., ALLEN, D., CHAPMAN, J., NORRIS, R. & WHITEOAK, J. 1989. *Nature,* **337**, 625
SOIFER, B.T., SANDERS, D., MADORE, B., NEUGEBAUER, G., DANIELSON, G., ELIAS, J., LONSDALE, C. & RICE, W. 1987. *ApJ* **320**, 238
WATSON, W.D. & WALLIN, B.K. 1994. *ApJ* **432**, L35
WILSON, A., BRAATZ, J. & HENKEL, C. 1995. *ApJ* **455**, L127

Weak radio spectral lines

HI and CO of the high-latitude cloud MBM7

By Y. C. MINH

Korea Astronomy Observatory, Hwaam, Yusong, Taejon 305-348, Korea

The HI 21-cm and CO (1-0) data reveal that the high-latitude cloud MBM7 consists of a cold dense molecular core of $5\,M_\odot$ surrounded by warm gas of about $25\,M_\odot$. We derive a total column density $N(HI + 2H_2)$ of 1 x 10^{21} cm^{-2} toward the centre and 1 x 10^{20} cm^{-2} toward the envelope of MBM7. The CO lines indicate the existence of dense cores ($n(H_2) \geq 2000$ cm^{-3}) of size (FWHM) ~ 0.1 pc. The morphology suggests shock compression. The HI cloud extends to the NE, and the velocity gradient appears to be about 2.8 km sec^{-1} pc^{-1} in this direction, which will disrupt the cloud in $\sim 10^6$ yr. Hydrodynamical turbulence may dominate the line broadening, and the dense cores appear not to be bound by gravity.

The distance to HLCs suggests that they belong to the galactic plane, since the scale height of the cloud is ~ 100 pc. Compared to the more familiar dense dark clouds, HLCs may differ only in their small mass and low density, with their proximity reducing the filling factor and enhancing the contrast of the core and envelope structure.

1. Introduction

The high-latitude clouds (HLCs) are usually distinguished from the dense clouds located in the Galactic plane by their low visual extinctions and small masses (10-100 $M_\odot$). HLCs, having typical $A_v \sim$1-2 mag, span a wide range of physical and chemical properties, and fall into the category known as translucent clouds (Magnani & de Vries 1986; van Dishoeck & Black 1988). CO and other molecules have been detected in some HLCs by their millimetre transitions. Such studies indicate that there exist cores of substantial density ($\leq 10^3$ cm^{-3}) in many HLCs. Turner and co-workers have examined the properties of the dense cores of HLCs and conclude that they may be in hydrostatic equilibrium with the surrounding gas (e.g. Turner *et al.* 1992). It is still controversial, however, whether these dense cores (some claimed as dense as $n(H_2) \approx 10^4$ - 10^5 cm^{-3}) are gravitationally bound and will form low-mass stars in the future. Most of the cores in HLCs appear to have much less gravitational energy than the internal turbulent energy (e.g. Magnani *et al.* 1985). If the large linewidths of the observed molecular lines indicate expansion, the clouds will be disrupted with time scales of about 10^5 - 10^6 yr.

The physical and chemical properties of HLCs have not been well defined. Many of the observed HLCs extend over several square degrees and show highly complex morphologies at scales down to the actual resolution of the telescopes. Filamentary and clumpy structures are seen on high-resolution optical images (e.g. Stark & van Dishoeck 1994), which may suggest that shocks are involved in their formation and evolution. The molecular abundances of HLCs are relatively poorly determined, and it remains controversial whether or not HLCs' physics and chemistry are similar to those of the dense clouds in the Galactic plane.

2. Observations and Results

The 21-cm line was observed with the Synthesis Telescope of the Dominion Radio Astrophysical Observatory, and CO and ^{13}CO J=1-0 lines using the 15-element focal plane array QUARRY with the 14-m Five College Radio Astronomy Observatory telescope. The observational parameters are summarized in Minh *et al.* (1996). Figs. 1 and 2 are the integrated intensity maps of the HI 21cm and the ^{12}CO J=1-0 lines, respectively.

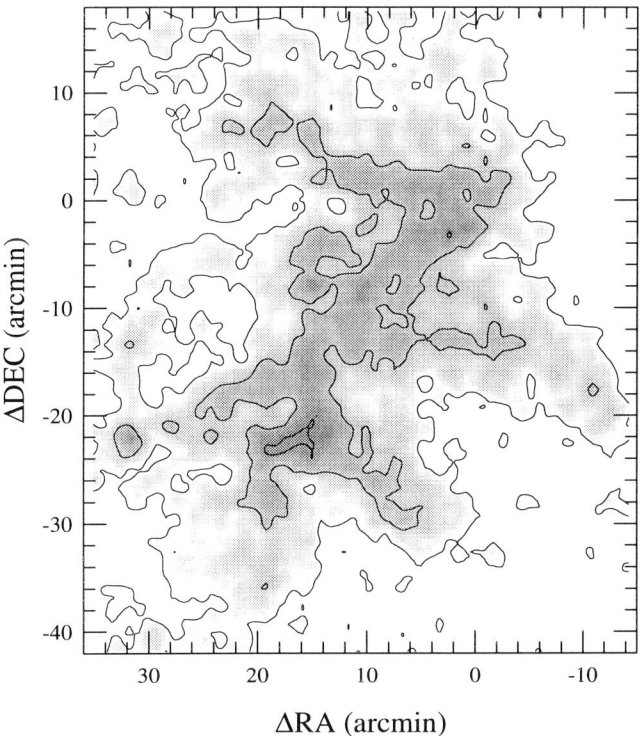

FIGURE 1. Integrated intensity maps of the H I emission. The (0, 0) position is RA(1950) = $2^h\ 19^m\ 36^s$; Decl.(1950) =$19°\ 40'\ 00''$. The lowest contour level is 330 K km sec^{-1}, and the increment is 40 K km sec^{-1}.

The molecular gas traced by the CO lines appears as localized clumps and forms an elongated structure along the NW to SE directions. Compared to the distribution of the H I emission, the molecular gas is located along a ridge perpendicular to the direction of the extended H I emission. The cloud parameters are included in Table 1 (see details in Minh et al. 1996).

A systematic velocity gradient of about 2.8 km sec^{-1}pc^{-1} appears in the NE direction, which may result from a systematic outward motion instead of rotation of the H I gas. The dynamical time scale for breaking up the H I cloud is estimated to be about $\sim 10^6$ yr. The observed linewidths are Δv_{obs}(H I) $\approx$ 7-8 km sec^{-1} and Δv_{obs}(CO) $\approx$ 1.5-2.5 km sec^{-1} which indicate that a substantial turbulent component exists in this region. Magnetic turbulence seems unlikely to be the major contributor to line broadening in this region. It is likely that the major contributor is non-magnetic hydrodynamic supersonic turbulence. Such bulk motions may result from external perturbations such as stellar wind or supernova explosions.

3. Discussion

The total extinction, A_v, is about 2 mag at the CO cores of MBM7, and it appears that molecules are more abundant than the atomic hydrogen by a factor of 10 in the lines of sight of bright CO emission. While the molecular gas is confined to a 2-3 pc^2 region, the

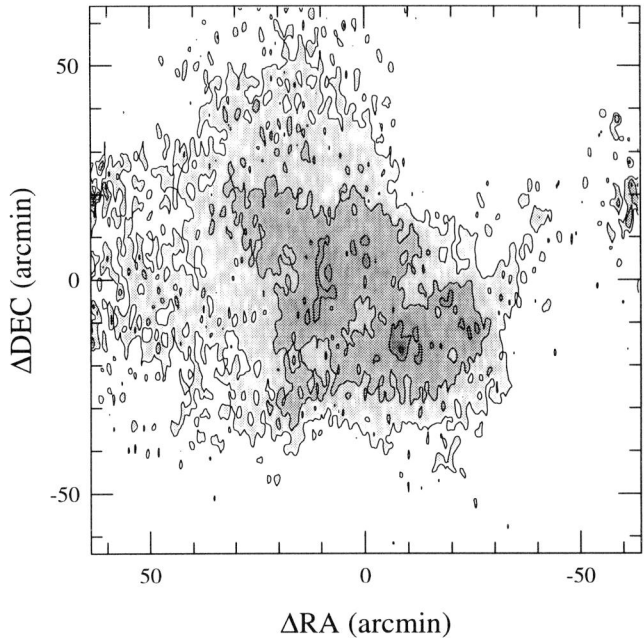

FIGURE 2. Integrated intensity map of the ^{12}CO (1-0) line. The contours increase by 3 K km sec^{-1} from 2 K km sec^{-1}.

	Size (FWHM) (pc)	N_{tot} (cm^{-2})	n_{tot} (cm^{-3})	T_k (K)	Mass (M$_\odot$)
Warm atomic/molecular cloud	1.5	3 10^{20}	1-2 10^2	30-300	22
Cold molecular cloud	0.4	1 10^{21}	1-2 10^3	10-15	5

TABLE 1. Cloud parameters (Minh et al. 1996)

H I is distributed over a much larger area ($\geq 10\,\text{pc}^2$) with a relatively uniform abundance of $\sim 10^{20}$ cm^{-2}. The peak CO and H I emission positions roughly coincide, so that the atomic and the molecular gas seem to be associated well with each other. The phase transition between atomic and molecular is thought to happen where the extinction of the radiation field is $A_v \sim 0.1$ mag for densities of ~ 500 cm^{-3}(van Dishoeck & Black 1988). Except for the CO-bright regions, $A_v \leq 1$ in the whole region. At the boundaries of the CO-emitting region we obtain a column density ratio of ~ 1 between the molecular and the atomic hydrogen, and the volume density is about 100 cm^{-3}. This may be the phase transition between H I and H$_2$.

The molecular cloud of MBM7 has a mass of ~ 5 M$_\odot$ and is surrounded by a warmer H I envelope of ~ 25 M$_\odot$. In the molecular cloud, several cold dense clumps, ~ 0.1 pc in size, $\sim$ a few 10^3 cm^{-3} in density, and ~ 1 M$_\odot$ in mass, exist. The velocity and pressure considerations in MBM7 seem to indicate that the gas cloud in MBM7 is not in equilibrium and the whole cloud is expanding. We believe that these MBM7 dense cores will finally break up or evaporate into H I gas with a disruption time scale of $\sim 10^6$ yr.

The morphology and velocity structure of the MBM7 H I emission suggest that shocks may have compressed and shaped the gas. Since a shock results in a density increase of the shocked region, and the cooling and recombination time scale is relatively short to 10^3 - 10^4 yr, the molecular cloud may be formed after the interaction. Without shock compression, it would be difficult to form molecules in a hot and diffuse H I cloud like that which still exists as an envelope of the molecular part in most HLCs. Turner (1995) has found that some sulphur-containing species are surprisingly abundant in MBM7 compared to typical interstellar clouds. This could be explained well by shocks, as long as the chemistry was equilibrated not too long ago, so that the subsequent cool-cloud ion-chemistry did not take over and erase the chemical memory of the shock.

Since the known HLCs are located relatively nearby (about 100 pc; Blitz *et al.* 1984), their actual scale heights from the galactic plane are less than 100 pc, although they are at high latitudes (b$\geq$40°). Since the scale height of H I of our galaxy from the galactic plane is $\sigma_Z \sim$100 pc and that of the molecular gas at the solar circle is $\sim$80 pc, most HLCs are still located inside the galactic plane. Therefore, it may not be proper to categorize the observed HLCs as different from those in the Galactic plane. In addition, the difference could also result from the large angular sizes due to the proximity to us, which reduces the filling factor and enhances the contrast of the core and envelope structure relative to the dense clouds. The study of HLCs will contribute very much to understanding the physical and chemical properties of interstellar clouds in general.

REFERENCES

BLITZ, L., MAGNANI, L., & MUNDY, L. 1984. *ApJ* **282**, L9.

MAGNANI, L., & DE VRIES, C. P. 1986. *A&A* **168**, 271.

MAGNANI, L., BLITZ, L., & MUNDY, L. 1985. *ApJ* **295**, 402.

MINH, Y. C., PARK, Y.-S., KIM, K.-T., IRVINE, W. M., BREWER, M., & TURNER, B. E. 1996 *ApJ* , in press.

STARK, R. & VAN DISHOECK, E. F. 1994. *A&A* **286**, L43.

TURNER, B. E., XU, LANPING, & RICKARD, L -J. 1992. *ApJ* **391**, 158.

TURNER, B. E. 1996, in preparation.

VAN DISHOECK, E. F. & BLACK, J. H. 1988. *ApJ* **334**, 771.

A large-scale, sensitive CO survey of the outer Galaxy towards Mon OB1.

By RICHARD J. OLIVER[1], M. R. W. MASHEDER[1], AND PATRICK THADDEUS[2].

[1] Astrophysics Group, H.H. Wills Physics Laboratory, Bristol University, Bristol, BS81TL, UK.

[2] Harvard-Smithsonian Center for Astrophysics, 60 Garden Street, Cambridge, MA02138, USA.

A new, fully-sampled, sensitive CO survey of 52.5 square degrees towards the Monoceros OB1 (Mon OB1) region has been completed using the CfA 1.2-m millimetre-wave radio telescope. This survey extends from $b = -1.5°$ to $b = +3.5°$ and from $\ell = 196.0°$ to $\ell = 206.5°$ on a uniformly spaced grid of $3'.75$ in ℓ and b, and has substantially better sensitivity than previous surveys of the region. CO is detected in 80% of the area, 60% of which is weak with integrated intensities less than 6 K km s^{-1}. This widespread weak CO emission reveals cold molecular gas tracing supernova remnants and expanding H II regions which are outside the discrete star-forming cloud cores previously mapped in less sensitive surveys. Weak emission at velocities greater than those expected from Local arm clouds are in a third quadrant extension of the Perseus arm. These abundant and cold outer Galaxy clouds have a mean kinematic distance of 3.5 kpc and are comparable in size to Local GMCs. The coincident positions of colour selected IRAS point sources indicates that these distant clouds are forming massive stars.

1. Introduction

Improvements in instrumentation used to observe the CO($J = 1 \to 0$) line, a tracer for molecular hydrogen, have improved dramatically since its first astronomical detection by Wilson et al. (1970). The CfA 1.2-m mm-wave radio telescope, the first to offer the area coverage necessary to map the large-scale distribution of CO in the Galaxy, is now used to make uniformly sensitive surveys of entire GMC complexes in reasonable time periods. This new CO survey of the Mon OB1 region has identical observational parameters to the new second-quadrant survey being undertaken with the CfA telescope and demonstrates the opportunity that these new surveys offer to improve our understanding of the large-scale distribution of molecular gas in the outer Galaxy.

The outer Galaxy offers an ideal laboratory in which to study molecular clouds. There is no kinematic distance ambiguity or velocity blending of features as experienced toward the inner Galaxy. Additionally, the Mon OB1 association lies towards one of the least obscured regions in the Galactic plane, offering excellent observational opportunities to study the large-scale distribution of molecular material and associated star formation. As a result of the poor sensitivity, spatial resolution and biased area coverage of earlier surveys, little insight was gained into the spiral arm structure of the outer Galaxy. For example, third-quadrant molecular clouds detected in CO at $R \leq 14$ kpc had been described as members of an "Outer Arm" which might be a separate spiral arm or an extension of the Perseus arm (Mead and Kutner 1988). Although the H I Perseus arm has been reported to extend into the third quadrant the justifications so far have been weak (Lindblad 1967, Henderson et al. 1982). Optical tracers are of little assistance since they are too infrequent and have a large dispersion in their estimated distances.

2. Observations and Data Presentation

The survey (Oliver et al. 1996) consists of 13,440 spectra obtained by position switching against reference positions previously shown by frequency switching to be free of CO emission to a level of 0.05 K (RMS). Two reference positions were chosen for each observation, higher and lower in elevation than the 'on' position. The time spent on each 'off' was adjusted to minimize the residual power in the difference spectrum. The angular separation between adjacent grid positions of $3'.75$ (0.4 beamwidth at FWHM) gives uniform sensitivity over area. In order to increase sensitivity the resulting data cube was convolved with a Gaussian of $4'.9$ FWHM resulting in an effective angular resolution of $10'$, and a mean RMS channel-to-channel noise of 0.115 K per beam.

When producing maps from a data cube, the signal-to-noise ratio is degraded by integration across emission-free channels. In order to minimise this sensitivity-reducing effect in the integrated maps we present the zeroth moment intensity. This technique has been successfully used with extragalactic CO data cubes that have emission spanning large velocity ranges (Adler et al. 1992).

3. The Large Scale Distribution of CO

Much more molecular gas is detected in the Mon OB1 region which is distributed in more complicated spatial and kinematic structures than previously known. Most of the emission is concentrated in two distinct velocity ranges. The strongest lies in the velocity interval from $-5 - +10$ km s^{-1} and is from the local spiral arm. However, identification of the high-velocity arm is subject to large uncertainties, owing to an absence of optical features outside of the local arm and to non-circular velocities.

A less ambiguous method of identifying spiral arms is to compare the kinematics of the CO to H I. The large-scale H I surveys provide the opportunity to trace spiral arms across a large Galactic longitude range towards the outer Galaxy, and allow a comparison to be made between the kinematics of the gas and the kinematics of a large number of optical features (Lindblad 1967). This analysis avoids the problems of localised non-circular velocity fields which induce considerable structure over five- to ten-degree ranges of Galactic longitude (Brand and Blitz 1993).

By comparing the CO (ℓ, v) structure with the H I from the Maryland-Green Bank survey we find that the weak arm of CO detected at velocities outside the range associated with the Local arm has a similar longitude-velocity structure to the high H I column densities of Lindblad's feature L, the Perseus arm.

The large-scale velocity structure is not, however, completely described by two spiral arms. With substantially improved sensitivity we detect not only the Perseus and Local arms, but also emission at velocities outside those of the two-arm model. Three notable departures are: (a) weak inter-arm emission at $\ell > 203°$, (b) two clouds with velocities that put them at kinematic distances beyond the Perseus arm (clouds 33 and 34 in Figure 1) and (c) emission between $\ell = 197.5°$ and $\ell = 201°$ at 'forbidden' negative velocities (clouds 4 and 9 in Figure 1), which cannot result from differential Galactic rotation in the third quadrant.

4. The Local Arm

A striking result of this new survey is the very large area over which weak CO emission ($\int T_R^* dv < 6$ K km s^{-1}) is now detected; 80% of the area surveyed has emission well above the noise ($> +3\sigma$), 60% of which is below the detection limit of Blitz's (1978) pioneering

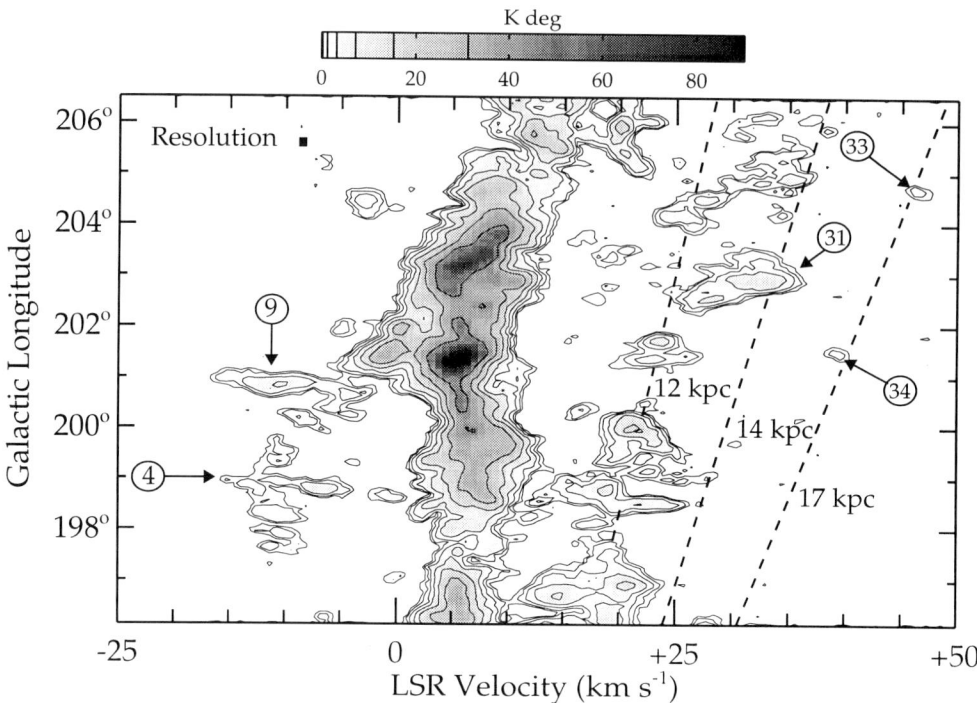

FIGURE 1. Longitude-velocity moment map of the present CO survey integrated over the latitude range $-1.75 < b < +3.5°$. The greyscale ranges from 0.0625 K deg (white) to 90.0 K deg (black). Contour levels are 0.0625, 0.125, 0.25, 0.5, 1, 2, 4, 8, 16, 32 K deg. The dotted lines indicate kinematic Galactocentric distances vs. longitude as labelled.

survey. This weak emission is spatially diffuse and fragmented, extending as halos around Lynds dark nebulae L1606, L1607, L1613, and the Cone Nebula.

5. Loops and Arcs

The improved sensitivity, area coverage and velocity resolution has resulted in detection of an abundant weak CO component, some of which appears to have kinematic and spatial relationships with large-scale expanding structures. All the loops and arcs have nearby physically and kinematically related optical or H I features which indicate associated SNRs or stellar wind blown shells. The forbidden velocity emission, which cannot be attributed to the large-scale spiral arm structure may be a result of these energetic sources.

6. The Perseus Arm Emission

The strong H I emission of the Perseus arm has a relatively weak molecular counterpart. However, the Perseus arm molecular clouds are comparable in mass and physical dimensions to local GMCs and similarly have extended cold halos. Their association with colour-selected (Richards *et al.* 1987) IRAS point sources provides evidence that they are forming massive stars.

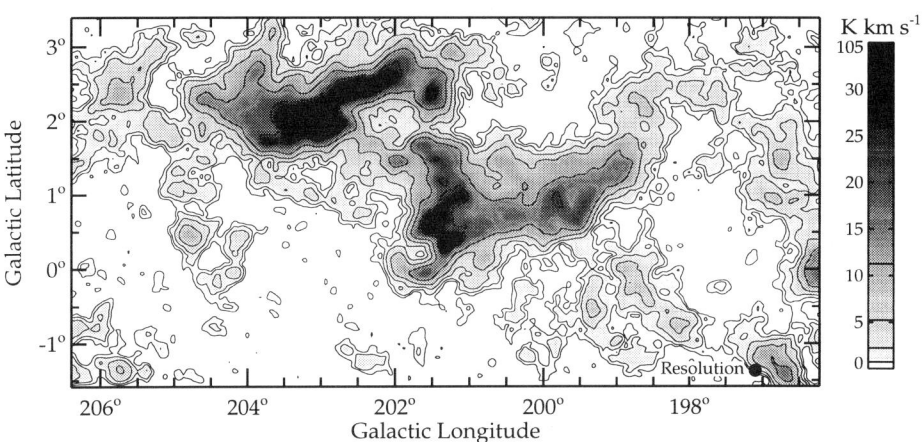

FIGURE 2. Moment map of the forbidden-velocity CO emission and the Local arm and inter-arm CO emission. The greyscale ranges from 0.75 K km s^{-1} (white) to 30.0 K km s^{-1} (black). Contour levels are 0.75, 1.5, 3.0, 6.0, 12.0, and 24.0 K km s^{-1}.

7. Conclusions

(a) This sensitive, unbiased CO survey has revealed far more molecular gas than previously detected; 80% of the spectra have emission $> 3\sigma$.

(b) Cold Perseus arm clouds extend into the 3rd Galactic Quadrant at $R \sim 12$ kpc.

(c) The Perseus arm GMCs are comparable in size and mass to Local arm GMCs and there is evidence that they are forming massive stars.

(d) The local weak emission is in complex structures which are spatially and kinematically associated with remnants of past star formation.

For a more detailed report of the survey and the results see Oliver et al. (1996).

REFERENCES

ADLER D.S., LO K.Y., WRIGHT W.C.H., RYDBECK G., PLANTE R.L., ALLEN R.J. 1992. *ApJ* **392**, 497

BLITZ L. 1978. PhD. Thesis, Columbia USA

BRAND J. AND BLITZ L. 1993. *A&A* **275**, 67

HENDERSON A.P., JACKSON P.D., KERR F.J. 1982. *ApJ* **263**, 116

LINDBLAD P.O., 1967. *Bull. Astr. Inst. Netherlands* **19**, 34

MEAD K. AND KUTNER M. 1988. *ApJ* **330**, 399

OLIVER, R. J., MASHEDER, M. R. W. & THADDEUS, P. 1996. *A&A*, accepted

RICHARDS, P. J., LITTLE, L. T. TORISEVA, M. AND HEATON, B. D., 1987. *MNRAS* **228**, 43

WILSON, R. W., JEFFERTS, K. B. & PENZIAS, A. A. 1970 *ApJ* **161**, L43

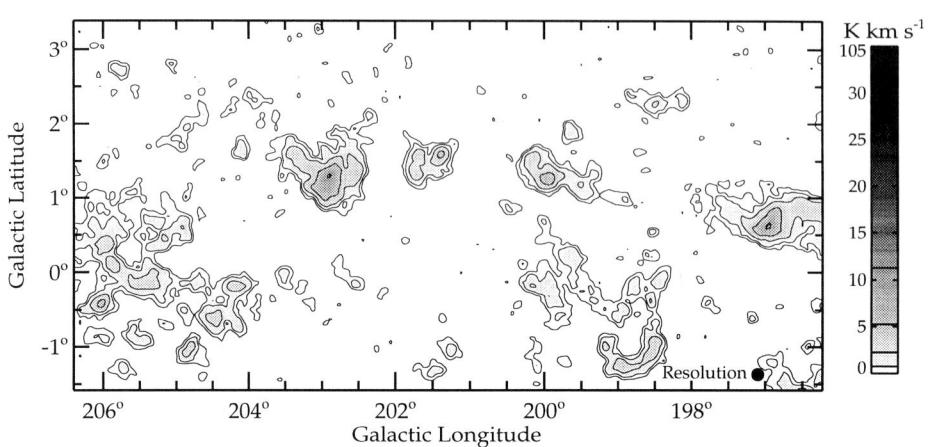

FIGURE 3. Moment map of the inter-arm and Perseus arm CO emission. The greyscale ranges from 0.75 K km s^{-1} (white) to 30.0 K km s^{-1} (black). Contour levels are 0.75, 1.5, 3.0, 6.0, 12.0, and 24.0 K km s^{-1}.

Quest for the 3-cm spectral limit: high sensitivity measurements of ^{3}He$^+$ emission from Galactic H II regions

By T. M. BANIA[1], D. S. BALSER[2], R. T. ROOD[3] AND T. L. WILSON[4]

[1]Department of Astronomy, Boston University, Boston, MA 02215, USA

[2]National Radio Astronomy Observatory, Green Bank, WV 24944, USA

[3]Astronomy Department, University of Virginia, Charlottesville, VA 22903, USA

[4]Max Planck Institut für Radioastronomie, 53121 Bonn, Germany

We are making precise determinations of the abundance of ^{3}He in H II regions distributed throughout the Milky Way. The abundance is derived using the 3.46 cm-wavelength hyperfine emission line of ^{3}He$^+$. For a sample of H II regions the abundances range between ^{3}He/H $\sim$ 1—5 $\times 10^{-5}$ by number. The abundance pattern observed across the Milky Way is difficult to reconcile with current models for the chemical evolution of the Galaxy. Moreover, our present limits on the ^{3}He abundance begin to challenge standard Big Bang nucleosynthesis models.

1. Introduction

The study of the origin and evolution of the elements is one of the cornerstones of modern astrophysics. For any given isotope of an element a crucial step is the observational determination of the abundance of that isotope and how that abundance varies temporally and spatially. We are making precise determinations of the abundance of the light isotope of helium, ^{3}He, in the interstellar medium of the Milky Way. The ^{3}He abundance is derived from measurements of the spin-flip line of ^{3}He$^+$ with a rest wavelength of 3.46 cm. Potentially observable sources of ionized gas include H II regions and planetary nebulae located throughout the Galaxy (e.g., Balser et al. (1994)). ^{3}He can serve both as a probe of cosmology and stellar/galactic evolution.

• The Big Bang theory for the origin of the Universe predicts that during the first $\sim$ 100 seconds significant amounts of the light elements (^{2}H, ^{3}He, ^{4}He, and ^{7}Li) were produced. The relative abundances of these light elements depend on various factors such as the average density of baryons in the Universe, the rate of expansion of the Universe, and details of particle physics. Studies of the abundances of these elements can provide information about the physical state of the Universe at these early times.

• Nuclear fusion reactions inside stars will change the relative amounts of the light elements from the primordial abundances produced by the Big Bang. Theory predicts not only that common solar-type stars are net producers of ^{3}He but also that the mass lost from winds generated at advanced stages of their evolution and the final planetary nebulae should be substantially enriched in ^{3}He. Planetary nebula ^{3}He abundances are therefore important tests of stellar evolution theory since these low-mass, evolved objects are expected to be significant sources of ^{3}He.

• The shedding and mixing of processed material by evolved stars occurs throughout the interstellar medium. New stars then form from this gas and subsequent evolution by this new stellar generation results in further processing of the elemental abundances. Measurement of the present ^{3}He abundance should therefore be an important diagnostic of chemical evolution in the Galaxy.

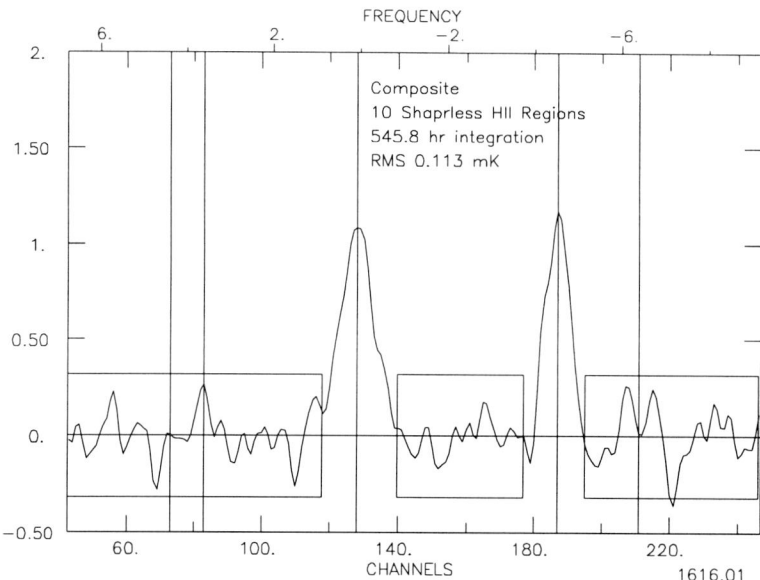

FIGURE 1. The observed ^{3}He$^+$ emission for a sample of ten Sharpless H II regions averaged together. The intensity scale is in milliKelvins of antenna temperature; the velocity resolution is 8.1 km sec^{-1}. Vertical lines flag from left to right the expected positions of the following recombination line transitions: C171η, He171η, H171η, ^{3}He$^+$, and H213ξ.

2. The Abundance of ^{3}He$^+$ in the Milky Way Interstellar Medium

The observational sample now includes ^{3}He$^+$ spectra taken toward 39 galactic sources which are H II regions and planetary nebulae. The H II regions are mostly within, but located throughout, the Milky Way's Population I disk, from the galactic centre to the far reaches of the outer Galaxy. Here we focus on the abundances we have determined for Galactic H II regions using the NRAO 140-ft telescope†. Our measurements of planetary nebulae and their astrophysical significance are discussed in this volume by Balser et al. (1996a). Figure 1 gives an example of the sensitivity we can achieve. Shown is the composite average spectrum for ten Sharpless H II regions. Note the "strong" H171η ($\Delta N=7$) hydrogen recombination line emission seen in this spectrum which has 0.113 mK RMS noise: both the ^{3}He$^+$ and H171η emission lines show comparable $\sim$1 mK intensities.

In recent years we have concentrated on measuring extremely accurate emission-line parameters for a few H II regions. Instrumental spectral baseline frequency structure limits the accuracy of the determination of these parameters. This structure, arising primarily from standing waves, is real; it *must* be removed to get accurate line measurements. We have developed observational protocols and baseline modelling procedures that can produce extremely accurate spectra.

These techniques: (1) allow weak (~ 10 mK) recombination lines to be measured to $\sim$15% accuracy within any single observational epoch; (2) give an inter-epoch calibration to within 10% for the spectral transitions of interest; and (3) produce results that

† The National Radio Astronomy Observatory is operated by Associated Universities Inc. under contract with the National Science Foundation.

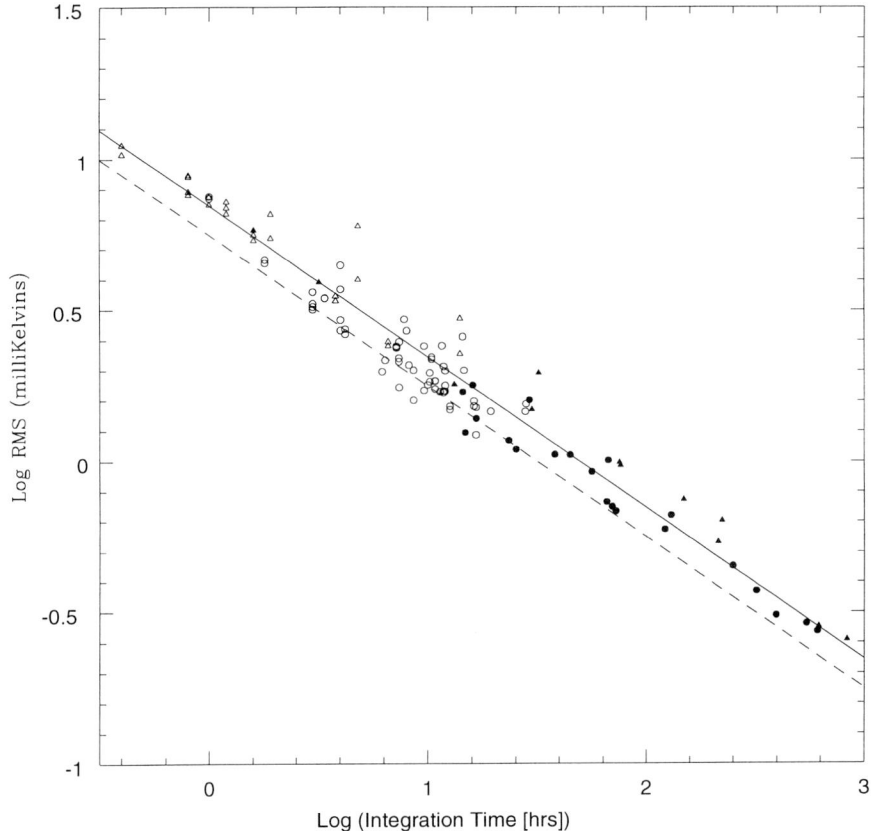

FIGURE 2. The empirical radiometer equation for the ^{3}He$^+$ experiment sample of HII regions. The lines indicate the noise performance expected for a perfect radiometer assuming system temperatures of 50 K (full) and 40 K (dashed). Note that the RMS noise measured by the 140 ft telescope spectrometer integrates down as expected for at least 1,000 hours.

make astrophysical sense. Here "astrophysical sense" means: the measured ^{4}He/H ratio does not vary with the order of the recombination line transition; the recombination line strengths and ^{4}He/H ratios do not vary with time; and the H recombination line strengths decrease smoothly with the order of the transition. Further, our source models (see below) can account in a *physically consistent way* for the intensities we observe in both the continuum and recombination lines. *We note that none of these things can be accomplished without modelling the spectral baselines to account for instrumental frequency structure.* In particular, the spectrometer performance shown in Figure 2 cannot be obtained unless these techniques are applied.

Deriving the abundance of any astrophysical species is a difficult multistep procedure. Recent ^{3}He/H abundance ratio determinations are summarized in Balser (1994) and Rood et al. (1995) where ^{3}He/H abundances are derived using models that account for source density and ionization structure as well as non-LTE effects and pressure broadening by electron impacts. The models are constrained by the observed ^{3}He$^+$ and recombination line parameters and the 3.5-cm continuum emission measured with the VLA D-array and the MPIfR 100-m telescope. Models derived in this manner not only account for the continuum flux from the sources but also reproduce the observed recombination line intensities.

3. Astrophysical Consequences

- HII regions sample the result of the chemical evolution of the interstellar medium since the formation of the Galaxy. The ^{3}He/H abundance ratio is expected to grow with time and to be higher in those parts of the Galaxy where there has been substantial stellar processing. The HII region abundances we derive range from ^{3}He/H $\sim 1-5 \times 10^{-5}$. The abundance differences are real, *i.e.*, the consequence neither of observational error nor of the source modelling (e.g., Rood *et al.* 1995). The observed abundance pattern across the Milky Way, however, is difficult to reconcile with current models for the chemical evolution of the Galaxy which predict a negative gradient in the ^{3}He/H abundance with galactocentric distance (e.g., Steigman & Tosi 1995).

- The best cosmological limits from ^{3}He currently are those based on the abundance derived for the galactic HII region S209. Balser *et al.* (1996b) use improved observations and source models for S209 to constrain the baryon-to-photon ratio, η. Using the other observed light elements and standard Big Bang nucleosynthesis (Wilson & Rood 1992), they find $\eta = 4.15 - 4.50 \times 10^{-10}$. These limits on the ^{3}He abundance begin to either challenge standard Big Bang nucleosynthesis models or call into question some of the standard assumptions concerning the chemical evolution of ^{3}He.

This research was supported by grant AST 91–21169 from the U.S. National Science Foundation.

REFERENCES

BALSER, D. S. 1994. Ph.D. thesis, Boston University

BALSER, D. S., BANIA, T. M., BROCKWAY, C. J., ROOD, R. T. & WILSON, T. L. 1994. *ApJ* **430**, 667

BALSER, D. S., BANIA, T. M., ROOD, R. T. & WILSON, T. L. 1996. This volume.

BALSER, D. S., BANIA, T. M., ROOD, R. T. & WILSON, T. L. 1996. In *Clusters, Lensing, and the Future of the Universe* (ed. V. Trimble and A. Reisenegger). A.S.P. Conf. Ser., vol. 88, p. 259. Astron. Soc. Pacific.

ROOD, R. T., BANIA, T. M., WILSON, T. L. & BALSER, D. S. 1995. In *ESO—EIPC Workshop of the Light Elements* (ed. P. Crane). p. 201. Springer.

STEIGMAN, G. & TOSI, M. 1995. *ApJ* **453**, 173

WILSON, T. L. & ROOD, R. T. 1992. *ARA&A* **32**, 191

High reliability galactic H I line observations

By P. M. W. KALBERLA[1], D. HARTMANN[2],
W. B. BURTON[3],
U. MEBOLD[1] AND G. WESTPHALEN[1]

[1]Radioastronomisches Institut der Universität Bonn (RAIUB), Auf dem Hügel 71, D-53121 Bonn, Germany

[2]Harvard-Smithsonian Center for Astrophysics, 60 Garden Street, Cambridge, MA 02138, U.S.A.

[3]Sterrewacht Leiden, Postbus 9513, 2300 RA Leiden, The Netherlands

Since 1962 it has been known that the accuracy of galactic H I observations, especially for weak lines at high galactic latitudes, is hampered by sidelobe contamination of the antenna. The Leiden/Dwingeloo Survey of galactic H I which becomes available soon is the first survey which has been corrected for stray radiation. These data offer new possibilities to analyze galactic H I with outstanding sensitivity. An analysis of the accuracy and limitations of the correction procedure is given. We report the discovery of very faint ($T_b < 0.1$ K) high velocity dispersion ($\sigma \sim 70$ km s^{-1}) H I emission probably originating in the galactic halo.

1. Introduction

The Leiden/Dwingeloo survey of neutral atomic hydrogen in our Galaxy provides high-quality line profiles of the entire sky north of $\delta = -30°$. Details of the observing and reduction procedures are given by Hartmann (1994). The survey will be published (Hartmann & Burton 1996) on CD-ROM. Compared to earlier large-scale H I surveys the new data are improved by an order of magnitude in either sensitivity, spectral resolution or spatial coverage.

The Dwingeloo 25-m telescope has an angular resolution of 36'. The observations were made on a grid of 30' both in galactic longitude and latitude. The spectral resolution of 1.03 km s^{-1} was obtained with an 1024-channel digital autocorrelator. Velocities between -450 km s^{-1} and 400 km s^{-1} (with respect to the Local Standard of Rest) are covered and allow the analysis of weak H I features exceeding the rms limit of 0.07 K.

21-cm line profiles observed by parabolic reflector antennas suffer from stray radiation originating from the sidelobes of the antenna diagram (van Woerden et al. 1962). Such contaminations which are variable with season due to the motion of the observer around the Sun are especially serious in high galactic latitudes as pointed out by Kalberla et al. (1980) and Lockman et al. (1986) and may account for more than 50% of the observed line radiation.

The Leiden/Dwingeloo survey is the first survey corrected for stray radiation. The main goal of the present paper is an analysis of the accuracy of this correction. We compare in section 3 the Dwingeloo data with the Bell Labs Survey (Stark et al. 1992). In section 4 we report the detection of low-level H I emission components with high velocity dispersion.

2. Reliability of the line profiles

The Dwingeloo observations have been carefully calibrated by using IAU standard positions for scale calibration and using regions of weak emission for baseline testing on a regular basis. All profiles have been corrected for stray radiation and baseline ripples.

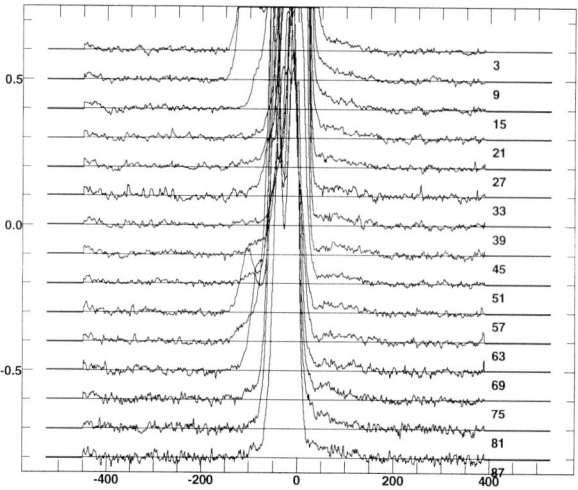

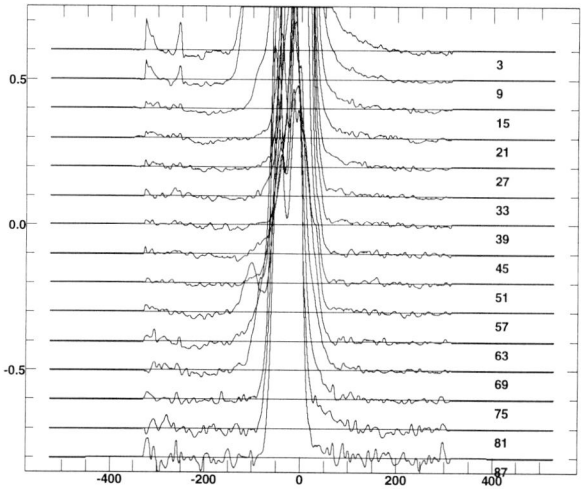

FIGURE 1. H I spectra in direction to the anticentre averaged over areas of $6° \times 6°$. The profiles are offset by 0.1 K, latitudes range from $b = 87°$ to $b = 3°$. Top panel: Leiden/Dwingeloo data, bottom panel: Bell Labs observations.

For details of the observing and reduction procedures the reader is referred to Hartmann (1994).

To test the reliability of the resulting profiles we have compared all profiles which have been observed independently at the same position. Calculating the variance of such profiles we conclude: i) The brightness temperature scale (determined in regions of strong H I emission) is accurate to 1%, ii) The accuracy of both baseline and correction for stray radiation (determined in regions of weak H I emission), is better than 0.1 K. iii)

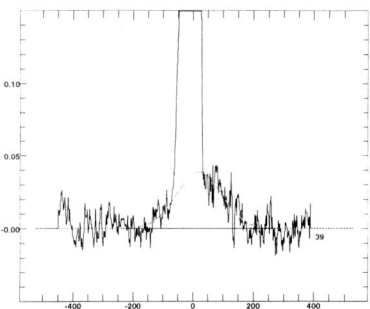 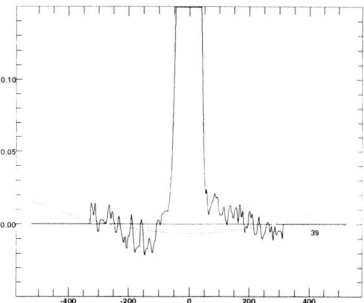

FIGURE 2. Left: The averaged Leiden/Dwingeloo spectrum at $b = 39°$ including a Gaussian fit to the extended wings. Right: Bell Labs data at the same position. A tentative second-order baseline fit is indicated.

Occasionally individual profiles are affected by interference, gain errors or other defects and have been eliminated.

The accuracy of the Dwingeloo profiles was found to be limited by reflections of H I radiation from the ground surrounding the telescope. From the survey data we extracted 2700 profiles showing such reflections which are unblended by the regular H I line emission. From these profiles we deduced an improved ground correction – our data differ with respect to this from those used by Hartmann et al. (1996).

The survey profiles have been smoothed over large areas of the sky aiming for improved sensitivity with respect to low-level H I emission. Such averages are most sensitive to systematic errors. We found a few problems which however can be overcome easily: a) the spectra are affected by a low level DC-offsets of the correlator which can be fixed by Hanning smoothing; b) the baseline partly may be still affected by ripples (features at velocity v=-300 km s^{-1}).

3. Comparison with Bell Labs Survey

To test the reliability of the Leiden/Dwingeloo data we have compiled comparisons to the Stark (1992) survey based on averages over $5° \times 5°$. The brightness temperatures of both surveys agree if one takes into account that the Bell Labs horn reflector data are affected by a) stray radiation from the near sidelobes and b) gain fluctuations up to 9%.

We found no indications for residual stray radiation problems in the Leiden/Dwingeloo Survey. The Bell Labs Survey profiles contain residual sidelobe contaminations originating from the near sidelobes. We conclude that the accuracy of the corrected Dwingeloo profiles is equivalent to, or taking the near sidelobes into account superior to, those of the Bell Labs Antenna. This is in particular due to the fact that the large velocity coverage of the Leiden/Dwingeloo data allows for a more reliable baseline correction than is possible for the Bell Labs data.

Fig. 1 displays in the upper panel profiles from the Leiden/Dwingeloo survey in direction to the anticentre averaged over areas of $6° \times 6°$. Latitudes are (from bottom to top) $87°$ to $3°$. The lower panel of Fig. 1 gives for comparison the average profiles derived from the Bell Labs Survey. We find in general a good agreement taking into account that the baseline of the Bell Labs Survey is uncertain due to the limited velocity coverage.

4. H I emission with high dispersion

Fig. 1 displays extended profile wings exceeding velocities of $|v| = 100$ km s^{-1}. We studied this phenomenon in more detail. In Fig. 2 (left) the profile at $b = 39°$ is displayed. A Gaussian analysis results in a low level H I emission component with a velocity dispersion of 75 km s^{-1}. Any attempts to explain this by residual contaminations from sidelobes fail. An analysis of the other profiles from Fig. 1 (top) gives similar results. A comparison of the Dwingeloo profile with the corresponding profile from the Bell Labs Survey is hampered by baseline effects. A tentative second-order baseline correction removes any discrepancies between both observations. It is obvious that the velocity coverage of the Bell Labs data is insufficient to recover emission with very high velocity dispersions reliably.

5. Summary and Discussion

It is mandatory to correct H I line observations in all cases where the analysis concentrates on low level ($T_b < 1$K) emission features. Alternatively the observations may be performed using a telescope with low sidelobe contamination. This requires that the main-beam efficiency of the telescope should be as close to 100% as possible, a requirement which needs special considerations with respect to the construction of a telescope. The new Green Bank Telescope (Lockman, this volume) is expected to become an ideal instrument for reliable galactic H I observations. At present only observations from the Bell Labs Telescope are available which unfortunately are hampered by the low spatial resolution ($\sim 2°$), low velocity resolution (5 km s^{-1}) and restricted velocity coverage.

The Leiden/Dwingeloo survey has been corrected for stray radiation using the algorithm given by Kalberla et al. (1980). Contaminations from sidelobes could be depressed below the detection limit. This way the significance of low level H I features could be improved by at least 1 order of magnitude. The survey data after correcting for residual effects of reflections from the ground allow for an analysis of extended profile wings.

We detect an ubiquitous low level H I emission with dispersions of $60 < \sigma < 80$ km s^{-1}, centre velocities of $-30 < v < +20$ km s^{-1} and column densities of $0.5 < N_H < 1.5\, 10^{19}$ cm^{-2}. Such components have been observed before, but the observational reliability of such detections was open to debate. According to Danly (1992) such components may originate in the halo at distances of more than 1 kpc above the galactic plane.

REFERENCES

Danly, L., Lockman, F.J., Meade, M.R., Savage, B.D. 1992. *ApJS* **81**, 125

Hartmann, D. 1994. PhD thesis, University of Leiden.

Hartmann, D., Kalberla, P.M.W., Burton, W.B., Mebold, U. 1996. *A&AS* in press

Hartmann, D., Burton, W.B. 1996, Atlas of Galactic Neutral Hydrogen, *Cambridge University Press*, in press

Kalberla, P.M.W., Mebold, U., Reich, W. 1980. *A&A* **82**, 275

Lockman, F.J., Jahoda, K., McCammon, D. 1986. *ApJ* **302**, 432

Stark, A.A., Gammie, C.F., Wilson, R.W., Bally, J., Linke, R.A., Heiles, C., Hurwitz, M., 1992. *ApJS* **79**, 77

van Woerden, H., Takakubo, K., Braes, L.L.E. 1962. *BAN* **16**, 321

Radio HI and optical absorption line studies of interstellar gas

By B. BATES,[1] C. R. SHAW,[1] S. N. KEMP,[1] F. P. KEENAN,[1] R. D. DAVIES[2] AND R. S. ROGER[3]

[1] The Queen's University of Belfast, Belfast BT7 1NN, Northern Ireland,

[2] University of Manchester, Nuffield Radio Astronomy Laboratories, Jodrell Bank, Cheshire SK11 9DL, England

[3] Dominion Radio Astrophysical Observatory, PO Box 248, Penticton, British Columbia V2A 6K3, Canada

An outline description is given of an investigation which uses optical spectroscopy, combined with HI observations obtained with the Lovell and DRAO Synthesis Radio Telecopes, to study an intermediate-velocity gas cloud in the foreground of the globular cluster M13.

1. Introduction

Using high-resolution spectroscopy we are investigating the structure and properties of the interstellar gas in the foreground of globular clusters and surrounding fields. These optical absorption line studies have been complemented with data from the IR and UV regions and, wherever possible, with λ 21-cm HI profiles obtained with the 76-m Lovell Telescope (LT). In particular, comparisons between the optical and radio data have been extremely important for deriving properties of different phases of the ISM. An outline description of results for several clusters is given in Bates (1995, and references therein).

Typically some 10 to 15 of the brightest cluster stars are used to 'map' the NaI column densities (and also KI for the higher column density sightlines) across the face of the cluster at $\approx$ arcmin resolution, according to the distribution of the stars. However, the HI profiles are obtained with much larger beamwidth ($\approx 12'$ FWHP for the LT) and these represent an average of gas properties over an angle which may be comparable with the angular size of the cluster. Ideally, the optical and radio data should be compared at a more similar spatial resolution.

Bates et al (1995) reported on the gas in the foreground of the globular cluster M13 (l=59°,b=41°; distance $\approx$ 7 kpc). In addition to the low-velocity gas components (observed at LSR velocities around +10 and -4 km s^{-1}) the HI spectra obtained over a field of some 1.5° revealed the patchy nature of an intermediate-velocity (IV) gas cloud observed at velocity ≈ -70 km s^{-1} and having low brightness temperature (T_B) ≈ 0.2 K. This IV gas was also detected for the first time in optical absorption line spectra with the observations suggesting significant column density variations across the cluster.

In 1994 we extended the investigation of this sightline by using the Dominion Radio Astrophysical Observatory (DRAO) Synthesis Radio Telescope (SRT) to map the HI gas at a resolution $\approx 2'$. Although the primary objective was to combine data to study the low-velocity gas components ($T_B \approx 5$ K), it proved feasible to use these observations to investigate the fine-scale properties of the low brightness temperature intermediate-velocity cloud (IVC). These new results complement the recent study of Kuntz & Danly (1996) who reported on the extensive distribution of IV gas in the northern Galactic hemisphere using the $\approx 2.5°$ FWHP Bell Laboratories HI survey. They may also be compared with studies of the high-velocity clouds (HVCs) which revealed the break-up

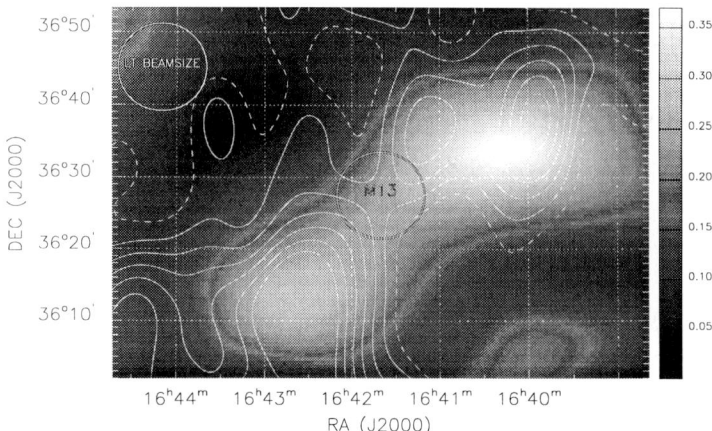

FIGURE 1. Brightness temperature map of the IV gas derived from LT observations. The contours are from DRAO SRT maps smoothed to $\approx 12'$ resolution. The dashed contour corresponds to 0 K and other contours are at 0.04 K intervals. The approximate size of the M13 cluster is indicated.

of HV gas structures into a collection of fine-scale concentrations when examined at the highest spatial resolution (e.g. Wakker & Schwarz 1991).

This report provides an outline description of our programme. A more detailed account of the observational methods and results will be given in Shaw et al (1996).

2. The large-scale structure of the M13 IV gas cloud

Figure 1 shows the brightness temperature maps obtained from the independent LT and SRT observations. The average T_B is obtained from spectra in the velocity range -64 to -80 km s^{-1} corresponding to those velocities close to the peak emission observed in the LT spectrum towards M13. The grey-scale is derived from LT spectra and the contours are from SRT observations in which the data have been smoothed to $\approx 12'$ to increase the signal-to-noise ratio (S/N) and to permit direct comparison of the independent observations. The main feature is the elongated ridge of gas having a typical column density $\approx 7 \times 10^{18}$ cm^{-2} which runs at an angle of some 70° to the Galactic plane and roughly across the face of the cluster. There are two smaller regions of higher T_B (≈ 0.3 to 0.5 K) which lie to the NW and SE of the cluster.

The LT spectra show the gas to have a broad velocity component ≈ 30 km s^{-1} FWHM. However, at a number of positions it is found necessary to include a narrow component (≈ 8 km s^{-1} FWHM) to give a satisfactory fit to the profiles. These positions occur where the LT beam was directed towards 'cloudlets', which are regions of smaller-scale structure of higher T_B observed in the higher resolution data. Such observations are similar to the two-component velocity structure found for HVCs using single dish observations of comparable beamwidth (Cram & Giovanelli 1976).

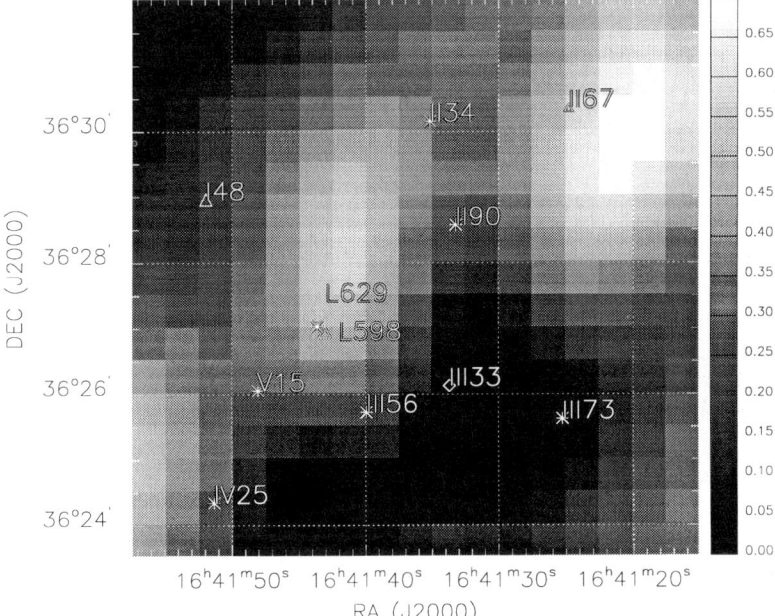

FIGURE 2. Brightness temperature map derived from LT and DRAO SRT observations for the cluster field. NaI absorption is measured towards the stars II-67 (the strongest detection), I-48 and L598. CaII absorption is measured towards star III-33.

3. Fine-scale structure and optical absorption spectra

In order to study the fine-scale distribution of the HI gas we constructed SRT maps at resolution $\approx 2.8' \times 2.0'$. These were obtained using the higher S/N data of the LT rather than the DRAO 26-m observations to fill in the lower-order spacings. An example is given in Figure 2 which shows the small region containing the M13 field. The positions of stars observed with the Utrecht Echelle Spectrograph at the William Herschel Telescope are indicated. The variation in T_B (averaged over the same velocity range as in Fig. 1) is evident. Detection in NaI absorption is observed for only the three stars II-67, I-48 and L598, whilst only upper limit column densities are obtained for the other stars. The strongest detection of the IV gas (NaI column density $\approx 8 \times 10^{11}$ cm^{-2}) is towards II-67 which lies close in direction to the region of highest T_B in this field. A particularly interesting observation is the detection in NaI absorption towards L598 but not towards L629. This implies a factor $\stackrel{\sim}{>} 2$ difference in gas column density for these two stars having an angular separation of only 10″. Kuntz & Danly (1996) have shown the IV Arch to lie at a z-height ≈ 1 kpc. Taking this distance for the M13 IV gas, then the corresponding spatial scale is only ≈ 0.05 pc. Over the several arcmin of the cluster the NaI column densities vary by a factor $\stackrel{\sim}{>} 10$.

The most southerly stars in the cluster lie in a region of lower T_B where the HI column density $\stackrel{\sim}{<} 9 \times 10^{18}$ cm^{-2}. There is no NaI absorption detected towards these stars but CaII absorption was observed towards III-33 (Bates et al 1995; this was the only star observed in CaII since it has much higher effective temperature than the other cluster stars). The Ca II/HI column density ratio $\stackrel{\sim}{>} 10^{-7}$ lies within the wide range reported for IV and HV gas clouds. The gas in this direction of the cluster appears to be associated with the broad velocity component. The low NaI/CaII ratio $\stackrel{\sim}{<}$ 0.1 observed for this gas would be consistent with a gas electron temperature $\approx 10^4$ K.

4. Concluding remarks

The combination of the HI observations obtained with the LT and the SRT, together with optical absorption line spectroscopy of cluster stars has provided the most detailed study of the fine-scale structure of an IVC. The cloud appears as a diffuse structure some 1.5° in size, but is likely to be part of a somewhat larger gas complex. Within this structure the SRT observations and the optical spectra reveal the presence of smaller-scale concentrations of higher column density. These features are unresolved at the resolution of the SRT data.

The HI spectra indicate a two-component velocity structure of the gas; a broad component associated with the more extensive and diffuse regions and narrow components identified with the fine-scale concentrations. Such properties resemble those described for the gas in HVCs. Also, our estimates of space densities for the small-scale concentrations which are based upon Na I/HI column densities and the physical scale of the structures (Shaw et al 1996) are consistent with values derived for HVC concentrations. These are of the order 75/D (cm^{-3}) where D is the gas distance in kpc (Wakker & Schwarz 1991).

The probing of the gas by optical absorption line spectroscopy reveals NaI column density variations of a factor $\stackrel{\sim}{>}$ 10 across the several arcmin face of the cluster and by a factor $\stackrel{\sim}{>}$ 2 on a scale $\approx$ 10 arcsec. In the light of such fine-scale variation, it may be difficult to define more precisely parameters such as relative abundance ratios derived from optical and radio observations. It is clear that these new data reinforce the well-known comments in the literature concerning the problem of using stellar and extragalactic sources as probes to detect, and to place distance limits upon, the HVCs and IVCs using absorption-line spectroscopy (e.g. Schwarz, Wakker & van Woerden 1995).

We would like to acknowledge the contribution made to the various observing programmes by our colleagues.

REFERENCES

BATES, B. 1995. In *The Formation of the Milky Way* (eds. J. Alfaro & A. J. Delgardo) pp. 68-71 Cambridge University Press.
BATES, B., SHAW, C. R., KEMP, S. N., KEENAN, F. P & DAVIES, R. D. 1995. *ApJ* **444**, 672
CRAM, T. R & GIOVANELLI, R. 1976. *ApJ* **48**, 39
KUNTZ, K. D & DANLY. 1996. *ApJ* **457**, 703
SCHWARZ, U. J., WAKKER, B. P & VAN WOERDEN, H. 1995. *A&A* **302**, 364
SHAW, C. R., BATES, B., KEMP, S. N., KEENAN, F. P., DAVIES, R. D & ROGER, R. S. 1996. *ApJ* submitted.
WAKKER, B. P & SCHWARZ, U. J. 1991. *Astron. Astrophys.* **250**, 484

Galaxies and cosmology

High resolution imaging of H I in galaxies

By J. M. van der HULST

Kapteyn Astronomical Institute, Postbus 800, NL-9700 AV Groningen, The Netherlands

High resolution imaging of galaxies with synthesis instruments has provided a wealth of information regarding the distribution of H I, the dynamics of galaxies and the properties of galaxies as a function of galaxy environment, galaxy type and galaxy luminosity. Resolved imaging has very often been crucial for interpreting the results from studies using merely global H I properties such as total H I mass and optical luminosity. This paper highlights several of the results to date, in particular to guide thinking about possibilities with future instrumentation.

1. Introduction

The earliest H I images of galaxies with a synthesis radio telescope were made nearly three decades ago with the Cambridge half-mile telescope (Emerson 1974, 1977; Newton & Emerson 1977; Newton 1980) and the Owens Valley Interferometer (Rogstad & Shostak 1971, 1973). The field developed really rapidly when faster and more sensitive instruments became available such as the Westerbork Synthesis Radio Telescope (WSRT), the Very Large Array (VLA) and, more recently, the Australia Telescope (ATNF). These instruments have imaged a few hundred galaxies with typical resolutions of $10''$ to $40''$ which at a distance of 10 Mpc corresponds to 0.5 to 2.0 kpc.

The resolved imaging of the H I in galaxies made it possible to study the detailed structure of H I in galaxies and to measure the kinematics of the gas disks out to radii which were inaccessible until then. This led to several important discoveries regarding the kinematics and distribution of the H I disks of galaxies. In the following sections I will briefly discuss some of these results, restricting myself to those aspects which have profited in particular from the increased sensitivity and resolution of modern synthesis arrays such as the WSRT, the VLA and the ATNF. For further information about the earlier work I would like to refer the reader to the review papers by Allen and van der Kruit (1978), Brinks (1990) and van der Hulst (1996).

2. HI Kinematics

2.1. *Global Kinematic Structure*

Studies of the kinematics of the H I in galaxies have shown the existence of both small scale and large-scale deviations from circular motion. These have been described in detail by Allen and van der Kruit (1978). The existence of streaming motions due to density waves received ample attention in the seventies (Visser 1980, Toomre 1981), but has not been actively pursued recently.

Large-scale streaming motions resulting from deviations of axisymmetry of the gravitational potential such as bar structures in galaxies remained an active line of research, though the results are still sparse (Sancisi *et al.* 1982, Ondrechen & van der Hulst 1989a, 1989b, England *et al.* 1990, Gottesman *et al.* 1984, Moore & Gottesman 1995, Jörsäter & van Moorsel 1996) and the theoretical calculations are still being improved (Sellwood & Sparke 1988, Athanassoula 1992).

Apparent large-scale deviations from circular motion arise from projection effects if the H I disk of a galaxy is not in one plane. This was first realized by Rogstad *et al.*(1974), who used a simple kinematic model consisting of concentric rings with slowly-changing

F568−1

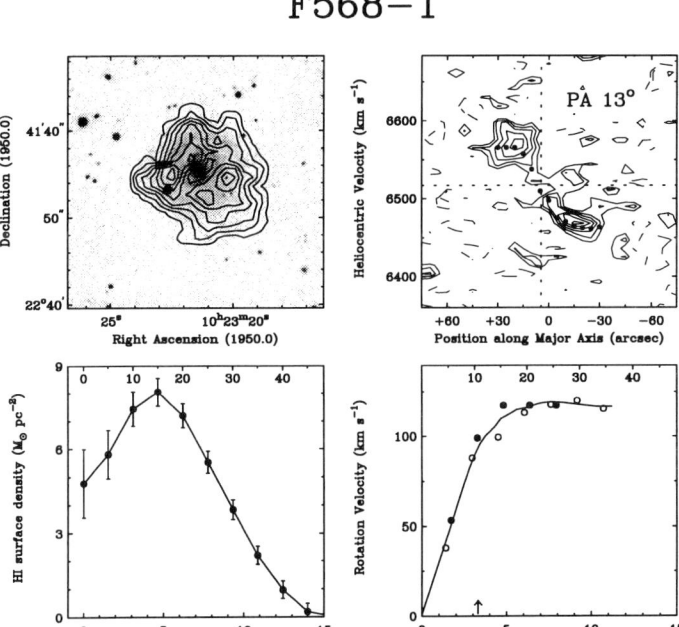

FIGURE 1. Optical image with H I distribution, major axis cut, radial H I surface density distribution and rotation curve of the low surface brightness galaxy F568-1 from de Blok et al. (1996)

orientations to describe the overall velocity field of M 83. Since many galaxies do show these effects in their velocity fields this description is adopted as a standard procedure for determining the rotation curves of galaxies from measured velocity fields (Bosma 1978, 1981, Begeman 1987).

2.2. Rotation Curves

The main result from studies of the rotation curves of galaxies can be summarized by the simple statement that rotation curves tend to be flat and do not show the near-Keplerian behaviour one expects on the basis of the distribution of light and gas. For a review on the subject of rotation curves see van Albada and Sancisi (1987) and Bosma (1989). The commonly accepted point of view is that galaxies have dark halos which provide the gravitational acceleration to keep the rotation speeds in the outer parts of galaxies high. A different view is that so-called modified Newtonian dynamics (MOND) keeps the rotation speeds high in the outer parts. In this picture (Milgrom 1983, 1988; Begeman et al. 1991) Newton's law has an additional acceleration term which causes the effective acceleration to become more like $1/r$ in the limit of low accelerations.

The existence of dark matter in galaxies being widely accepted, modern studies concentrate on determining the precise shapes of the dark halos in galaxies. This can be done for example by measuring the kinematics of H I in polar ring galaxies (Sackett & Sparke 1990, Arnaboldi et al. 1993).

Most of the rotation curve studies to date concern bright, nearby galaxies for the obvious reason that these offered the best opportunity to measure rotation curves out to

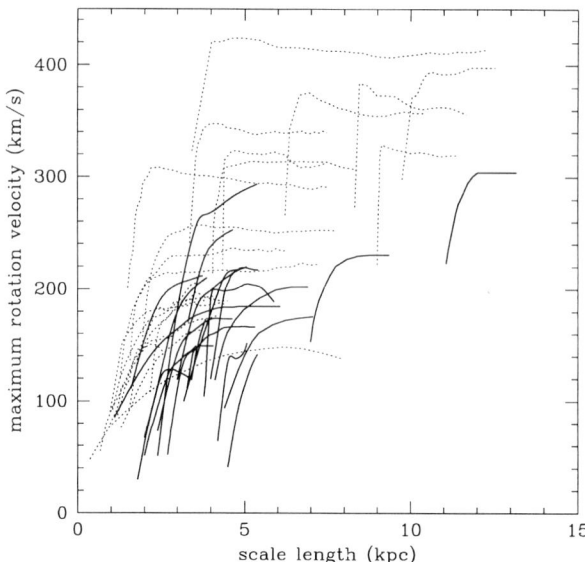

FIGURE 2. Rotation curves for galaxies with exponential disks ranging in blue central surface brightness from 23.5 to 20.6 mag arcsec^{-2}. The dashed curves are from Broeils (1992), the solid curves from van der Hulst et al. 1993 and de Blok et al. 1996. The lower left point of each curve is placed at the position given by the maximum rotation velocity and scale length of the optical disk.

a large number of optical disk scale lengths. With the improvement of the sensitivity of synthesis arrays studies of fainter and more distant galaxies can be pursued and are being carried out. Two examples are a study of all spiral galaxies in the Ursa Major cluster by Verheijen, Sancisi and Tully (priv. comm.) and studies of the HI in galaxies (mostly of late Hubble type) with low-surface-brightness disks (van der Hulst et al. 1993, de Blok et al. 1996). An example of the typical quality and resolution of such data is shown in figure 1.

Though the latter rotation curves do not have the same quality and linear resolution of those in nearby, bright galaxies, they do illustrate that there exists a large variety of rotation curves. This is illustrated in figure 2 which shows rotation curves for a sample of galaxies with a large range in optical central surface brightness of the exponential disk. The rotation curves have been placed in this diagram following the method of Casertano and van Gorkom (1991): the lower left point of each rotation curve has been plotted at a position indicating the maximum rotational velocity of the rotation curve and the scale length of the optical disk.

It is quite clear from figure 2 that the galaxies with low-surface-brightness disks (which in this case happen to be the later-type galaxies as well) have rotation curves which are very shallow and hardly reach a prominent plateau as do the rotation curves of many bright, nearby galaxies. Analysis of the mass distribution in low surface brightness galaxies shows them to be more dark-matter dominated and less dense than the high-surface-brightness galaxies studied thus far (de Blok and McGaugh 1996, de Blok et al. 1996).

Future radio telescopes with a tenfold increase in collecting area and sensitivity will provide data of similar quality as now obtained for galaxies out to redshifts of $z = 0.02$, for galaxies out to redshifts as far as $z = 0.5$. This has been demonstrated by Braun

(1996) who investigated the capabilities of synthesis radio telescope with a collecting area of one square kilometre. The capabilities of present instruments are dramatically illustrated by a deep (100-hour) integration with the VLA by Van Gorkom et al. (1996) of the cD cluster A2670 at z = 0.0767. This observation has a 5-σ limit of $5 \times 10^8 M_\odot$ and some 20 galaxies have already been detected.

In addition, of course, the increased sensitivity of such an instrument will enable tracing the rotation curves in nearby galaxies out to larger radii, unless the HI surface density drops to dramatically low levels as suggested by the present observations (see Maloney 1993).

3. HI distributions

3.1. Large-Scale Distribution

The large-scale HI distributions of galaxies can briefly be characterized as follows:

(i) most galaxies have HI sizes (measured at the current limits of sensitivity with synthesis instruments which typically are $\leq 10^{20}$ atoms cm^{-2} HI column density) which are at least as large and often exceed the optical sizes as measured by the isophotal de Vaucouleurs' size, R_{25}, (Bosma 1978, 1981, Allen & van der Kruit 1978);

(ii) the HI sizes depend on the environment (Warmels 1986, Cayatte et al.1990, 1993);

(iii) HI distributions are often not quite symmetric but lopsided (Baldwin et al.1980);

(iv) the HI distributions are much flatter than the light distributions of galaxies and in early-type galaxies the HI column densities are often lower in the central parts than in the main disk; in addition the HI disks show a wealth of structure ranging from clear spiral arms (e.g. M 101 Allen et al. 1973, M 51 Rots et al. 1989) to more irregular structures (e.g. M 33 Deul and van der Hulst 1987, Ho II Puche et al. 1992);

(v) many HI disks show large-scale warping of the plane, in general outside the optical disk (Sancisi 1976, Bosma 1978, 1981, 1991);

(vi) the edge of the star-forming disk in bright galaxies always appears to occur at a critical gas surface density level (Kennicutt 1989);

(vii) HI imaging of galaxies with disks with low optical surface brightnesses (Van der Hulst et al. 1993, De Blok et al. 1996) shows that these objects have low surface density HI disks. The HI surface densities are close to or below this threshold for star formation, consistent with the low present and past star-formation rate in these galaxies;

3.2. Small Scale Distribution

Surveys of the HI in the Galaxy (Hartmann 1994, Burton 1988, and references therein) show a wealth of small scale structure. Some aspects, in particular the filamentary and shell shape features have been discussed in detail by Heiles (1979, 1984, 1990) and are thought to be related to star formation in the disk through stellar winds and supernova explosions (MacLow & McCray 1988, MacLow et al. 1989, Elmegreen & Chiang 1982, Tomisaka 1992, Tomisaka & Ikeuchi 1986, Norman & Ikeuchi 1989).

Synthesis imaging of the HI in nearby galaxies has revealed similar features in nearby galaxies. The first such studies were those of Brinks & Bajaja (1986) and Deul & den Hartog (1990) of the Local Group galaxies M 31 and M 33, respectively. These observations demonstrated clearly that the HI disk in galaxies has a very percolated structure with many holes and filaments: a sort of tunnelling network as proposed by Miller & Cox (1993) based on the ideas of Cox & Smith (1974) and McKee & Ostriker (1977). Some 150 holes have been identified in the HI disks of these galaxies with sizes ranging from 40 pc (the resolution limit) to 1 kpc.

Since the work in M 31 and M 33, a few more galaxies have been studied in detail.

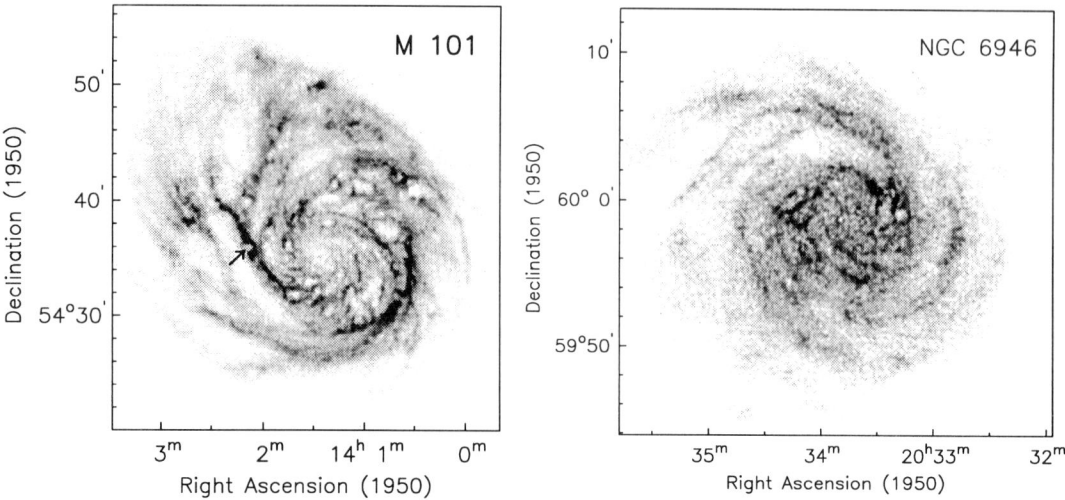

FIGURE 3. HI Distribution of the spiral galaxies M 101 (left panel) and NGC 6946 (right panel)

Large holes have been discovered in the irregular galaxies IC 10 (Shostak & Skillman 1989) and Ho II (Puche et al. 1992). The holes in these galaxies have sizes ranging from 0.1 to 1.5 kpc. The energy input derived is $10^{50} - 10^{52}$ ergs. The largest holes correspond in size to the supershells found in our Galaxy (Heiles 1979).

More recently, Kamphuis (1993) studied two nearly face-on, large galaxies in HI: M 101 and NGC 6946. These ScI galaxies both have several very large star-forming regions along well-defined, massive spiral arms. The HI extends farther than the Holmberg radius in both galaxies and shows spiral structure even outside the optical image. The HI column density distributions of these galaxies, illustrating the percolated structure of the HI disk, are shown in figure 3. The resolution is 15″ or 500 and 700 pc at the distances of M 101 and NGC 6946 respectively.

It is obvious from figure 3 that the HI disks of M 101 and NGC 6946 have tens of large holes with sizes of 1 to more than 5 kpc. Some of the large holes are ambiguous and could equally well be considered as inter-arm regions. The large, well-defined holes occupy about 10% of the disk and are located throughout the whole HI disk, even in the outer regions where the spiral arms are barely visible in the optical. No clear holes could be found beyond the Holmberg radius of the optical disk.

Because of the larger distance to M 101 and NGC 6946, holes smaller than 500 pc can not be distinguished, and we have no information about whether a population of smaller holes such as found in M 33 exists in these giant galaxies. The associated HI masses range from $1 - 5 \times 10^7$ $M_\odot$ and kinematic ages are $20 - 50 \times 10^6$ years. The typical kinetic energy associated with individual holes is 10^{53} ergs.

Kamphuis (1993) found no strong correlation between large holes, and OB associations and HII regions. There is some relation to the star formation and the holes in that the HII regions are often located close to HI concentrations and near the boundaries of the holes.

There is, however, one particular good example of the effect of star formation on the

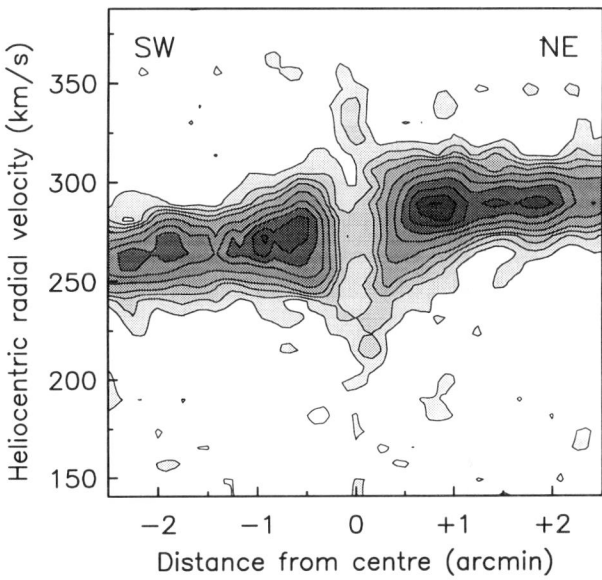

FIGURE 4. Position – velocity diagram of the HI superbubble near the HII region NGC 5462 in M 101

surrounding HI. This is the expanding superbubble found by Kamphuis et al.(1991) around the HII region NGC 5462 at $\alpha = 14^h\ 2^m\ 9^s$, $\delta = 54°\ 36'\ 2.5''$. Figure 4 best shows the kinematic behavior of the HI around the HII region. This Figure is a position – velocity map showing the HI profiles at each position along a line of position angle 30° centred on the superbubble. The high-intensity ridge running from the SW to the NE along the position axis (centred around a velocity of ≈ 274 km s^{-1}) is the HI in the spiral arm in which NGC 5462 is located. When coming close to the centre of the bubble one clearly sees a slight increase in intensity and in velocity width of the HI profiles. At the centre position the intensity drops and the profiles become very broad with a full width of ≈ 150 km s^{-1}. This signature is precisely what one expects to see in the case of an expanding HI bubble. At the position of the centre of the bubble one observes the full expansion velocity along the line of sight, both the blueshifted front and the redshifted backside of the bubble. As one moves away from the centre the component along the line of sight gradually becomes smaller and one observes a gradually smaller velocity width. The amount of HI blown out is about 3×10^7 M$_\odot$ representing a kinetic energy of about 3×10^{53} ergs. It would require many hundreds of supernovae to provide this much energy. This is not inconceivable since NGC 5462 alone requires a few hundred OV stars for its ionization.

With this example at hand it becomes quite clear one needs to examine in detail both the spatial distribution of the HI and the velocity structure of the HI to further investigate the interaction between the star formation and the interstellar medium.

3.3. Small Scale Kinematic Structure

Kamphuis (1993) examined the velocity structure of M 101 and NGC 6946 on a smaller scale (a few kpc and motions between 30 and 80 km s^{-1}), the regime as shown by the superbubble in NGC 5462. In M 101 there are some additional features like the superbubble, but never as symmetric and as clear as the case of NGC 5462. The features often are at the limit of the sensitivity and resolution. The gas moving at velocities in

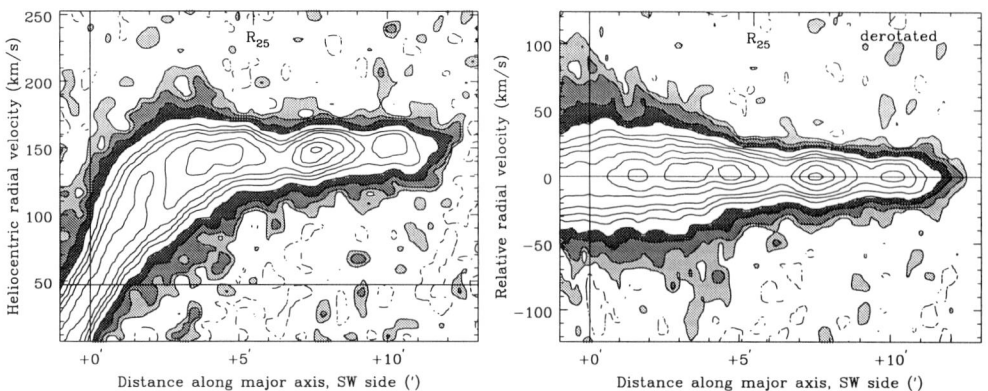

FIGURE 5. Left panel: Position − velocity diagram along the major axis of NGC 6946 (position angle 60°), strip integrated over 6′ in the perpendicular direction. The de Vaucouleurs radius (R_{25}) is indicated in the top of the figure. Right panel: same data but derotated.

excess of the local rotation is often associated with the H I holes, but not always. The situation in NGC 6946 is quite similar.

To get a better idea of how widespread such high-velocity gas features are, Kamphuis & Sancisi (1993) averaged position − velocity diagrams in a strip 6′ (18 kpc) wide along the major axis of the galaxy. This was only done for the receding half of the galaxy because the approaching half is partly confused with foreground Galactic H I emission. The averaging was performed on the smoothed 30″ resolution data. Figure 5 (left panel) shows the strip-averaged position − velocity diagram. It clearly shows the presence of low-intensity wings in the H I profiles (vertical slices through the position − velocity diagram), especially at radii from 0′ to 5′. These wings are indicated in figure 5 by the three levels of grey scale.

The faint profile wings are much more obvious if one corrects the data for the regular differential rotation of the galaxies by shifting the individual H I profiles to the same reference velocity. This way one removes the rotation and other large scale systematic motions to get profiles centred at 0 km s^{-1}. The right panel in figure 5 shows the same data as shown in the left panel after such a de-rotation.

A very obvious effect is that the low-intensity extensions to the H I profiles are only present within the de Vaucouleurs radius (R_{25}). This strongly suggests a relation between the star-formation activity in the disk and the presence of this fast-moving, redshifted and blueshifted H I gas. The amount of H I in this excess velocity component is 5×10^8 M$_\odot$ or 2% of the total amount of H I in NGC 6946. The total kinetic energy in this extra component is about 10^{55} ergs.

This global relation to the star-formation activity in the disk is also seen in the edge-on galaxy NGC 891 (Swaters et al. 1995, 1996). Figure 6 shows a H I column density map of the gas in the inner parts of the galaxy. The gas in the inner parts has been isolated by only integrating H I within the areas shown in the major axis position − velocity diagram (right panel of figure 6). If NGC 891 exhibits normal differential rotation, the velocities within the trapezoidal areas can only arise from H I on orbits in the inner regions of the galaxy; gas in the outer regions would have radial velocities much closer to the systemic velocity of the galaxy because of projection effects.

It is quite obvious from figure 6 that the H I is extending up to to z − distances of 1′

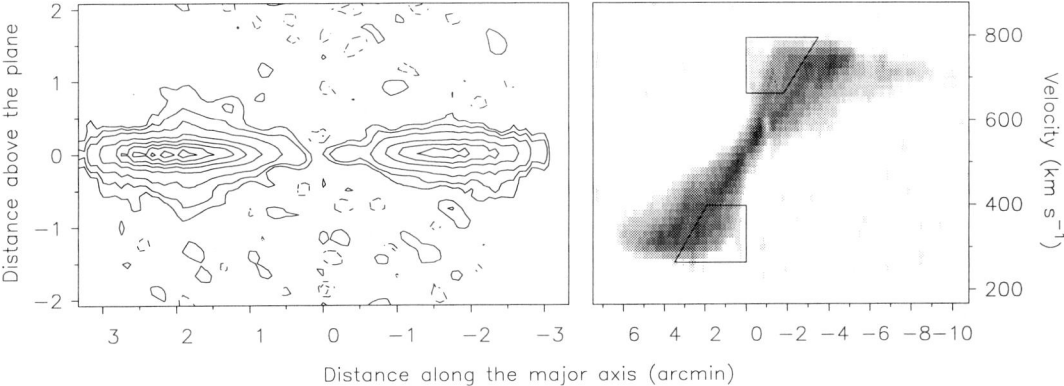

FIGURE 6. Major axis position − velocity diagram (right panel) of NGC 891 and H I column density map (left panel) constructed by integrating over the velocities within the shaded area of the position − velocity diagram in order to isolate H I in the inner parts of the galaxy.

or 2.5 kpc between 1′ and 2.5′ the NE part (left panel) of the galaxy. The H I in the SW part of the galaxy does not extend as far from the plane as the H I in the NE. This is in remarkably good agreement with the Hα structure which also is much more extended in the NE than in the SW (Rand et al. 1990 and Dettmar 1990). Also the non-thermal radio emission shows the same asymmetry (Dahlem et al. 1994). The distribution of star formation in the disk, as estimated from the Hα luminosity distribution is also asymmetric with much more vigorous star formation in the NE half than in the SW half of the galaxy. This suggests a direct relation between the star formation in the disk and the H I at high z distances.

A detailed study of high-velocity features on a smaller scale can at present only be done in very nearby galaxies. Such a study of M 33 is underway (Kolkman & van der Hulst, in preparation), using the WSRT H I observations of Deul & van der Hulst (1987). An example of a high velocity feature in M 33 is shown in figure 7. The general structure in figure 7 is very similar to the velocity structure seen in the hydrodynamical calculations of Rosen & Bregman (1995). The H I masses of such features are a few 10^5 M$_\odot$ and have kinetic energies of a few 10^{51} ergs, an amount that a few supernovae can easily produce.

4. Future Instrumentation

The data presented in the previous sections demonstrated clearly that fine structure in the H I in nearby galaxies is rather complex, much like we are used to see in detailed images of the H I distribution in our own Milky Way. Once sufficient angular resolution and sensitivity are available each galaxy examined reveals the presence of a rather frothy H I structure with evidence for H I shells, filaments and structures with deviations from the local differential rotation. The amounts of H I involved typically are a few percent of the total amount of H I in the disk. For large galaxies such as M 101, NGC 891 and NGC 6946 this amounts to several 10^8 M$_\odot$ and represents a total bulk kinetic energy of 10^{54-55} erg s^{-1}.

The evidence for a direct relation to the star-formation activity in the disk is circumstantial and global. No detailed correlation with the star-forming regions in the disk of a galaxy has yet been established. The main results in this respect are the high-z gas found

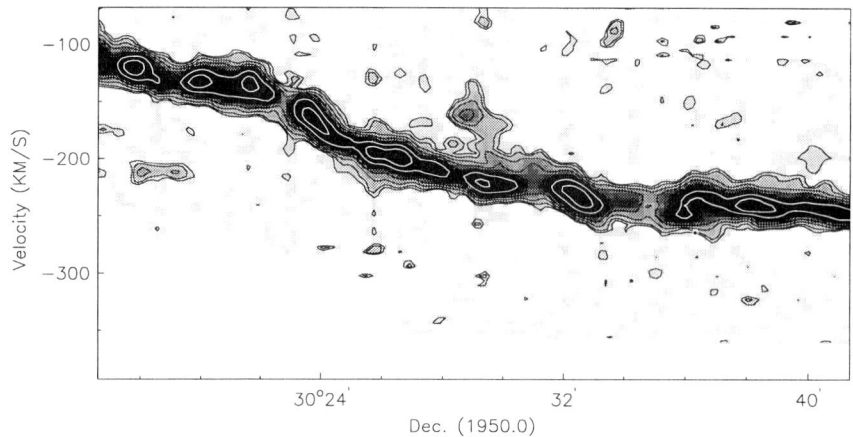

FIGURE 7. Example position − velocity diagram in M 33.

in the inner parts of NGC 891 whose scale height so beautifully reflects the north-south asymmetry in the distribution of star formation in the disk and the declining velocity dispersion in face-on galaxies with as prime example the extra broad velocity component found within the optical disk of NGC 6946.

Here one caveat should, however, be mentioned: in face-on galaxies we only observe the radial velocity of the HI and have no way of knowing whether the gas we observe is in front of or behind the main disk. We therefore cannot directly distinguish between gas falling into and gas moving out of the plane of a galaxy. The interpretation in terms of gas moving out of the disk into the halo is based on such nice and symmetric cases as NGC 5462, on the connection in velocity with the gas in the disk and on the (subjective) notion that the current picture of the effects of stellar winds and supernovae suggests that outflows of gas are more readily expected. To really make sure one sees outflows one needs to perform an HI absorption measurement against a continuum source located in the disk of a galaxy. These sources exist but are too faint for performing an HI absorption experiment with the present instrumental capabilities. HI absorption measurements will in addition provide a means of measuring more widely the spin temperature of the HI in galaxies.

With future, more sensitive instrumentation and observations it will be possible to search for small scale structure at faint levels of emission in a larger number of galaxies out to greater distances and get a better idea of the systematics of the relation between the star formation in the disks of galaxies and the presumably associated flow of HI out of the disk, the effects of the environment etc. With new data in other spectral domains such as Hα imaging, imaging in CO and in the IR, observations of the [CII] 158-μm line with ISO and from the KAO (and in the future SOFIA) it will be possible to assess in detail the heating and cooling of the ISM and study the relation between the energy balance in the ISM and the star-formation activity. So far, relations between star formation and other phenomena such as the ones discussed in this chapter have been very loose and phenomenological. A detailed assessment of the real physics involved is required and future observations such of the kind mentioned above will make this possible.

I would like to thank Olaf Kolkman, Rob Swaters and Jurjen Kamphuis for letting me use unpublished material.

REFERENCES

ALLEN, R.J., VAN WOERDEN, H. & GOSS, W.M. 1973. *A&A* **29**, 447.
ALLEN, R.J. & VAN DER KRUIT, P.C. 1978. *ARA&A* **16**, 103.
ARNABOLDI, M., CAPACCIOLI, M., CAPPELLARO, E., HELD, E.V. & SPARKE, L.S. 1993. *A&A* **267**, 21.
ATHANASSOULA, E. 1992. *MNRAS* **259**, 328.
BALDWIN, J.E., LYNDEN-BELL, D. & SANCISI, R. 1980. *MNRAS* **193**, 313
BEGEMAN, K. 1987. Ph.D. Dissertation, University of Groningen
BEGEMAN, K., BROEILS, A.H. & SANDERS, R.H. 1991. *MNRAS* **249**, 523.
BOSMA, A. 1978. Ph.D. Dissertation, University of Groningen
BOSMA, A. 1981. *AJ* **86**, 1825
BOSMA, A. 1989. In: *Large Scale Structure and Motions in the Universe* ed. Mezetti, M., Kluwer Academic Publishers, p. 65.
BOSMA, A. 1991. In: *Warped Disks and Inclined Rings around Galaxies* ed. S. Casertano, P. Sackett & F. Briggs, Cambridge University Press, p. 181.
BRAUN, R. 1996. In: *Cold Gas at High Redshift* eds. M. Bremer, Paul P. van der Werf, H.J.A. Rottgering, & C.L. Carilli, Kluwer Acad. Press, in press.
BRINKS, E. & BAJAJA, E. 1986. *A&A* **169**, 14
BRINKS, E. 1990. In: *The Interstellar Medium in Galaxies* ed. H. A. Thronson & J. M. Shull, Kluwer Ac. Publ., p. 39.
BROEILS, A. 1992. Ph.D. Dissertation, University of Groningen
BURTON, W.B. 1988. In: *Galactic and Extragalactic Astronomy* ed. G.L. Verschuur and K.I. Kellermann, Springer Verlag, N.Y., p. 245.
CASERTANO, S. & VAN GORKOM, J.H. 1991. *AJ* **101**, 1231
CAYATTE, V., BALKOWSKI, C., VAN GORKOM, J. H., & KOTANYI, C. 1990. *AJ* **100**, 604.
CAYATTE, V., KOTANYI, C., BALKOWSKI, C., & VAN GORKOM, J. H. 1993. *AJ* **107**, 1003.
COX, D.P. & SMITH, B.W. 1974. *ApJ* **189**, L105
DAHLEM, M., DETTMAR, R.J. & HUMMEL, E. 1994. *A&A* **290**, 384
DETTMAR, R.J. 1990. *A&A* **232**, L15
DEUL, E.R. & VAN DER HULST, J.M. 1987. *A&A* **67**, 509
DEUL, E.R. & DEN HARTOG, R.H. 1990. *A&A* **229**, 362
DE BLOK, W.J.G., MCGAUGH, S.S. & VAN DER HULST, J.M. 1996. *MNRAS* in press.
ELMEGREEN B.G. & CHIANG, W.H. 1982. *ApJ* **253**, 666
EMERSON, D.E. 1974. *MNRAS* **169**, 607
EMERSON, D.E. 1977. *MNRAS* **176**, 321
ENGLAND, M.N., GOTTESMAN, S.T. & HUNTER, J. H. 1990. *ApJ* **348**, 456.
GOTTESMAN, S.T., BALL, R., HUNTER, J.H. & HUNTLEY, J. M. 1984. *ApJ* **286**, 471.
HARTMANN, D. 1994. Ph. D. Dissertation, University of Leiden
HEILES, C. 1979. *ApJ* **229**, 533
HEILES, C. 1984. *ApJ Suppl.* **55**, 585
HEILES, C. 1990. *ApJ* **354**, 483
JÖRSÄTER, S. & VAN MOORSEL, G.A. 1995. *AJ* **110**, 2037.
KAMPHUIS, J.J. 1993. Ph.D. Dissertation, University of Groningen
KAMPHUIS, J.J., SANCISI, R. & VAN DER HULST, J.M. 1991. *A&A* **224**, L29
KAMPHUIS, J.J., & SANCISI, R. 1993. *A&A* **273**, L1
KENNICUTT, R.C. 1989. *ApJ* **344**, 685.
MALONEY, P. 1993. *ApJ* **414**, 41.

MacLow, M. & McCray, R. 1988. *ApJ* **324**, 776

MacLow, M., McCray, R. & Norman, M.L. 1989. *ApJ* **337**, 141

McKee, & Ostriker, J.P. 1977. *ApJ* **218**, 148

Milgrom, M. 1983. *ApJ* **270**, 365

Milgrom, M. 1988. *ApJ* **333**, 689

Moore, E.M. & Gottesman, S.T. 1995. *ApJ* **447**, 159.

Newton, K. 1980. *MNRAS* **190**, 689

Newton, K. 1980. *MNRAS* **191**, 169

Newton, K. 1980. *MNRAS* **191**, 615

Newton, K. & Emerson, D.T. 1977. *MNRAS* **181**, 573

Norman, C.A. & Ikeuchi, S. 1989. *ApJ* **345**, 372

Ondrechen, M.P. & van der Hulst, J.M. 1989a. *ApJ* **342**, 39.

Ondrechen, M.P. & van der Hulst, J.M. 1989b. *ApJ* **342**, 29.

Puche, D., Westphal, D., Brinks, E. & Roy, J. 1992. *AJ*, **103**, 1841

Rand, R. J., Kulkarni, S. R. & Hester J. J. 1990. *ApJ* **352**, L1

Rogstad, D.H, Lockart, I.A. & Wright, M.C.H. 1974. *ApJ* **193**, 309

Rogstad D.H & Shostak, G.S. 1971. *A&A* **13**, 99

Rogstad D.H & Shostak, G.S. 1973. *A&A* **22**, 111

Rosen, A. & Bregman, J.N. 1995. *ApJ* **440**, 634

Rots, A.H., Crane, P.C. Bosma, A., Athanassoula, E. & van der Hulst, J.M. 1990. *AJ* **100**, 387

Sancisi, R. 1976. *A&A* **53**, 159.

Sancisi, R., Allen, R.J. & Sullivan, W.T. 1979. *A&A* **78**, 217.

Sackett, P.D. & Sparke, L.S. 1990. *ApJ* **361**, 408

Sellwood, J.A. & Sparke L.S. 1988. *MNRAS* **231**, 25.

Swaters, R. Sancisi, R. & van der Hulst, J.M. 1995. *Astro. Lett. and Communications* **31**, 161

Swaters, R. Sancisi, R. & van der Hulst, J.M. 1996. *ApJ* in preparation

Tomisaka, K. 1992. *Pub. Ast. Soc. of Japan* **44**, 177.

Tomisaka, K.& Ikeuchi, S. 1986. *Pub. Ast. Soc. of Japan* **39**, 109.

Toomre, A. 1981. In: *The Structure and Evolution of Normal Galaxies* ed. S.M. Fall & D. Lynden-Bell, Cambridge University Press, 111.

van Albada, T.S. & Sancisi R. 1987. *IAU Symposium 117* ed. J. Kormendy & G.R. Knapp, p. 67.

van der Hulst, J.M. 1996. 'Bubbles and Holes in the Interstellar Medium' in *HI in the Universe*, Minn. Lecture Series no. 5, ed. E.D. Skillman, ASP Conference series, in press.

van der Hulst, J.M., Skillman, E.D., Smith, T.R., Bothun, G.D., McGaugh, S.S., & de Blok, W.J.G. 1993. *AJ* **106**, 548.

van Gorkom, J.H., Dwarakanat, K.S. & Guhathakurta, P. 1996 In: *Cold Gas at High Redshift* eds. M. Bremer, Paul P. van der Werf, H.J.A. Rottgering, & C.L. Carilli, Kluwer Acad. Press, in press.

Visser, H.C.D. 1980. *A&A* **88**, 149.

Warmels, R. 1986. Ph.D. Dissertation, University of Groningen

Radio emission from galactic disks

By RAINER BECK

Max-Planck-Institut für Radioastronomie, Auf dem Hügel 69, D-53121 Bonn, Germany

Galactic radio continuum emission emerges from a thin disk (supernova remnants, HII regions, gas and magnetic field concentrations) and a thick disk filled by extended magnetic fields and cosmic-ray electrons. Their energy spectra, derived from the synchrotron spectra of normal galaxies, are similar to the spectrum of the local Galactic cosmic-ray electrons, supporting a universal acceleration mechanism. Spectral steepening beyond $\simeq 1$ GHz is observed in galaxies interacting with the intergalactic medium which leads to magnetic field amplification and thus higher synchrotron losses of the electrons.

The tightness and non-linearity of the radio–far infrared correlation can be explained by a model where interstellar magnetic fields are coupled to gas clouds. This is supported by observations of field enhancements in regions with large-scale concentrations of dust or cool gas.

Interstellar magnetic fields, as deduced from multi-frequency polarization observations, show a well-ordered structure largely following the spiral arms. In some galaxies an axisymmetric spiral pattern dominates (the field being directed *inwards*), while others host a dominating bisymmetric spiral field or "mixed modes", as predicted from non-linear dynamo theory.

As long as star-formation activity is low, the magnetic fields are rather regular. Strong star formation leads to turbulent cloud motions and supernova explosions, which tangle the field, so that the radio emission is only weakly polarized. As a consequence the highest fractional polarizations and polarized intensities at centimetre wavelengths are found in *interarm* regions. At decimetre wavelengths, galactic disks become "optically thick" for polarized emission.

In NGC 6946 the regular field is concentrated in narrow "magnetic arms" located *in between* the optical spiral arms. The field cannot simply be frozen into the gas and orientated by a density-wave flow. A galactic dynamo may provide a stable spiral pattern of the field, but non-axisymmetric models are still being developed.

---

1. Introduction

The observation of radio disks requires both single-dish and synthesis telescopes. Synthesis telescopes provide high resolution, but are insensitive to spatial structures larger than some angular limit. This reduces the visibility of the diffuse disk. Missing large-scale structures in maps of Stokes parameters Q and U may systematically distort the polarization angles. Single-dish telescopes provide the missing spatial information.

Fig.2 shows an example of the maximum-entropy combination of synthesis and single-dish data in total power and linear polarization (Beck & Hoernes 1996). The maps with VLA data alone (Fig.1) contain only 27% of the total flux density and 40% of the polarized flux density visible in the combined maps.

Galactic radio continuum emission has three components:
- the nuclear region
- the thin disk (HII regions, gas/magnetic field concentrations, supernova remnants)
- the thick disk (or "radio halo" if exceptionally extended)

This paper only deals with thin and thick disks.

In edge-on galaxies, the thin and thick disks are clearly separated due to their different vertical extents (Fig.3). The contribution from the thick disk to the total radio emission dominates in most edge-on galaxies (Dumke et al. 1995), but can be small in galaxies with low star-forming activity like M31 (Beck 1982) or NGC 4565 (Sukumar & Allen 1991). For our own Galaxy, Beuermann et al. (1985) found that 90% of the total emission at

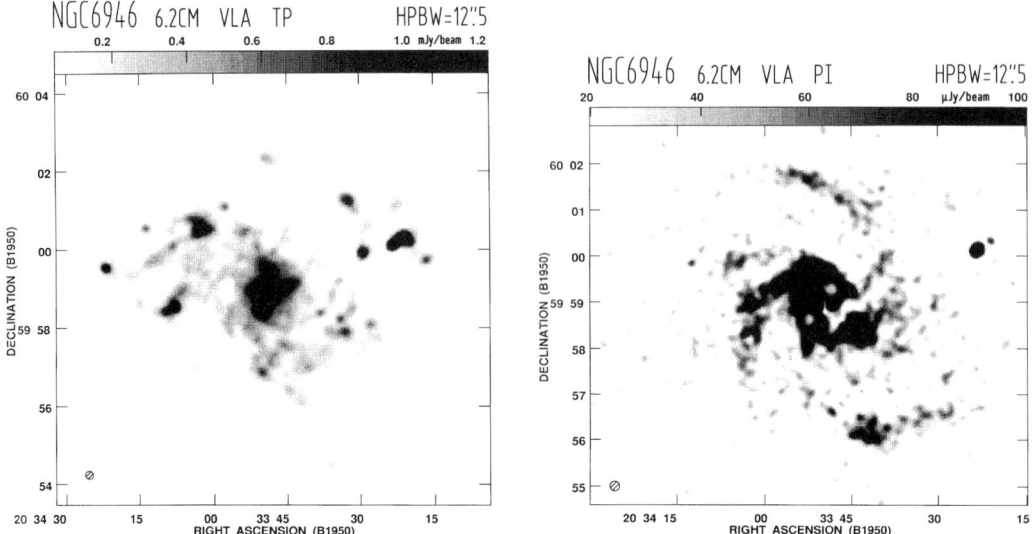

FIGURE 1. Total and polarized emission of NGC 6946, observed at λ6.2cm with the VLA (12.5″ synthesized beam). From Beck & Hoernes (1996)

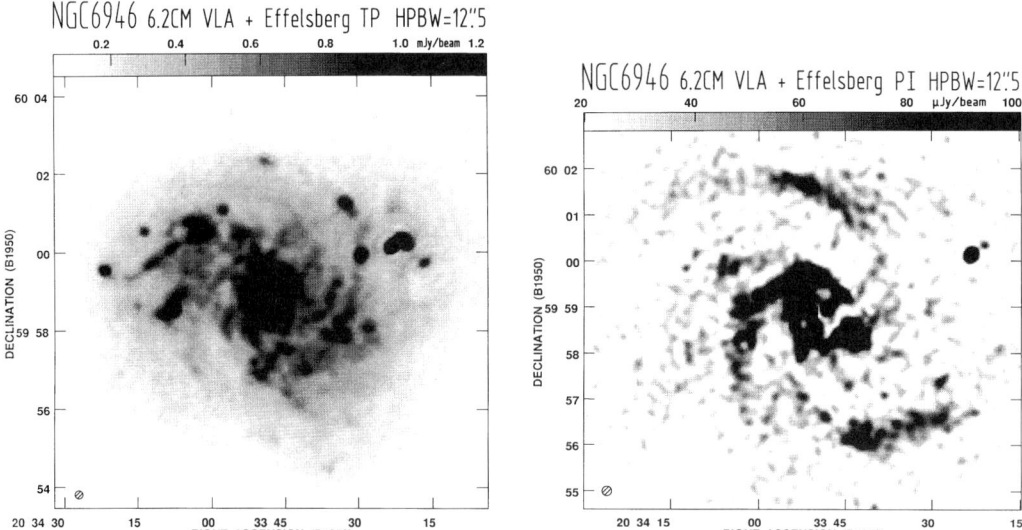

FIGURE 2. Total and polarized emission of NGC 6946, observed at λ6.2cm with the VLA (12.5″ synthesized beam) and combined with the extended emission observed with the Effelsberg 100-m telescope (2.5' resolution). From Beck & Hoernes (1996)

408 MHz is from the thick disk with a scale height of $\simeq 1.5$ kpc near the Sun (scaled to 8.5 kpc galactocentric distance), compared with $\simeq 0.15$ kpc for the thin disk.

In face-on galaxies, the thin disk can be resolved into spiral arms and other regions related to star-forming activity, whereas the thick disk is much smoother and also fills interarm regions. In NGC 6946 the spiral-arm component of the thin disk (Fig.1a)

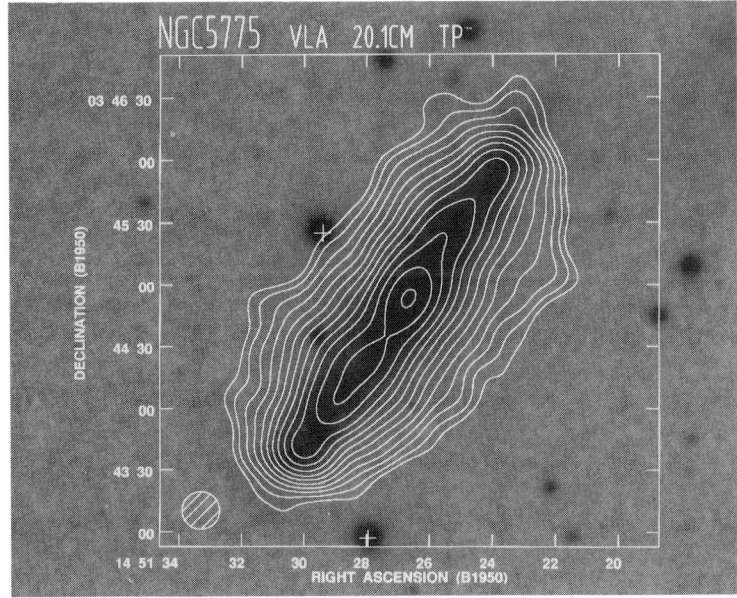

FIGURE 3. Total emission of NGC 5775, observed at λ20.1cm with the VLA (18″ synthesized beam). (Golla, unpublished)

contains only $\simeq 30\%$ of the total flux density, while e.g. in M31 the spiral-arm component dominates (Beck 1982).

Supernova remnants account for less than 10% of the total radio luminosity of spiral galaxies (Condon 1992). Thermal emission from HII regions and diffuse ionized gas contributes, on average, only 7% at λ20cm (Niklas 1995). Thermal radio and optical H_α emission generally coincide in spiral arms.

The spiral-arm component of the nonthermal radio emission is enhanced due to magnetic field concentrations and has a somewhat broader distribution, organized in longer structures (Tilanus et al. 1988). Even in star-forming regions the radio emission is mostly nonthermal at $\lambda \geq 3$cm (Deutsch & Allen 1993; Ehle et al. 1996a). In density-wave galaxies, the nonthermal emission is offset from the thermal emission and mostly follows the dust lanes (Tilanus et al. 1988; Deutsch & Allen 1993; see also Fig.6).

Synchrotron emission bears information on cosmic-ray electrons and magnetic fields in the interstellar medium. In the following sections, implications from observations for these components are discussed.

2. Spectrum and propagation of cosmic-ray electrons

Niklas et al. (1995a) observed a sample of 214 Shapley-Ames galaxies. The spectra of the integrated radio flux density from spiral galaxies are generally straight between $\simeq 400$ MHz and $\simeq 5$ GHz (Niklas 1995). Below $\simeq 400$ MHz, flattening due to energy losses of the cosmic-ray electrons occurs. Beyond $\simeq 5$ GHz, the thermal contribution flattens the spectrum. After subtraction of the thermal emission (possible for 74 galaxies of the sample), the average synchrotron spectral index is $\alpha_{nth} = -0.85$ (where $S_\nu \propto \nu^{\alpha_{nth}}$), with a standard deviation of only 0.13.

Steady-state models for cosmic rays predict a gradual steepening of the electron spectrum with increasing energy E_e due to various energy losses (e.g. Pohl 1993a). As $\nu_{obs} \propto B_t E_e^2$, the steepening of the synchrotron spectrum is smoother than that of the energy spectrum, and it is further smeared out across a galaxy by variations in the strength of the total magnetic field B_t. Although the steady-state models could successfully be fitted to the data of some nearby spiral galaxies (Pohl et al. 1991), most synchrotron spectra compiled by Niklas (1995) allow a spectral steepening of only $\Delta\alpha \leq 0.05$ over the available frequency range which is significantly smaller than predicted. Furthermore, the spectral index does not correlate with the equipartition field strength as expected in case of synchrotron losses (Niklas & Beck 1996). Better low-frequency data are needed.

The energy spectrum of the local Galactic cosmic-ray electrons shows a steepening beyond 10 GeV (e.g. Golden et al. 1994). However, in the range below 10 GeV relevant for GHz synchrotron emission (assuming $B_t \simeq 9\mu G$, see Section 4), the energy spectrum is consistent with an almost straight power-law with a slope of $s \simeq -2.7$ (Golden et al. 1994) or $s = -2.83 \pm 0.11$ (Basini et al. 1991), yielding $\alpha_{nth} = (s-1)/2 = -0.85...0.90$, in good agreement with the extragalactic data. An injection spectrum for the electrons with $s_0 \simeq -2.2$, as obtained from acceleration models in supernova remnants (Völk et al. 1988) and from local cosmic-ray observations (Engelmann et al. 1990), plus leakage out of the galaxy $\propto E_e^{0.6}$ (Engelmann et al. 1990) can explain the extragalactic spectra.

Significant spectral steepening beyond $\simeq 1$ GHz was observed in galaxies interacting with the intergalactic gas, e.g. in NGC 2276 (Hummel & Beck 1995) and in some galaxies of the Virgo cluster (Niklas et al. 1995b). These galaxies have exceptionally strong magnetic fields due to the interaction, and thus synchrotron losses of the cosmic-ray electrons are significant already at energies of a few GeV.

Cosmic rays are believed to be accelerated in supernova remnants (e.g. Völk et al. 1988), diffuse away and fill the thick disk. The diffusion lengths are a few kpc in the disk (Bicay & Helou 1990; Duric 1991), while the diffusion in the vertical direction (derived from typical synchrotron scale heights of edge-on galaxies) is about 1 kpc (Dumke et al. 1995). This explains the smooth appearance of the thick radio disks. Most variations of the spectral index of the total emission can be explained by variations of the relative thermal contribution (Klein et al. 1982; Hummel & Gräve 1990). Outside the star-forming radius of galaxies the synchrotron spectrum steepens due to aging of the cosmic-ray electrons.

It is possible that large variations of synchrotron spectral indices may occur on small scales which have not been resolved in most observations. However, high angular resolution requires observations with synthesis telescopes which often miss large-scale structures so that spectral variations are amplified. Combined data from single-dish and synthesis telescopes, or from synthesis telescopes in several configurations, are required. In NGC 3556, Bloemen et al. (1993) found abrupt spectral changes on scales of ≤ 1 kpc. Future observations should show whether this occurs also in other galaxies.

The radio spectrum of the thin disk can be determined only in edge-on galaxies. For the prototype NGC 891, Hummel et al. (1991) found an almost constant spectral index of $\alpha \simeq -0.65$ until 8-kpc distance from the centre, compared with $\alpha \simeq -1.0$ in the thick disk which supports a higher thermal fraction in the thin disk. In the halo the spectrum further steepens.

3. The radio–far infrared correlation

The correlation between radio and far-infrared luminosities of galaxies is one of the tightest correlations known in astrophysics. It is not just a distance or volume effect, and it holds also within galaxies down to scales of $\simeq 1$ kpc (e.g. Bicay & Helou 1990). From their radio survey of Shapley-Ames galaxies, Niklas et al. (1995a) showed that the correlation is non-linear with a slope of 1.25 ± 0.08.

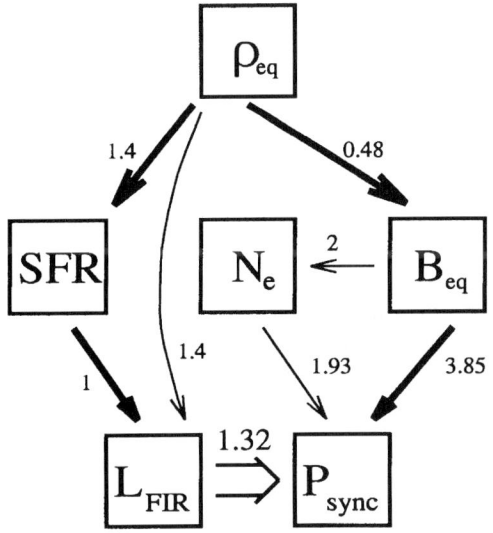

FIGURE 4. The relations responsible for the correlation between synchrotron luminosity P_{sync} and far infrared luminosity L_{FIR}. ρ_{eq}: equivalent gas density (derived from HI and H_2 masses), SFR: star-formation rate, B_{eq}: equipartition magnetic field strength, N_e: number density of cosmic-ray electrons. The numbers give the exponents in the relationships, e.g. $P_{sync} \propto B_{eq}^{3.85}$. From Niklas & Beck (1996)

Most attempts to explain the correlation tried to couple the radio and far-infrared emissivities to the star-formation rate (e.g. Hummel et al. 1988; Völk 1989; Condon 1992; Lisenfeld et al. 1996). Finite escape probabilities of cosmic-ray electrons and UV photons further complicate the picture (Helou & Bicay 1993). Niklas & Beck (1996) explained the correlation, globally and locally, and with the correct slope, by a close coupling of magnetic fields to gas clouds ($B_t \propto \rho^{0.5}$), a connection of the star-formation rate to gas density, and equipartition between the cosmic-ray and magnetic field energy densities (Fig.4).

A detailed comparison between the total synchrotron intensity and the cool gas (HI + 2 H_2) as observed in the HI and CO lines confirmed a coupling of the magnetic field to the gas in a spiral arm of M31 (Berkhuijsen et al. 1993). The total field strength B_t is higher in spiral arms because the number density of clouds and ρ of the diffuse gas increase. The observed synchrotron–CO correlation in galactic disks on larger scales (Adler et al. 1991) is one aspect of the relation between magnetic field and gas.

4. Polarization observations

Linearly polarized emission is detected if the magnetic field component perpendicular to the line of sight has a preferential orientation $B_{reg,\perp}$ within the telescope beam. In a completely regular magnetic field the degree of polarization of the synchrotron emission takes values of $P_0 = 70 - 75\%$, depending on the nonthermal spectral index.

In a "magneto-ionic" medium containing magnetic fields and ionized gas, the plane of polarization of a linearly polarized radio wave is rotated by an angle $\Delta \psi$ which increases with the integral of $n_e B_{reg,\|} \lambda^2$ along the line of sight (where n_e is the thermal electron density and $B_{reg,\|}$ is the component of the regular magnetic field along the line of sight).

The quantity $\Delta\psi/\Delta\lambda^2$ is called the rotation measure, RM. The sign of RM allows the two opposite directions of $B_{reg,\parallel}$ to be distinguished.

Polarized emission is reduced by various depolarization effects:
- unresolved magnetic field structures across the telescope beam
- variations in magnetic field orientation along the line of sight within the synchrotron-emitting layer
- differential Faraday rotation along the line of sight within the synchrotron-emitting layer
- Faraday dispersion (along the line of sight and across the beam) by cells with random magnetic field orientations within the synchrotron-emitting layer or in the foreground
- gradients in Faraday rotation across the telescope beam.

The first two effects are independent of wavelength, magnetic field strength and electron density and dominate at centimetre wavelengths. The other effects only occur in magneto-ionic regions and increase as different functions of λ, field strength and electron density (Burn 1966; Ehle & Beck 1993).

As Faraday effects are large at *long* wavelengths ($\lambda \geq 20$cm), galactic disks become "optically thick" for polarized emission so that only polarized emission from a nearby layer can be detected (Sukumar & Allen 1991; Horellou et al. 1992; Ehle & Beck 1993; Neininger et al. 1993; Beck 1993). Rotation angles observed in the decimetre range are no longer proportional to λ^2, and RM may reverse its sign even without magnetic field reversals (Burn 1966). At certain wavelengths "anomalous polarization" may appear (Horellou et al. 1992).

Fig.5 demonstrates the effect of Faraday depolarization as observed in the spiral galaxy NGC 6946. At λ20cm, some regions are almost completely depolarized, whereas at λ6cm there is significant polarized emission.

At *short* wavelengths ($\lambda \leq 6$cm) Faraday effects become weak and the radio astronomer gets a clear view of galactic magnetic fields. Typical fractional polarizations are only 1–10% in spiral arms, but 20–40% in between the spiral arms and in outer regions of galaxies. The observed $B_{reg,\perp}$ vectors trace the field orientations in the sky plane (see Sect.6).

5. Magnetic field strengths

The average "equipartition" strength of the total field $\langle B_{t,\perp} \rangle$ can be derived from the total radio synchrotron intensity if energy-density equipartition (or minimum total energy) between cosmic rays and magnetic fields is assumed. Furthermore, the ratio K between the energy densities of cosmic-ray protons and electrons, the synchrotron spectral index α_{nth}, the extent of the radio emission along the line of sight, and the volume filling factor of the magnetic fields have to be known. As the derived equipartition field strength depends only on the power $1/(\alpha_{nth} + 3)$ of any of these parameters, even large uncertainties lead to moderate errors in field strength. An estimate of the regular field strength $B_{reg,\perp}$ can be obtained by using the observed intensity of the polarized emission.

The standard minimum-energy formula uses a fixed integration interval in *frequency* to determine the total energy density of cosmic-ray electrons. This procedure makes it difficult to compare minimum-energy field strengths between galaxies because a fixed frequency interval corresponds to different electron energy intervals, depending on the field strength itself. When, correctly, a fixed integration interval in *electron energy* is used, the minimum-energy and energy-equipartition estimates give almost the same values for $\langle B_t^2 B_{t,\perp}^{1+\alpha_{nth}} \rangle \simeq \langle B_{t,\perp}^{3+\alpha_{nth}} \rangle$. The resulting estimate $\langle B_{t,\perp}^{3+\alpha_{nth}} \rangle^{1/(3+\alpha_{nth})}$ is larger than the

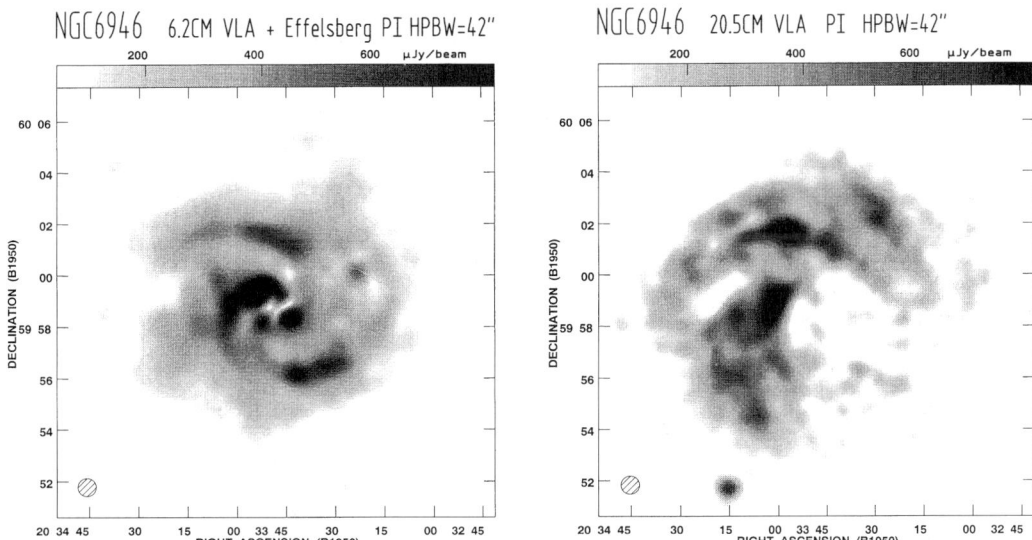

FIGURE 5. Polarized emission of NGC 6946, observed at λ6.2cm (as Fig.2b, but smoothed to a 42″ beam) and at λ20cm, observed with the VLA (42″ synthesized beam, from Beck (1991)).

mean field $\langle B_{t,\perp} \rangle$ if the field strength varies along the path length, since $\langle B_{t,\perp}\rangle^{3+\alpha_{nth}} \leq \langle B_{t,\perp}^{3+\alpha_{nth}}\rangle$.

In nearby galaxies the average total field strengths range between $\simeq 4\,\mu$G in M33 (Buczilowski & Beck 1991) and $\simeq 19\mu$G in NGC 2276 (Hummel & Beck 1995). In spiral arms they can reach $20\,\mu$G as in NGC 6946 (Beck 1991) and NGC 1566 (Ehle et al. 1996a). The strengths of the regular fields are typically 1–5 μG, but $\simeq 10\,\mu$G in the interarm regions of NGC 6946 (Fig.2b). The observed ratio B_{turb}/B_{reg} is $\simeq 2 - 3$ in spiral galaxies (Buczilowski & Beck 1991), but decreases with increasing resolution.

For the Galactic interarm field near the Sun, Heiles (1996) derived $B_{turb} \simeq 3.6\,\mu$G, $B_{reg} \simeq 2.2\,\mu$G and hence $B_t \simeq 4.2\,\mu$G, whereas $B_t \simeq 6\,\mu$G in spiral arms, mainly due to a stronger B_{turb}.

The mean magnetic field strength for the sample of 74 spiral galaxies by Niklas (1995) is $9\,\mu$G with a standard deviation of $3\,\mu$G, using $K = 100$. For his sample of 88 Sbc galaxies Hummel (1986) gave a mean minimum-energy field of $\simeq 8\,\mu$G. For the sample of 146 late-type galaxies by Fitt & Alexander (1993), the mean total minimum-energy field strength for $K = 100$ is $10 \pm 4\,\mu$G.

The "equipartition" assumption has been frequently questioned. Using γ-ray observations to obtain indirect data about the distribution of cosmic-ray electrons in the Magellanic Clouds and comparing with radio data, Chi & Wolfendale (1993) claimed that energy equipartition is not valid. Pohl (1993b) replied that the data do not require deviations from equipartition, but only a smaller ratio K than in our Galaxy. In addition, γ and radio emission may originate from different regions and thus may trace different electron spectra.

Arguments in favour of equipartition conditions have been presented by Duric (1990). Furthermore, the observed Faraday rotation measures are consistent with equipartition strengths of the regular field and electron densities in the diffuse ionized medium (M. Krause et al. 1989a; Buczilowski & Beck 1991).

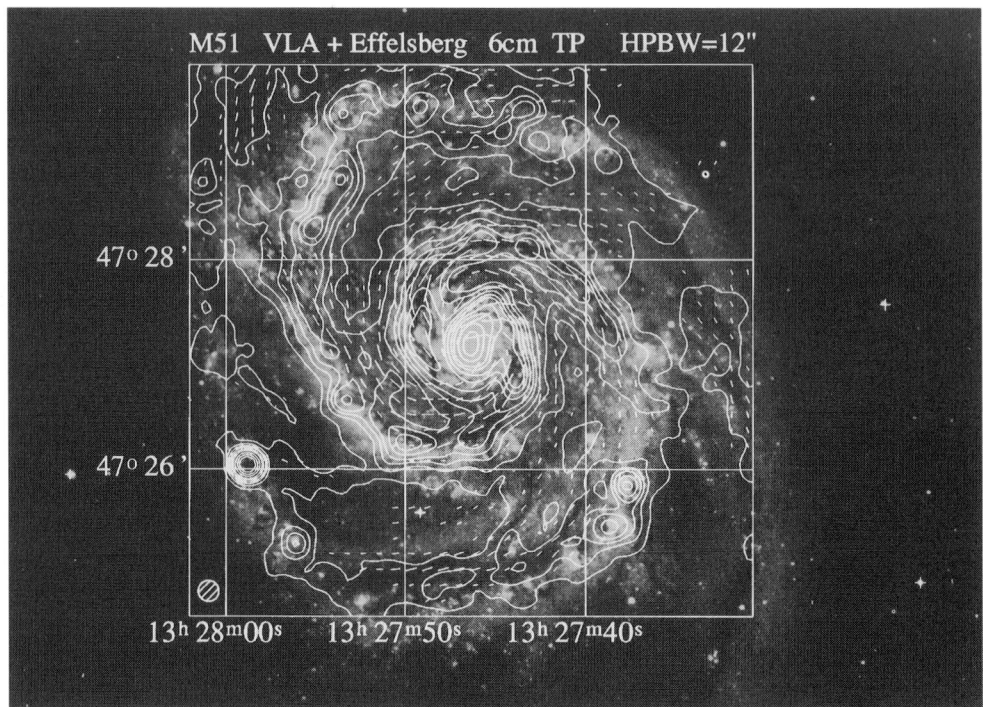

FIGURE 6. Total and polarized radio continuum emission of M51, observed at λ6.2cm with the VLA (12″ synthesized beam) and combined with Effelsberg λ6.3cm data. From Neininger & Horellou (1996)

6. Magnetic field structures

Alignment of magnetic fields along dust lanes has been found in many galaxies, e.g. in M31 (Beck et al. 1989), in M51 (Fig.6) and even in galaxies lacking grand-design spiral structure, such as the "flocculent" galaxy NGC 5055 (Soida et al., in prep.).

The nearby spiral galaxies M51 and M81 exhibit strong density waves. In the density-wave picture the magnetic field is frozen into the gas and is transported by the gas flow. Thus the field orientation should reflect the streamlines of the gas, not the structure of the spiral wave itself. The orientations of the streamlines are roughly aligned with the arms, but show large variations in interarm regions (Otmianowska-Mazur & Chiba 1995). Radio polarization observations show that the pattern of the regular magnetic field follows closely the optical spiral pattern in M51 (Neininger 1992; Neininger & Horellou 1996), M81 (M. Krause et al. 1989b), M83 (Neininger et al. 1993; Ehle et al. 1996b) and NGC 1566 (Ehle et al. 1996a), but *not* the streamlines of the gas.

Strong shocks should compress the magnetic field and should lead to degrees of radio polarization of 40–70% (Beck 1982) at the inner edges of the spiral arms. The strongest total fields in M51 are indeed found at the positions of the prominent dust lanes at the inner edges of the optical spiral arms (Mathewson et al. 1972). The regular fields are also strongest in dust lanes (Fig.6), although some polarized features extend far into the interarm regions and are only 10–30% polarized at λ6 cm. Remarkably, one dust lane crosses the eastern spiral arm of M51, and so does the field (Fig.6). Hence, the regular fields in M51 are more closely coupled to the cool gas than to the density waves.

In M81, M83 and NGC 1566 the strongest polarized intensities and hence strongest regular fields occur in *interarm regions*, while the total synchrotron intensity (showing the total field) is highest on the spiral arms. Field tangling in the spiral arms due to increased turbulent motions of gas clouds and supernova shock fronts may explain this result (Sukumar & Allen 1989; Beck 1991). High-resolution observations of M81 (Schoofs 1992) confirmed that the regular fields extend across almost the entire interarm region, but are somewhat stronger near the inner edge of the prominent western spiral arm, where some dust clouds are visible.

The distribution of the magnetic field orientations is much smoother between spiral arms and interarm regions than that of the regular magnetic field strengths. Soida *et al.* (1996) showed that strength and orientations of the regular fields in NGC 4254 reveal much less arm-interarm variations than expected from density-wave compression in its two major arms, and regular fields even exist in regions of chaotic optical pattern.

In summary, density waves cannot be the only agent acting on the field structure. The galactic dynamo (Sect.8) has to preserve the large-scale spiral pattern. On small scales, field anchoring in gas clouds (Beck 1991) can explain some increased field alignment in the interarm space, but not the small variation of field orientations.

7. Magnetic spiral arms

Arms of polarized emission in IC 342 are located *between* the inner optical arms (M. Krause 1993), but the outer polarized arms cross the optical arms (M. Krause *et al.* 1989a). High-resolution observations of another gas-rich galaxy, NGC 6946 (Beck & Hoernes 1996), revealed a much more regular distribution of polarized intensity (Fig.2b) with long magnetic arms located in *interarm* regions, running perfectly parallel to the adjacent optical spiral arms. The magnetic structure is more symmetric than that of the total field, the gas and the young stars, all of which show a quite irregular pattern.

The magnetic arms in NGC 6946 do not fill the entire interarm spaces like in M81, but are only $\simeq$ 500–1000 pc wide. As their shapes are very similar at λ6 cm and λ3 cm, they cannot be due to wavelength-dependent Faraday depolarization effects. They are also visible as small ridges in total emission so that wavelength-independent depolarization can also be excluded.

After subtraction of the (unpolarized) background the maximum degree of polarization in the northern magnetic arm becomes $\simeq$ 65%. Compared with 75% as the theoretical upper limit, the fields in the magnetic arms must be *almost totally aligned*. The peak strength of the regular field is $\simeq 13\,\mu$G in the northeastern magnetic arm. No present-day model can explain how such strong and regular magnetic arms could be generated.

8. Dynamos and their radio signatures

The linear mean-field dynamo (e.g. Ruzmaikin *et al.* 1988; Wielebinski & F. Krause 1993) generates magnetic field modes which are of spiral structure due to their azimuthal and radial field components. The pitch angle depends on the amount of differential rotation and on the thickness of the gas disk (e.g. Elstner *et al.* 1992). Non-linear dynamo theory predicts mixed modes or even irregular field structures (Beck *et al.* 1996).

We distinguish between spiral structures that can be considered as basically axisymmetric and basically antisymmetric (bisymmetric) with respect to rotation by 180°: ASS ($m = 0$) and BSS ($m = 1$), respectively. Higher azimuthal Fourier modes m may be superimposed on these dominant ones. Fields containing several Fourier components of significant amplitude have mixed spiral structure, MSS.

An ASS (BSS) field produces a 2π-periodic (π-periodic) distribution of RM along azimuthal angle ϕ (M. Krause 1990; Wielebinski & F. Krause 1993). An observational check is difficult if the data suffer from Faraday depolarization, if the regular field is not parallel to the plane of the galaxy, if its pitch angle in the disk is not constant, or if the disk is surrounded by a halo with magnetic fields of comparable strengths. A more direct method of analysis considers polarization angles ψ without converting them into Faraday rotation measures (Sokoloff et al. 1992; Berkhuijsen et al. 1996).

Singly-periodic RM variations indicative of ASS fields have been detected in the disks of M31 (Sofue & Takano 1981; Beck 1982) and IC 342 (Gräve & Beck 1988; Krause et al. 1989a; Sokoloff et al. 1992). Here the phase of the azimuthal variation of RM is equal to the magnetic pitch angle. In NGC 6946 the phase of the 2π-periodic RM variation differs significantly from the value of the mean magnetic pitch angle (Ehle & Beck 1993), in conflict with a pure ASS field. The two main magnetic spiral arms have RMs of opposite sign, indicative of a superposition of the $m=0$ and the $m=2$ mode.

The only clear candidate for a BSS symmetry is M81 (M. Krause et al. 1989b; Sokoloff et al. 1992). In M33 the weak polarized emission leads to large uncertainties in RM, and a bisymmetric field can be claimed only with some caution (Buczilowski & Beck 1991). The same is valid for NGC 2276 (Hummel & Beck 1995). The galaxies M33, M81 and NGC 2276 show signs of gravitational interaction, which may be important in producing non-axisymmetric dynamo fields (Moss 1995).

The strongly interacting galaxy M51 is a special case. Analyzing all available polarization angle data, the field in M51 can be described as MSS, with axisymmetric and bisymmetric components having about equal weights in the disk, together with a horizontal axisymmetric halo field with opposite direction (Berkhuijsen et al. 1996).

In all other galaxies observed so far the data are still insufficient to allow a firm conclusion. Any statistical analysis sorting galaxies with respect to single dynamo modes should be regarded with great caution.

By comparing the signs of the RM distribution and the velocity field, inward and outward directions of the radial component of the spiral magnetic field can be distinguished. Surprisingly, all ASS fields (M31, IC 342, NGC 253) and the MSS field in NGC 6946 point *inwards*. Dynamo action does not prefer one direction. This indicates some asymmetry in the initial seed field (F. Krause, in prep.).

A dynamo spiral field interacts with optical spiral arms in various ways. Firstly, streaming of gas clouds and their collision rates are modified. Secondly, magnetic fields are essential for the onset of star formation as they allow angular momentum to be removed from the protostellar cloud during its collapse. Thirdly, strong fields shift the stellar mass spectrum towards the more massive stars (Mestel 1994). On the other hand, the conditions leading to efficient star formation may oppose dynamo action (Nozakura 1993). Dynamo models taking into account basic non-axisymmetric effects, e.g. in the velocity field and distribution of gas in a density-wave potential (Subramanian & Mestel 1993; Hanasz & Chiba 1994) or in a bar potential (Chiba & Lesch 1994), are being developed.

9. Future observations

Future studies of galactic disks in radio continuum need higher resolution and higher sensitivity. This calls for more collecting area for synthesis telescopes and for multi-beam systems at highest frequencies for single dishes. High-resolution, low-frequency data are required to obtain better synchrotron spectra and to search for "anomalous" polarization. Intermediate frequencies ($\lambda \simeq 13$cm, available at the Effelsberg and ATCA telescopes, but not yet at the VLA) allow one to observe layers at various depths in polarization.

Finally, ultracompact synthesis arrays are desired to give better sensitivity for diffuse emission.

REFERENCES

Adler, D.S., Allen, R.J. & Lo, K.Y. 1991. *ApJ* **382**, 475

Basini, G., Bongiorno, B., Brunetti, M.T., et al. 1991. In *22nd Int. Cosmic Ray Conf.*, vol. 2, pp. 137

Beck, R. 1982. *A&A* **106**, 121

Beck, R. 1991. *A&A* **251**, 15

Beck, R. 1993. In *IAU Symp. 157. The Cosmic Dynamo* (ed. F. Krause et al.), p. 283. Kluwer, Dordrecht.

Beck, R. & Hoernes, P. 1996. *Nature* **379**, 47

Beck, R., Loiseau, N., Hummel, E., Berkhuijsen, E.M., Gräve, R. & Wielebinski, R. 1989. *A&A* **222**, 58

Beck, R., Brandenburg, A., Moss, D., Shukurov, A. & Sokoloff, D. 1996. *Ann. Rev. A&A*, in press

Berkhuijsen, E.M., Bajaja, E. & Beck, R. 1993. *A&A* **279**, 359

Berkhuijsen, E.M., Horellou, C., Krause, M., Neininger, N., Poezd, A., Shukurov, A. & Sokoloff, D. 1996. *A&A*, submitted

Beuermann, K., Kanbach, G. & Berkhuijsen, E.M. 1985 *A&A* **153**, 17

Bicay, M.D. & Helou, G. 1990 *ApJ* **362**, 59

Bloemen, H., Duric, N. & Irwin, J. 1993. In *23rd Int. Cosmic Ray Conf.*, vol. 2, p. 279

Buczilowski, U.R. & Beck, R. 1991. *A&A* **241**, 47

Burn, B.J. 1966 *MNRAS* **133**, 67

Chi, X. & Wolfendale, A.W. 1993. *Nature* **362**, 610

Chiba, M. & Lesch, H. 1994. *A&A* **284**, 731

Condon, J.J. 1992. *Ann. Rev. A&A* **30**, 575

Deutsch, E.W. & Allen, R.J. 1993. *Astron. J.* **106**, 1812

Dumke, M., Krause, M., Wielebinski, R. & Klein, U. 1995. *A&A* **302**, 691

Duric, N. 1990. In *IAU Symp. 140. Galactic and Intergalactic Magnetic Fields* (ed. R. Beck et al.), pp. 235–236. Kluwer, Dordrecht.

Duric, N. 1991. In *The Interpretation of Modern Synthesis Observations of Spiral Galaxies* (ed. N. Duric & P.C. Crane). ASP Conf. Ser. vol. 18, pp. 17–26

Ehle, M. & Beck, R. 1993. *A&A* **273**, 45

Ehle, M., Beck, R., Haynes, R.F., Vogler, A., Pietsch, W., Elmouttie, M. & Ryder, S. 1996a *A&A* **306**, 73

Ehle, M., Beck, R., Haynes, R.F. & Sukumar, S. 1996b *A&A*, submitted

Elstner, D., Meinel, R. & Beck, R. 1992. *A&A Suppl.* **94**, 587

Engelmann, J.J., Ferrando, P., Soutoul, A., et al. 1990 *A&A* **233**, 96

Fitt, A.J. & Alexander, P. 1993. *Mon. Not. Roy. Astron. Soc.* **261**, 445

Golden, R.L., Grimani, C., Kimbell, B.L., et al. 1994. *ApJ* **436**, 769

Gräve, R. & Beck, R. 1988. *A&A* **192**, 66

Hanasz, M. & Chiba, M. 1994. *MNRAS* **266**, 545

Heiles, C. 1996. In *Polarimetry of the Interstellar Medium* (ed. W. Roberge & D. Whittet). ASP Conf. Ser., in press

Helou, G. & Bicay, M.D. 1993. *ApJ* **415**, 93

Horellou, C., Beck, R., Berkhuijsen, E., Krause, M. & Klein, U. 1992. *A&A* **265**, 417

HUMMEL, E. 1986 *A&A* **160**, L4

HUMMEL, E. & BECK, R. 1995. *A&A* **303**, 691

HUMMEL, E. & GRÄVE, R. 1990 *A&A* **228**, 315

HUMMEL, E., DAVIES, R.D., WOLSTENCROFT, R.D., VAN DER HULST, J.M. & PEDLAR, A. 1988. *A&A* **199**, 91

HUMMEL, E., DAHLEM, M., VAN DER HULST, J.M. & SUKUMAR, S. 1991. *A&A* **246**, 10

KLEIN, U., BECK, R., BUCZILOWSKI, U.R. & WIELEBINSKI, R. 1982 *A&A* **108**, 176

KRAUSE, M. 1990. In *IAU Symp. 140. Galactic and Intergalactic Magnetic Fields* (ed. R. Beck et al.), pp. 187–196. Kluwer, Dordrecht.

KRAUSE, M. 1993. In *IAU Symp. 157. The Cosmic Dynamo* (ed. F. Krause et al.), pp. 305–310. Kluwer, Dordrecht.

KRAUSE, M., HUMMEL, E. & BECK, R. 1989a *A&A* **217**, 4

KRAUSE, M., BECK, R. & HUMMEL, E. 1989b *A&A* **217**, 17

LISENFELD, U., VÖLK, H.J. & XU, C. 1996. *A&A* **306**, 677

MATHEWSON, D.S., VAN DER KRUIT, P.C. & BROUW, W.N. 1972 *A&A* **17**, 468

MESTEL, L. 1994. In *Cosmical Magnetism* (ed. D. Lynden-Bell), pp. 181–211. Kluwer, Dordrecht.

MOSS, D. 1995. *MNRAS* **275**, 191

NEININGER, N. 1992. *A&A* **263**, 30

NEININGER, N. & HORELLOU, C. 1996. In *Polarimetry of the Interstellar Medium* (ed. W. Roberge & D. Whittet). ASP Conf. Ser., in press

NEININGER, N., BECK, R., SUKUMAR, S. & ALLEN, R.J. 1993. *A&A* **274**, 687

NIKLAS, S. 1995. PhD Thesis, University of Bonn.

NIKLAS, S. & BECK, R. 1996. *A&A*, submitted

NIKLAS, S., KLEIN, U., BRAINE, J. & WIELEBINSKI, R. 1995a *A&A Suppl.* **114**, 21

NIKLAS, S., KLEIN, U. & WIELEBINSKI, R. 1995b *A&A* **293**, 56

NOZAKURA, T. 1993. *MNRAS* **260**, 861

OTMIANOWSKA-MAZUR, K. & CHIBA, M. 1995. *A&A* **301**, 41

POHL, M. 1993a *A&A* **270**, 91

POHL, M. 1993b *A&A* **279**, L17

POHL, M., SCHLICKEISER, R. & HUMMEL, E. 1991. *A&A* **250**, 302

RUZMAIKIN, A., SHUKUROV, A. & SOKOLOFF, D. 1988. *Magnetic fields of Galaxies*. Kluwer, Dordrecht.

SCHOOFS, S. 1992. Diploma Thesis, University of Bonn.

SOFUE, Y. & TAKANO, T. 1981 *Publ. Astr. Soc. Japan* **33**, 47

SOIDA, M., URBANIK, M. & BECK, R. 1996. *A&A*, in press

SOKOLOFF, D., SHUKUROV, A. & KRAUSE, M. 1992. *A&A* **264**, 396

SUBRAMANIAN, K. & MESTEL, L. 1993. *MNRAS* **265**, 649

SUKUMAR, S. & ALLEN, R.J. 1989. *Nature* **340**, 537

SUKUMAR, S. & ALLEN, R.J. 1991. *ApJ* **382**, 100

TILANUS, R.P.J., ALLEN, R.J., VAN DER HULST, J.M., CRANE, P.C. & KENNICUTT, R.C. 1988. *ApJ* **330**, 667

VÖLK, H.J. 1989. *A&A* **218**, 67

VÖLK, H.J., ZANK, L.A. & ZANK, G.P. 1988. *A&A* **198**, 274

WIELEBINSKI, R. & KRAUSE, F. 1993. *A&A Rev.* **4**, 449

Weak radio emission from Active Galactic Nuclei

By MAREK J. KUKULA[1], J. S. DUNLOP[1] AND A. PEDLAR[2]

[1]Institute for Astronomy, University of Edinburgh, Edinburgh, EH9 3HJ, UK

[2]N.R.A.L., University of Manchester, Jodrell Bank, Macclesfield, SK11 9DL, UK

The majority of Active Galactic Nuclei do not exhibit powerful radio emission. However, with modern instruments many radio-quiet objects are now easily detectable and their radio emission, though weak, can provide us with a great deal of information about the processes occurring in AGN. This article gives an overview of the history of radio observations of 'radio-quiet' AGN, and discusses some of the most recent results.

1. Introduction

Radio astronomy provides us with an extremely effective tool for studying Active Galactic Nuclei (AGN). Optical depths are low at radio wavelengths, allowing us to bypass intervening gas and dust and peer directly into the nuclei of distant galaxies. In addition, the sub-arcsecond resolution and micro-Jansky sensitivities of modern aperture synthesis telescopes allow us to obtain images on a spatial scale and with a level of detail that cannot be matched at shorter wavelengths.

Galaxies with active nuclei were first identified optically over fifty years ago but it was not until the 1950s and 60s, when some of the most powerful radio sources in the sky were identified with galaxies and distant quasars, that the radio properties of AGN began to receive significant attention. However, only about 1% of optically identified AGN exhibit strong radio emission (typically several orders of magnitude more powerful than the optical emission) and it is now clear that there are two distinct populations of AGN: the small fraction of objects which are 'radio loud' – Radio Galaxies (RGs) and radio-loud quasars (RLQs); and the majority which are 'radio quiet' – Seyfert galaxies and radio-quiet quasars (RQQs).

Subsequently much has been learned about the radio properties of RLQs and RGs and our understanding of the origins and nature of the radio emission in these objects, whilst still incomplete, has advanced considerably (Antonucci 1993, Urry & Padovani 1995). By comparison, the radio properties of radio-quiet AGN have remained largely unexplored. Initially this neglect was a natural consequence of practical constraints – sensitivities were simply not high enough to detect the majority of radio-quiet objects – but later the perception that weak radio emission was intrinsically unimportant and could tell us little about the workings of the AGN also played a rôle.

The last fifteen years have seen a change in attitudes and attention has increasingly focussed on what can be learned from radio observations of radio-quiet AGN. However, our knowledge of the radio properties of these objects still lags some way behind that of their radio-loud counterparts.

Why study the radio properties of radio-quiet AGN?

The mechanism by which a small fraction of AGN can produce powerful radio sources whilst the majority either cannot or do not is still one of the most pressing mysteries in the field of AGN research. However, it is clear that powerful radio emission is the exception rather than the rule and we therefore need to ensure that we are fully conversant with the radio properties of 'normal' AGN before we can be confident that we understand why a small fraction of objects are so very different. It is important to realise that the term 'radio quiet' does not imply radio silence. Although the radio luminosities of RQQs and Seyferts are orders of magnitude smaller than those of radio-loud objects, with modern high-sensitivity instruments most radio-quiet objects are easily detectable at radio wavelengths.

It should also be remembered that the *nearest* AGN are all radio quiet. The closest powerful radio galaxy is Cygnus A at a redshift of 0.056, but there are many Seyfert galaxies with redshifts $z < 0.005$. These nearby radio-quiet objects offer the opportunity to study the structure of AGN on physical scales that are more than an order of magnitude smaller than those which can be resolved in any radio galaxy.

The aim of this article is to present a brief overview of what is known about the radio properties of AGN which are *not* radio loud and to show what can be learned from radio studies of these objects. As well as the 'classical' radio-quiet AGN – the Seyfert galaxies and RQQs – we will also discuss studies which show that weak radio emission may be associated with low-level activity in the nuclei of many apparently normal galaxies, and that sensitive radio surveys are an effective way of identifying such activity.

2. The origin of the radio emission in radio-quiet AGN

In order to explain the powerful radio lobes, highly-collimated jets and apparently superluminal motions which are observed in radio-loud AGN, the most plausible model seems to be that of a central black hole fed by an accretion disc. However, although the optical properties of radio-quiet AGN are similar to those of radio-loud objects, the lack of *powerful* collimated outflows of radio-emitting plasma removes one of the principal arguments for the presence of a central 'monster' to power the activity in radio-quiet objects.

One alternative to the monster hypothesis is the idea that radio-quiet activity is the result of a central starburst. Circumnuclear starbursts are a relatively common feature of nearby Seyfert galaxies (*eg* NGC 1068: Balick & Heckman 1985) and it seems likely that the galaxy mergers and interactions which are thought to play a part in triggering nuclear activity and fuelling the AGN would also lead to enhanced rates of star formation in the vicinity of the nucleus (Hernquist 1989).

The total radio luminosities of radio-quiet AGN do not differ greatly from those of starburst galaxies, but the spectroscopic characteristics of Seyferts/RQQs and starburst regions are quite distinct. However, luminous supernovae exploding in HII regions can mimic some of the spectral features of Seyfert nuclei, leading to claims that supernova remnants in a dense *nuclear* starburst might be responsible for the activity in radio-quiet AGN (Terlevich *et al.* 1992 and references therein).

Both thermal and non-thermal radio emission would be expected from a starburst region, the former from HII regions associated with massive ($M > 10^7 M_\odot$) stars and the latter from supernovae (SN) and supernova remnants (SNR). Various authors have suggested ways in which such emission would differ from that produced by a central monster (*e.g.* Condon 1982, Ulvestad 1982, Carral *et al.* 1990) and Condon (1992) describes several observational tests with which to distinguish between starburst- and monster-related radio sources:

(i) Brightness temperature. The radio sources in starbursts have brightness temperatures $T_b \lesssim 10^5$ K above $\nu \sim 1$ GHz, and become optically thick when T_b significantly exceeds 10^4 K. Brightness temperatures in the sources produced by monsters can be considerably higher, and their spectra will remain steep until synchrotron self-absorption occurs at $T_b \sim 10^{10}$ K.

(ii) Morphology. The radio brightness distribution of a starburst must follow the distribution of massive stars in the host galaxy. Starbursts cannot produce well-collimated or precessing radio jets.

(iii) FIR/radio correlation. Starbursts obey a very tight FIR/radio luminosity correlation. A FIR/radio ratio which deviates significantly from this relation is a strong indicator of the presence of a monster.

Judged by these criteria, the majority of the radio-quiet AGN discussed in this article appear to require a monster origin for their radio sources. For this reason, along with the similarity in the optical properties of radio-quiet and radio-loud objects, it seems justifiable to assume that the radio emission in radio-quiet AGN is produced in a similar way to that in radio-loud objects. However, if a compact nuclear starburst such as that proposed by Terlevich et al. were to coexist with a central monster it would be very difficult to separate the two components observationally, and thus starburst-related emission might still make a significant contribution to the total radio flux of radio-quiet AGN.

3. Seyfert galaxies

Seyfert galaxies were the first type of active galaxy to be identified and are defined optically as galaxies with bright 'semi-stellar' nuclei and prominent, broad emission lines (Seyfert 1943). However, unlike radio galaxies, whose nuclei often exhibit similar optical properties, Seyferts were found not to possess strong radio sources. Their radio properties remained largely unexplored until the 1970s, when improvements in sensitivity meant that many Seyferts were detectable in the radio for the first time. Although a handful of objects are known with flux densities of several hundred mJy, the majority have fluxes S≤ 10 mJy.

The advent of multi-element aperture synthesis instruments, combining high sensitivity with high resolution, allowed real progress to be made. The late 1970s and 1980s saw several major radio surveys of Seyfert galaxies including the 1.4 GHz WSRT survey (de Bruyn & Wilson 1976, Meurs & Wilson 1981, Wilson & Meurs 1984), the 5 GHz VLA studies of Ulvestad & Wilson (1984a,b), and the MERLIN observations of Unger et al. (1986). The principal results to emerge from these studies were as follows:

• Flux densities are typically a few milliJanskys at radio wavelengths, corresponding to luminosities L< 10^{24} WHz^{-1}.

• The radio emission is generally confined to the central regions of the host galaxy but can exhibit structures on a range of scales from ~10 kpc down to ~ 10 pc.

• When the radio structure can be resolved it is often linear or 'jet-like', i.e. elongated or consisting of a series of aligned radio knots.

• Typical spectral indices of $\alpha = 0.7$ (where $S \propto \nu^{-\alpha}$) are measured, indicating a synchrotron origin for the radio emission. However, compact radio knots have been found with flatter spectra ($\alpha \sim 0.3$), often associated with the optical nucleus and presumably analogous to the synchrotron self-absorbed cores seen in radio-loud AGN. In at least one object, NGC4151, there is evidence for an additional, thermal source of radio emission - possibly the ionised gas in the optical line-emitting region (Pedlar et al. 1993). Brightness temperatures $T_b \gg 10^4$K are estimated for the radio knots.

Thus, the radio sources in Seyfert nuclei resemble the core-jet-lobe structures found in Radio Galaxies and RLQs, but on a much smaller (sub-galactic) physical scale and several orders of magnitude less luminous (to date there is no evidence for bulk relativistic

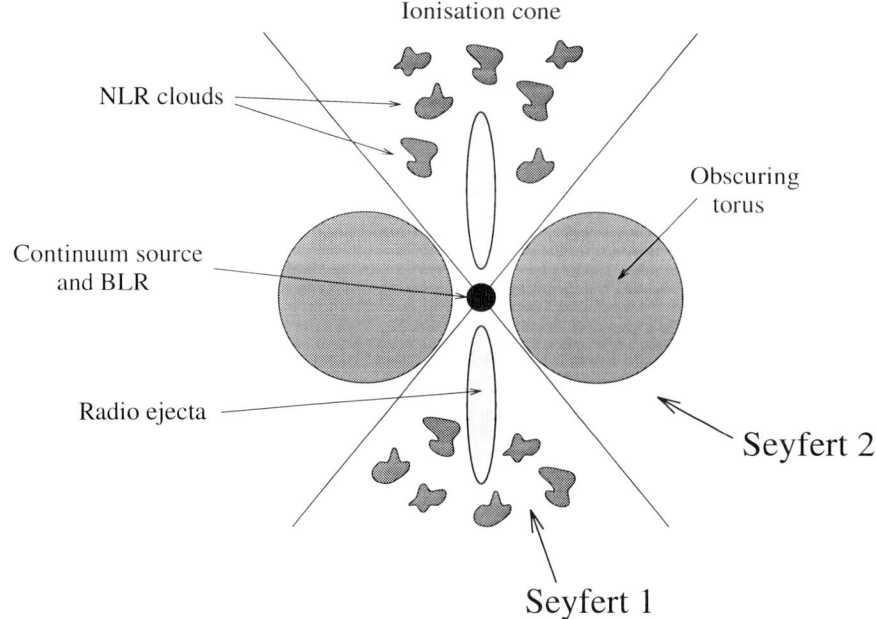

FIGURE 1. Schematic representation of a Seyfert nucleus, showing the major constituents of the unified model and the effect of viewing angle on the appearance of the object.

motion in Seyfert nuclei, but long-term VLBI programmes will be required before we can be sure that they do not occur).

Testing Seyfert unification schemes using radio properties

Radio observations provide an independent means for testing models of anisotropy in Seyfert nuclei. Figure 1 shows a sketch of the current 'unified scheme' for the two main types of Seyfert nucleus; Seyfert 1s, with very broad ($\sim 10^4$ km s^{-1}) permitted lines and narrower (~ 500 km s^{-1}) forbidden lines, and Seyfert 2s, in which both the permitted and forbidden lines have similar widths (~ 500 km s^{-1}). The essential ingredients of the Seyfert unified scheme are as follows. An opaque torus surrounds the optical continuum source and broad-line region (BLR), collimating the radiation from the nucleus into twin cones. Gas clouds inside the cones are ionised by the high energy photons to form the optical narrow-line region (NLR), and radio-emitting material is ejected along the axis of the cones either as a series of discrete plasmons or as a continuous jet. When our line of sight lies inside the cones we have an unobstructed view of the continuum source and BLR and the object is classified as a Seyfert 1. If our line of sight lies outside the cones the BLR and continuum source are obscured and we call the object a Seyfert 2.

This model was constructed in order to explain the optical and infrared properties of the two types, but it also makes two easily testable predictions about the radio properties of Seyfert nuclei. First, because the torus is optically thin at radio wavelengths the radio luminosity should be independent of viewing angle and thus the luminosity distributions of the two types should be identical. Second, if Seyfert 1s are viewed 'head on' then their radio structures should be foreshortened relative to those of Seyfert 2s.

Results from the early radio surveys appeared to indicate that Type 2 Seyferts were both larger and more luminous than Type 1s. However the samples were derived from the Markarian lists of galaxies with an ultraviolet excess, a selection criterion which biassed against weak Type 2 objects. Attempts to correct for this bias by adding non-Markarian Seyferts into the samples greatly reduced the significance of the luminosity difference (Ulvestad & Wilson 1989) and the first radio survey of the spectroscopically-selected CfA Seyferts (Edelson 1987) found no difference between the luminosity distributions of the two types. Thus, recent measurements of Seyfert radio luminosities appear to be consistent with unification of the two types.

Results from two very recent studies

The large (15″) beam in Edelson's VLA D-array survey of the CfA Seyferts meant that no structural information was obtained. In an attempt to rectify this Kukula et al. (1995) re-observed these objects at 8.4 GHz with A-array, giving a 0.25″ beam. The sample contains 48 objects, with approximately equal numbers of each type (26 type 1s, 22 type 2s) and covers a redshift range $0.01 < z < 0.06$.

Similar detection rates of $\sim 80\%$ were achieved for both types of Seyfert but their radio structures were found to be vastly different. 44% of the detected Seyfert 2s were resolved and contained multiple radio components in double, triple or linear structures. However, resolved structure was found in only one (5%) of the Seyfert 1s and the remaining objects were unresolved point sources. At first sight this difference appears to be consistent with the unified scheme – in the Seyfert 1s the radio axis lies close to the line of sight and so we see the structure foreshortened and in some cases multiple radio components are superimposed to form a single unresolved source. In fact the difference is too extreme to be due simply to orientation effects. Observed opening angles of ionisation cones and relative number densities of the two types imply that an object could be misaligned with the line of sight by $\sim 40°$ and yet still be classified as a Seyfert 1. For 95% of the Seyfert 1s in the CfA sample to be intrinsically the same size as the Seyfert 2s and yet to be unresolved, they would all have to be aligned to within 5° of the line of sight. Thus, orientation is unlikely to be the only effect at work in the CfA sample.

Roy et al. (1994) took a different approach to determining the radio structures of Seyfert nuclei. Observing at 2.3 GHz with the single-baseline Parkes-Tidbinbilla Interferometer (PTI) their survey was sensitive to the presence only of structures smaller than 0.1″ (corresponding to 20 – 200 pc over a redshift range of $0.01 \leq z \leq 0.1$) and with brightness temperatures $T_b > 10^5$ K (thus excluding any cases of starburst-related radio emission). 102 objects (35 type 1s and 67 type 2s), selected on the basis of FIR properties, were observed and compact radio components were detected in 50% of Seyfert 2s but in only 25% of Seyfert 1s – a statistically significant difference.

This result does not agree with the simple unified scheme, which predicts equal detection rates for both types (and equal numbers of compact radio components in each type). Roy et al. suggest that free-free absorption of the radio emission in the ionised NLR clouds may be at least partly responsible. In Seyfert 1s we see the nucleus through the full depth of ionised gas in the NLR whereas in Seyfert 2s the column density of ionised gas between us and the radio structure is much smaller (the dusty torus is neutral). It is very difficult to estimate how much free-free absorption one might expect from the NLR since densities and filling factors are not accurately known. However, with the right set of conditions, optical depths rather larger than unity might be possible in the NLR at 2.3 GHz.

Whether or not this mechanism is viable, the results of the PTI survey and the findings of Kukula et al. appear to be in direct conflict. Kukula et al. achieved equal detection rates for the two types but found that Seyfert 2s were more likely to contain *multiple* radio components. Roy et al. found that Seyfert 2s were more likely to contain a compact radio core than Seyfert 1s. Optical depths due to free-free absorption will be smaller at

8.4 GHz than at 2.3 GHz, but probably not by enough to cause such a marked difference in the detection rates of the two studies.

However, there are ways in which the two studies might be reconciled, both with each other and with the unified model. The CfA survey was an *imaging* survey, able to detect radio emission on scales of up to several arcseconds, whereas the PTI survey was sensitive only to components less than 0.1 arcseconds in size. If compact radio components are always associated with the Seyfert core, with larger components further from the nucleus, then free-free absorption in Seyfert 1s might preferentially screen out the compact core (seen through the full depth of the NLR) whilst leaving the diffuse radio 'lobes' on the nearside of the nucleus relatively unaffected. These lobes would still be visible in the VLA maps but would not be detected in the PTI measurements. This would also explain the apparent dearth of Seyfert 1s with multiple radio components in the VLA images; the difference this would cause in the measured radio luminosities of the two Seyfert types might be lost in the scatter due to the relatively small number of objects.

Perhaps more importantly the CfA and PTI Seyfert samples have different selection criteria, cover different redshift ranges and are not strictly comparable. Significantly, when one considers only the 16 CfA Seyferts which are also included in the PTI sample *both* surveys give approximately equal detection rates for Seyfert 1s and 2s.

A number of other factors may also be at work. Relativistic beaming, if it occurs in Seyferts, could significantly affect the composition of Seyfert samples by allowing intrinsically faint but closely aligned Seyfert 1s to be included by virtue of their boosted flux. One interesting hypothesis is that there are actually two types of Seyfert: 'extended' objects with radio jets and lobes and 'compact' objects with only a radio core. Either sort could have the optical characteristics of a Seyfert 1 or a Seyfert 2 depending on orientation, but only those objects with jets would have resolveable radio structure (recall that over half of the Seyfert 2s in the CfA sample were also unresolved). The two types of Seyfert might be evolutionary stages of one another, or the presence of jets might be related to environmental factors such as the Hubble type of the host galaxy. Finally we cannot rule out the possibility of intrinsic differences between Seyfert 1s and 2s.

Thus, although the results to date are not inconsistent with Seyfert unification they strongly suggest that orientation alone is not the answer. Detailed radio studies of individual Seyfert nuclei and radio surveys of new, larger samples will doubtless help to clarify these issues.

4. Radio-quiet quasars

Soon after the discovery in the 1960s of quasars with radio fluxes of several Janskys it became apparent that there were many more objects with similar redshifts, spectra and optical luminosities but with no detectable radio emission. As sensitivities improved it was found that the radio detection rate increased only very slowly for fluxes below 100 mJy and it became clear that there is a pronounced gap in the quasar radio luminosity function: a few objects (RLQs) have radio powers $\geq 10^{25}\,\mathrm{W\,Hz^{-1}}$, but the majority are radio quiet with powers $P < 10^{24}\,\mathrm{W\,Hz^{-1}}$ (see for example Kellermann *et al.* 1989 and Miller, Peacock & Mead 1990).

Miller, Rawlings & Saunders (1993) have demonstrated that this gap is too large for the RLQs to be the result of Doppler boosting in an underlying population of RQQs (though there are a handful of objects with intermediate radio luminosities which could conceivably be beamed radio-quiet objects). It therefore seems inescapable that there really is a fundamental difference between the radio-loud and radio-quiet quasars.

Because the optical properties of RQQs appear to form a continuous sequence with those of Seyfert 1 nuclei it is generally thought that RQQs and Seyferts represent the high- and low-luminosity ends of a physically similar series of objects. Lending support to this idea, Kellermann et al. (1989) found that the radio luminosity function of RQQs in the Bright Quasar Survey (BQS; Schmidt & Green 1983) dovetails quite neatly with the radio luminosity function of nearby Seyferts. However, the combined effects of their large nuclear luminosities and high redshifts has made it very difficult to determine the host galaxy types of RQQs from the ground. By analogy with Seyferts, it has generally been assumed that RQQs lie in disc systems but recent results indicate that this assumption may not be justified. Dunlop et al. (1993) found no significant differences in the near-infrared (K-band:2.2μm) luminosities of RQQ and RLQ hosts at a level which suggests that both may be found in giant ellipticals and Lacy, Rawlings & Hill (1992) showed that the most optically-luminous low-redshift RQQ, E1821+643, is associated with a cD elliptical galaxy. More recently Taylor et al. (1996) found that for half of the RQQ hosts in their sample the best fit to the luminosity profile was given by a de Vaucouleurs law rather than an exponential disc.

Optical HST images of quasar hosts confirm the findings of these ground-based studies (eg Disney et al. 1995) and it now seems certain that *not all* radio-quiet quasars lie in disc systems. Since galaxy type seems to be one of the key factors determining the radio loudness of AGN at low redshifts one would very much like to know whether RQQs in disc and elliptical galaxies can be distinguished in terms of their radio emission. Unfortunately, at present the majority of radio data on RQQs consists only of upper limits on the flux density and consequently their radio properties are even less well understood than those of Seyfert galaxies.

The nature of the radio emission in RQQs

Recent studies have reached very different conclusions about the cause of the radio emission in RQQs. Sopp & Alexander (1991) argued that, since the RQQs in the BQS adhere to the far-infrared – radio luminosity correlation defined by spiral and starbursting IRAS galaxies, the bulk of the radio emission originates in the supernova remnants of a starburst region. However, more recently Miller, Rawlings & Saunders (1993) used the correlation between the 5GHz radio and [OIII]λ5007 luminosities of the same sample to argue that the radio emission comes from jets, similar to those in RLQs of comparable optical luminosity, but far less powerful. Even so, their study does still require some starburst emission in many objects to explain correlations between non-nuclear radio and [OIII] emission.

In order to address the origins of RQQ radio emission in a systematic fashion, Kukula et al. (1996) carried out a multi-frequency, high-resolution VLA survey of 14 RQQs in the redshift range $0.1 \leq z \leq 0.3$. Observations were made in A-configuration at 1.4, 5 and 8.4 GHz, with angular resolutions of 1.4, 0.4 and 0.24 arcseconds respectively.

8.4-GHz radio luminosities were found to be in the range 10^{22} to 10^{23} W Hz^{-1}, an order of magnitude larger than those of typical Seyferts, and the radio spectra were generally steep with typical spectral indices of ~ 0.7. There was no obvious tendency for radio luminosity to correlate with the host galaxy type, as determined by Taylor et al. (1996), although the small size of the quasar sample rules out any definitive interpretation.

In the majority of cases the VLA beam was too large to reveal significant details of the radio structure, but some images revealed multiple and/or extended radio components. Because the majority of objects were unresolved only lower limits could be placed on the brightness temperatures of individual radio sources, which were found to range from 10^2 to 10^5 K.

VLBI observations will place more stringent constraints on the value of T_b, and will hopefully allow a distinction to be made between starburst- and monster-related radio emission, but overall the features revealed in the VLA maps closely resemble those seen in early low-resolution maps of Seyfert nuclei, suggesting that similar processes are at work in RQQs. In future VLBI studies we might expect to find small-scale jets and compact radio components.

5. Other active nuclei with weak radio emission

Spiral galaxies. With sufficient angular resolution the radio emission from normal spiral galaxies can be separated into two components (Hummel 1981): a diffuse component associated with the disc and extending over 10-20 kpc, and a compact central component on a scale of 1 kpc or less. The nuclear component appears to be distinct from the disc (and is sometimes clearly present even when the disc component is very weak or undetected) but it is not clear whether it signals the presence of non-stellar activity or of a nuclear starburst.

In a 1.4-GHz VLA survey Hummel et al. (1985) detected 36/100 optically bright Sbc galaxies with fluxes S>5mJy. These 36 objects subsequently formed the basis of a 5-GHz VLA survey with 2-arcsec resolution, the findings of which were published by Vila et al (1990). 21 galaxies harbour a compact ($\sim$ 100pc) central component and a number also have 'jet-like' extended structures, i.e. they closely resemble Seyfert galaxies albeit 1-2 orders of magnitude less powerful.

The radio properties of these 21 objects form a continuum with those of Seyfert nuclei, but the data of Vila et al. cannot distinguish between starburst- and monster-related emission, either in terms of morphology or spectral index and brightness temperature. Sensitive, high-resolution VLBI measurements would help to resolve these issues, and careful spectroscopy might provide optical evidence for non-stellar activity (the problem being to subtract the underlying starlight with sufficient accuracy). Indeed, recent studies have shown that many apparently normal galaxies display spectroscopic evidence for weak 'dwarf Seyfert' activity (Filippenko & Sargent 1985 and Ho et al. 1995), and many more have LINER-type spectra which can also be taken as evidence for non-stellar nuclear activity (Ho et al. 1993).

Elliptical galaxies. The host galaxies of radio loud AGN are invariably ellipticals so it seems likely that the ability of an AGN to produce large, powerful radio sources depends very much upon its having an elliptical host. However, it does not follow that all AGN in elliptical galaxies are radio loud; as we have already seen there is now convincing evidence that several RQQs have elliptical hosts, and in this section we will discuss several radio surveys which suggest that weak, radio-quiet AGN are relatively common in nearby elliptical galaxies.

The usual method for finding extragalactic radio sources involves surveying large regions of the sky for discrete sources and then searching for optical counterparts. This has proved a very successful way of discovering radio-loud AGN but, because it is often impractical to carry out such surveys with high sensitivity, they are a very inefficient means of finding low-luminosity radio sources.

An alternative method is to make high-sensitivity observations of a pre-selected sample of galaxies. Early radio studies of low-redshift elliptical galaxies yielded some very interesting results. As might be expected, the space density of E/S0 galaxies *detected* as radio sources was found to rise rapidly at low powers (Dressel & Condon 1978, Hummel 1980). In some cases this radio emission was found to arise in compact (< 200pc) cores, some with flat or inverted spectra. These cores have radio powers in the range $10^{21} - 10^{23}$ W Hz^{-1}, and some are observed to be time variable. VLBI observations reveal that the mas-scale structure is elongated and/or asymmetric, resembling the one-sided jets seen in RLQs, with high brightness temperatures on parsec scales. Meanwhile, arcsec-scale (kpc) structures resembling jets and lobes with low radio powers ($10^{21} - -10^{22} \mathrm{W\,Hz^{-1}}$) are found in some objects. In other words, they resemble the radio sources in quasars and powerful radio galaxies, but with radio luminosities more reminiscent of Seyfert nuclei.

A more systematic search, specifically for weak, AGN-related radio emission in nearby ellipticals, was carried out by Wrobel (1991) and Wrobel & Heeschen (1991). They observed ~ 200 nearby E/S0 galaxies with the VLA at 5 GHz, achieving a detection threshold of ~ 0.5mJy. $\sim 25\%$ of the galaxies were detected and the vast majority of these were unresolved (< 580pc) nuclear sources with powers as low as $10^{19} \mathrm{W\,Hz^{-1}}$ (cf Sagittarius A in our own galaxy which has $L_{5GHz} \sim 10^{18} \mathrm{W\,Hz^{-1}}$). From these data it is impossible to distinguish between a monster or a starburst origin for the radio using the criteria of Condon (1992), and Wrobel & Heeschen speculate that a starburst origin is more likely in an S0 host than an elliptical (unfortunately in practice it is often impossible to distinguish between the two galaxy types). Further deep imaging will be required before we can be sure how much of this emission is associated with a nuclear monster and how these weak radio sources are related to RGs and RLQs in similar galaxies.

Conceivably some of these nearby objects are the relics of former radio-loud AGN which were active in earlier epochs but have now largely exhausted their fuel supply. Accurate number counts and luminosity functions will be necessary to determine whether the local population of weakly active elliptical galaxies is consistent with the number of radio-loud ellipticals known to exist at higher redshifts. Estimates of the central mass concentrations would place stringent limits on the number of these objects which could have been powerful AGN in the past. However, the small-scale radio structures in many of the nuclei suggest that, like Seyferts, the majority were never radio loud, and that the usual form of activity in elliptical galaxies is radio quiet, just as it is in disc systems. Perhaps we should be asking not 'Why are AGN in elliptical galaxies radio loud?', but 'Why are only some AGN in elliptical galaxies radio loud?'.

6. Summary

Radio-quiet AGN are not 'radio silent'. Their radio emission, though weak, is now easily detectable with modern instruments and can tell us a great deal about the conditions in what are both the nearest and by far the most numerous class of active galaxy. It is likely that the radio sources in radio-quiet AGN are powered by the same mechanism which produces the relativistic jets and radio lobes observed in radio-loud objects. However, the AGN in spiral galaxies always produce weak radio sources, and it seems that this may also be the most common form of activity in elliptical galaxies. At present we do not understand why this should be the case.

REFERENCES

ANTONUCCI, R. 1993. *ARA&A* **31**, 473.
BALICK, B., & HECKMAN, T. M. 1985. *ApJ* **419**, 553.
CARRAL, P., TURNER, J. L., & HO, P. T. P. 1990. *ApJ* **362**, 434.
CONDON, J. J., CONDON, M. A. GISLER, G., & PUSCHELL, J. J. 1982. *ApJ* **242**, 102.

CONDON, J. J. 1992. in 'Testing the AGN Paradigm', eds. Holt, Neff & Urry *AIP Conference Proceedings* **254**, p629.
DE BRUYN, A. G., WILSON, A. S. 1976. *A&A* **53**, 93.
DISNEY, M. J. et al. 1995. *Nat.* **376**, 150.
DRESSEL, L. L., & CONDON, J. J. 1978. *APJS* **36**, 53.
DUNLOP, J. S., TAYLOR, G. L., HUGHES, D. H., & ROBSON, E. I. 1993. *MNRAS* **264**, 455.
EDELSON, R. A. 1987. *ApJ* **313**, 651.
FILIPPENKO, A. V., & SARGENT, W. L. W. 1985.. *ApJS* **57**, 503.
HERNQUIST, L. 1989. *Nat.* **340**, 687.
HO, L. C., FILIPPENKO, A. V. & SARGENT, W. L. W. 1993. *ApJ* **417**, 63.
HO, L. C., FILIPPENKO, A. V. & SARGENT, W. L. W. 1995. *ApJS* **98**, 477.
HUMMEL, E. 1980. *A&AS* **41**, 151.
HUMMEL, E. 1981. *A&A* **93**, 93.
HUMMEL, E., PEDLAR, A., VAN DER HULST, J. M., & DAVIES, R. D. 1985. *A&AS* **60**, 293.
KELLERMANN, K. I., et al. 1989. *AJ* **98**, 1195.
KUKULA, M. J., PEDLAR, A., BAUM, S. A., & O'DEA, C. P. 1995. *MNRAS* **276**, 1262.
KUKULA, M. J., DUNLOP, J. S., HUGHES, D. H., & RAWLINGS, S. 1996, in preparation.
LACY, M., RAWLINGS, S. & HILL, G. J. 1992. *MNRAS* **258**, 828.
MEURS, E. J. A., & WILSON, A. S. 1981. *A&AS* **45**, 99.
MILLER, L., PEACOCK, J. A., & MEAD, A. R. G. 1990. *MNRAS* **244**, 207.
MILLER, P., RAWLINGS, S., & SAUNDERS, R. 1993. *MNRAS* **263**, 425.
PEDLAR, A., et al. 1993. *MNRAS* **263**, 471.
ROY, A. L., et al. 1994. *ApJ* **432**, 496.
SCHMIDT, M. & GREEN, R. F. 1983. *ApJ* **269**, 352.
SEYFERT, C. K. 1943. *ApJ* **97**, 28.
SOPP, H. M., & ALEXANDER, P. 1991. *MNRAS* **251**, 14P.
TAYLOR, G. L., DUNLOP, J. S., HUGHES, D. H., & ROBSON E. I. 1996. *MNRAS in press*.
TERLEVICH, R., TENORIO-TAGLE, G., FRANCO, J., & MELNICK, J. 1992. *MNRAS* **255**, 713.
ULVESTAD, J. S. 1982. *ApJ* **259**, 96.
ULVESTAD, J. S. & WILSON, A. S. 1984a. *ApJ* **278**, 544.
ULVESTAD, J. S. & WILSON, A. S. 1984b. *ApJ* **285**, 439.
ULVESTAD, J. S., & WILSON, A. S. 1989. *ApJ* **343**, 659.
UNGER, S. W., PEDLAR, A., BOOLER, R. V., & HARRISON, B. A. 1986. *MNRAS* **219**, 387.
URRY, C. M., & PADOVANI, P. 1995. *PASP* **107**, 803.
VILA, M. B., et al. 1990. *MNRAS* **242**, 379.
MEURS, E. J. A., & WILSON, A. S. 1984. *A&A* **136**, 206.
WROBEL, J. M. 1991. *AJ* **101**, 127.
WROBEL, J. M., & HEESCHEN, D. S. 1991. *AJ* **101**, 148.

High sensitivity decimetre observations of Seyfert nuclei

By A. PEDLAR[1], N.G. HAMILTON[1] AND M. J. KUKULA[2]

[1]NRAL, University of Manchester, Jodrell Bank, Macclesfield, UK

[2]Institute for Astronomy, Blackford Hill, University of Edinburgh, UK

We discuss the need for high sensitivity when observing the radio continuum emission from Seyfert nuclei. Studies of extended, low-brightness regions of non-thermal emission associated with Seyferts are underway. At present this weak emission may be attributed either to extended collimated flows or starburst-driven superwinds and high-sensitivity measurements are required at low frequencies to distinguish between the models. We also consider the detection of thermal gas in Seyferts by free-free absorption and report on the detection of localised free-free absorption in NGC 4151 at 73 cm. We suggest that the ionised gas may be closely associated with a neutral molecular torus.

1. Introduction

Almost all Seyfert nuclei are 'radio quiet'. For example, ~60% of the CfA sample of Seyferts have flux densities less than 10 mJy at 8 GHz (Kukula et al. 1995). Hence even the detection of radio emission from these nuclei requires high sensitivity. Most of the radio emission is non-thermal synchrotron with a steep spectral index, and hence there are advantages in working at low frequencies. However, given the small angular size of much of the emission (~100 pc corresponding to <1 arcsec for all but the nearest Seyferts) only instruments with subarcsecond angular resolution can make detailed studies of the Narrow Line Regions (NLR) of these objects. In this paper we will briefly consider in which areas higher sensitivity at low frequencies would benefit Seyfert studies and report on some recent MERLIN observations at low frequencies.

2. Extended Lobes associated with Seyfert Nuclei

At present the specification for the Square Kilometre Array (SKA) would result in microJy sensitivity with an angular resolution of order an arcsecond at decimetre wavelengths. Although the instrument would marginally resolve the radio emission associated with the NLR, one of its main applications in the Seyfert area would be the study of weak radio emission, extended over several kpc, which has recently been found to be associated with Seyfert nuclei (Baum et al 1993).

As a case study we report on briefly on sensitive observations of Markarian 6. Baum et al. (1993) presented WSRT observations of weak radio lobes with a total extent of 15 kpc. These lobes, in common with those of several other Seyferts, were seen to be extended along the minor axis of the host galaxy, and were interpreted by Baum et al. as evidence for a starburst-driven wind. Another possibility is that the lobes are a continuation of the collimated ejection seen as small radio jets in many Seyferts, and are analogous to similar structures seen in Fanaroff–Riley 1 radio galaxies. Kukula et al (1996) have used MERLIN to image the 500-pc radio jet (Fig 1a) and have detected 1 kpc shell-like structures orthogonal to the radio jet (Fig 1b). The 15-kpc extended structure is also shell-like (Fig 1c) and hence these observations favour the starburst interpretation.

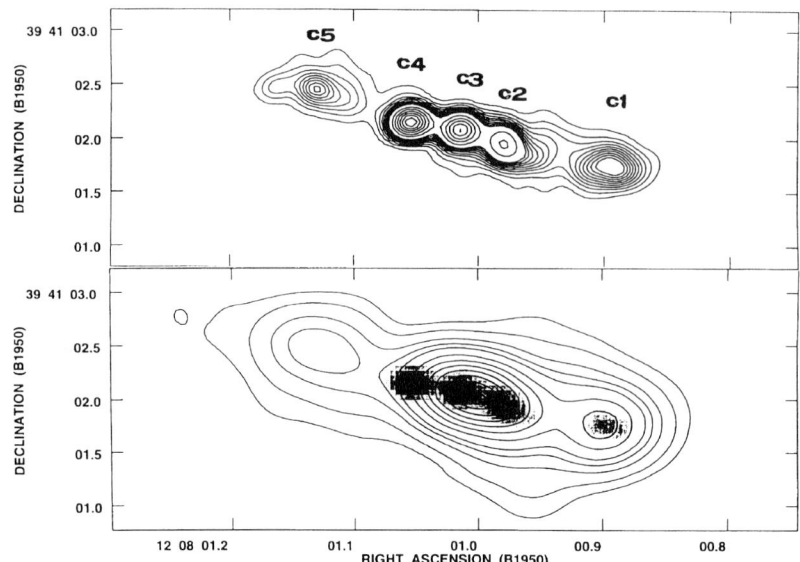

FIGURE 1. (a) A 5-GHz MERLIN contour map of Markarian 6. (b) A 1.6-GHz MERLIN image of Markarian 6 and (c) A WSRT 5-GHz image of Markarian 6. Full details of these images are given in Kukula et al. 1996

If detailed modelling of such flows is to be attempted it is essential that the radio emission be directly comparable to optical studies of the associated ionised gas (e.g. Meaburn et al 1989). However even though Markarian 6 is one of the stronger radio Seyferts, much of the extended emission is below current detection limits at 1-arcsec resolution. It is clear that detailed studies of the weak extended emission will require the sensitivity of the SKA at low frequencies.

3. Detection of Free-Free absorption

Optical studies have shown that Seyfert nuclei contain significant quantities of ionized gas in their Narrow Line Regions. The ionized gas should emit via free-free radiation in the radio, although the expected emission is weak (Pedlar et al 1993) and should be preferentially observed at higher frequencies possibly using the enhanced VLA (Perley 1996) or MERLIN Phase 3 (Cohen 1996). At the relatively low frequencies envisaged for the SKA, the increasing strength of non-thermal synchrotron emission would make free-free emission studies difficult, requiring careful separation of thermal and non-thermal components. However the increasing strength of the non-thermal emission at low frequencies ($\propto \nu^{-0.7}$), coupled with the increasing optical depth of the free-free emission ($\propto \nu^{-2.1}$), would enable the detection of relatively small quantities of ionized gas by free-free absorption. Absorption studies have the additional advantage over emission studies in that they can be used to constrain the relative positions of thermal and non-thermal gas in the line of sight (i.e. the absorbing thermal gas must be in front of the non-thermal).

Using the old MERLIN at 73 cm we pioneered low-frequency studies in Seyferts and found evidence for free-free absorption in NGC 1068 and Markarian 3 (Pedlar et al. 1983, 1984). Recently, using the new MERLIN with improved angular resolution (~ 0.5 arcsec) and making use of phase referencing techniques, we have observed more Seyferts at 73 cm. In the case of the compact triple in NGC 5929 we find that only the NE "lobe" shows

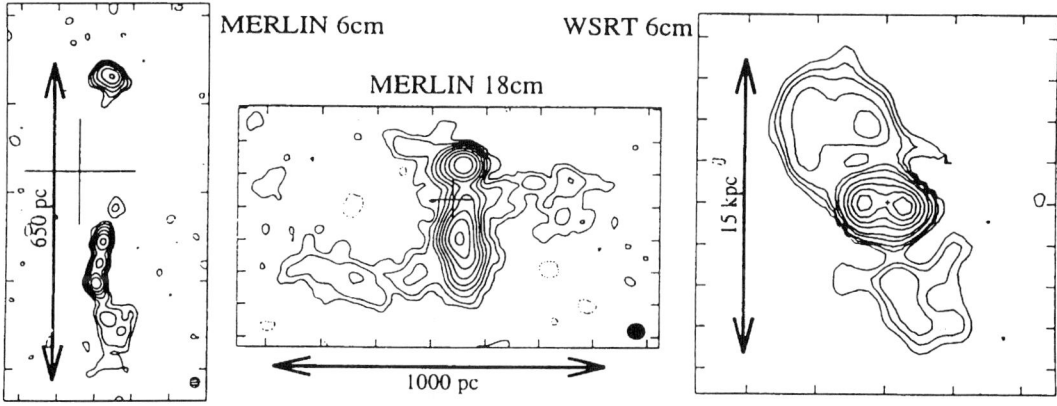

FIGURE 2. (a) MERLIN 18 cm contour map of NGC 4151 observed with an angular resolution of 0.24 × 0.21 arcsec. The components are labeled C1 to C5 following Carral et al (1990). Contour levels are 1,2,3,4,5,6,7,8,9,10,15,20,25,30,35,40 mJy/beam. (b) MERLIN 73 cm contour map of NGC 4151 observed with an angular resolution of 0.56 × 0.51 arcsec, and the levels are -5,5,10,20,30,40,60,80,100,120,160,180 mJy/beam. The 18 cm image is superimposed as a greyscale.

a low-frequency turnover (Su et al 1996). This is consistent with free-free absorption by the NLR region with an emission measure of 10^5 pc cm^{-6} and suggests that the NE lobe is on the far side of the active nucleus.

4. free-free absorption in NGC 4151 — a photoionized torus?

A MERLIN low-frequency study of NGC 4151 has produced evidence for highly localised free-free absorption. In Figure 2a we show a MERLIN 18-cm image, and in Figure 2b we show the MERLIN 73-cm image with the 18-cm image superimposed as a greyscale. Despite the difference in angular resolution, there is no doubt that component C4 is relatively much weaker at 73 cm than at 18 cm. This component appears to be closely associated with the active nucleus and has a spectral index of −0.4 between 5 and 15 GHz (Pedlar et al 1993). Our 18 cm measurements have a flux density of 55 mJy, consistent with a continuation of this power law. However at 73 cm we can only set an upper limit to the flux density of C4 of < 60 mJy, approximately a factor of 2 less than the flux density extrapolated from the power law determined from higher frequency measurements. Thus the spectrum of component C4 shows a clear low-frequency turnover (Fig. 3) consistent with an optical depth >0.5 at 73 cm. The rest of the radio emission shows no evidence for a low-frequency turnover. For example the combined flux density of components C2+C3 increases from ∼110 mJy at 18 cm to ∼270 mJy at 73 cm, consistent with higher-frequency measurements (Pedlar et al 1993), and giving a spectral index of ∼ −0.9 (Fig. 3).

Component C4 is known to be extended over at least 20 mas (Pedlar et al 1993, Harrison et al 1986) and hence the low-frequency turnover cannot be explained by synchrotron self absorption. The most likely explanation for the turnover is free-free absorption, corresponding to an emission measure $> 3 \times 10^5$ pc cm^{-6} (assuming an electron temperature of 10^4 K). However if the emission measure was significantly greater than 10^6 pc cm^{-6} this would result in a turnover at 18 cm which we do not observe.

Mundell et al. (1995) have determined that only component C4 in NGC 4151 shows

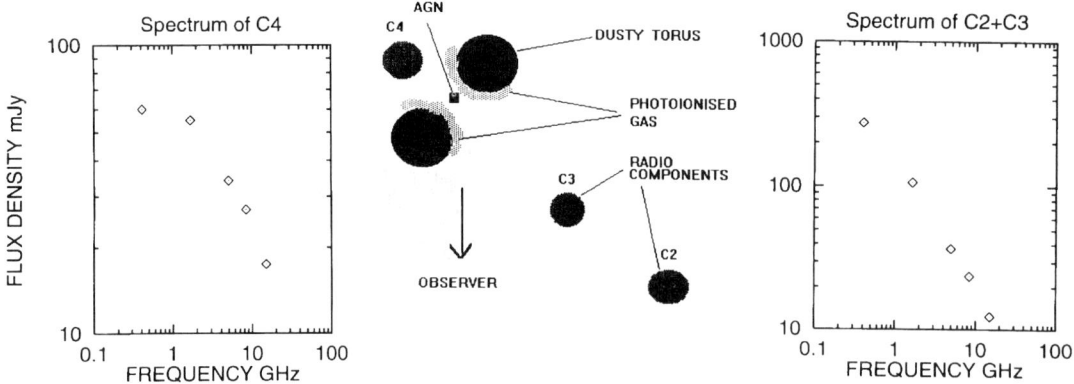

FIGURE 3. A sketch showing the possible relation between the neutral torus (see Mundell et al 1995) and the ionised gas responsible for free-free absorption reported in this paper. The radio spectra of component C4 and the combined spectrum of C2+C3 are inset

measurable neutral hydrogen absorption, which they tentatively identify with a dusty molecular torus approximately 50 pc in extent. It seems more than a coincidence that only component C4 should show low frequency absorption and we suggest that the two phenomena are related. Given the proximity of the neutral torus to the intense UV field of the AGN, it is difficult to see how the surface of the torus can avoid being photo-ionised. In fact if the electron densities are 10^3 cm^{-3}, then the inferred emission measures only require pathlengths of order a parsec which could correspond to an ionisation front at the surface of the torus (Fig. 3). Hence we can speculate that that the neutral molecular torus, as well as hiding the broadline region and collimating the nuclear UV, may also be the source of the ionized gas which forms the Narrow Line Region.

REFERENCES

BAUM ET AL 1993. *Ap.J.* **419**, 533
CARRAL ET AL 1990. *Ap.J.* **362**, 434
COHEN 1996. This meeting.
HARRISON ET AL 1986. *MNRAS* **218**, 775.
KUKULA ET AL 1995. *MNRAS* **276**, 1262
KUKULA ET AL 1996. *MNRAS* In Press
MEABURN ET AL 1989. *MNRAS* **241**, 1P
MUNDELL ET AL 1995. *MNRAS* **272**, 355
PEDLAR ET AL 1983. *MNRAS* **202**, 647.
PEDLAR ET AL 1984. *MNRAS* **214**, 463.
PEDLAR ET AL 1993. *MNRAS* **263**, 471.
PERLEY 1996. This meeting.
SU ET AL. 1996. *MNRAS* In Press

Neutral hydrogen in Seyfert galaxies

By C. G. MUNDELL

NRAL, University of Manchester, Jodrell Bank, Macclesfield, UK

Seyfert galaxies are the closest and most common type of Active Galactic Nuclei and as such, represent ideal sites for the study of the AGN phenomenon and its relationship to the host galaxy environment, which is unobservable in more distant and powerful AGN. Neutral hydrogen (HI) is a valuable tracer of galactic structure and dynamics and so detailed study of the distribution and kinematics of HI on all size scales is vital to determine the relationship between AGN and their host galaxies.

Here, I present high-sensitivity, high-resolution observations of neutral hydrogen in the archetypal Seyfert galaxy, NGC4151. The emission observations (resolution $5'' = 300$ pc for $H_0 = 75$ kms^{-1} Mpc^{-1}) reveal shocks along the leading edges of the gas-rich galactic bar. Velocity profiles across the shocks show the clearest evidence yet of streaming motions in a galactic bar. Dissipation in the shocks may be responsible for channelling the gas towards the nucleus. A new rotation curve is presented showing a turnover at a radius of $\sim 35''$ which was undetected in lower resolution studies.

On a much smaller scale ($0.15'' = 10$ pc), the MERLIN HI absorption measurements show that absorption is seen against only one of the 5 radio components in NGC4151. That component is thought to contain the active nucleus and if the HI cloud responsible for the absorption is part of a circumnuclear torus (invoked in Unified Schemes) we deduce the torus to be ~ 40 pc thick and ~ 70 pc in extent.

1. Introduction

The physical conditions in the host galaxy of an Active Galactic Nucleus (AGN) are likely to be intimately related to the nuclear activity, with HI playing an important role in triggering and fuelling the nuclear activity. However, to date, there have been few detailed synthesis studies of HI in AGN host galaxies†.

Seyfert galaxies are the best candidates among AGN hosts for HI studies since, although their active nuclei are relatively weak, they exhibit many of the properties of their more luminous counterparts (in particular quasars), and they are the closest and most common type of AGN. Sensitivity limitations have typically constrained such studies to extended structure in nearby galaxies. However, recent increases in sensitivity of both MERLIN and the VLA have resulted in the detection of HI structure on the smallest size-scales yet. In fact, it is now possible, with sensitive high resolution λ21-cm observations, to study HI *emission* on sub-kiloparsec scales in nearby galaxies revealing previously unseen detail in structure and kinematics of the gas.

On a much smaller scale there is often evidence for significant quantities of neutral hydrogen in the vicinity of the active nucleus. Unfortunately, HI *emission* studies are limited to angular resolutions of order 5 arcsec ($\sim$ few hundred pc) because the spin temperature of the hydrogen is typically a few hundred °K, and hence, even when optically thick, its arcsecond structure falls below the sensitivity of present-day instruments. Neutral hydrogen *absorption* studies, however, are sensitivity limited mainly by the brightness of the background continuum source and so are eminently suited for detailed investigations

† (e.g. Bosma, Ekers & Lequeux, 1977; Bosma, 1981; Simkin *et al.*, 1987; Irwin & Seaquist, 1991; Pedlar *et al.*, 1992; Brinks et al, 1993; Prieto & Freudling, 1993; Liszt & Dickey, 1995; Mundell *et al.*, 1995a,b,T, Thean, 1995)

of the nuclear environment and its dynamics on scales of a few parsecs. Absorption measurements also discriminate between infall and outflow of gas.

H I observations allow us to investigate models for AGN and their hosts on a wide range of scales, from the outer-most regions, where the gas may be affected by tidal interactions, down to the circumnuclear regions, where it may play a role in the fuelling of AGN and in turn be affected by the nuclear activity. In particular, *bars* may be an efficient fuelling mechanism for the active nucleus (Shlosman et al., 1990, Larson, 1994) and, on nuclear scales, molecular *tori* are thought to surround the AGN and account for many of the observed nuclear properties of Seyferts. Both of these components are amenable to study with high resolution H I emission and absorption studies. The famous, nearby Seyfert NGC4151 is a prime example.

1.1. *Gas in Galactic Bars*

At least 30% of all galaxies and roughly half of all disk galaxies are thought to contain bars (Binney & Tremaine, 1987). Since neutral gas may respond in a highly non-linear way to even small deviations from axial symmetry, it is a very good tracer of the underlying gravitational potential of a barred galaxy (Teuben et al., 1986). λ21-cm observations therefore provide a useful way to investigate the kinematics of H I in barred galaxies.

1.2. *The Nuclear 'Obscuring Torus'*

A small molecular torus, invoked by unification schemes, is thought to surround the active nucleus in Seyferts, obscuring our view of the BLR in type 1 Seyferts and collimating the nuclear UV radiation. The existence of the molecular torus has been inferred from molecular line imaging (e.g., CO emission, Tacconi 1994) and the detection of large molecular gas concentrations in many Seyfert nuclei (e.g., Meixner et al. 1990), but angular resolutions at these wavelengths are, at present, insufficient to resolve the torus.

Significant quantities of neutral material are also expected to be present in the molecular torus due to dissociation of some of the molecules by the nuclear UV radiation (Pedlar et al., 1994). Similar dissociation fronts are proposed to precede the ionisation fronts of galactic H II regions embedded in dense molecular clouds (Hill & Hollenbach, 1978). Therefore, whilst general properties of the torus may be indirectly inferred from optical and infra-red observations, at present, only high-angular-resolution MERLIN absorption observations ($\sim 0.15''$ = ~ 10 pc in nearby Seyferts), are capable of direct investigation of the torus and its kinematics.

2. Streaming Shocks in the Bar of NGC4151

NGC4151 has a Seyfert type 1.5 nucleus (Osterbrock and Koski, 1976) in an SABab host galaxy (de Vaucouleurs, 1991). The host galaxy has a well-defined two-arm outer spiral pattern but, out to a radius of ~ 5 kpc, NGC4151 is dominated by an oval or 'fat' bar-like structure. The bar has dimensions of $2' \times 3'$, is elongated along PA $\sim 130°$ (Pedlar et al., 1992), and possesses two small regions of enhanced emission in the NW and SE corners (Fig.1). These two bright regions are short and curved, with their concave edges pointing towards the major axis of the bar. Observed H II regions (Perez-Fournon & Wilson, 1990) coincide with these H I emission arcs and the velocity dispersion in the arcs is significantly higher than in the rest of the galaxy (Mundell, 1995T). The arcs closely resemble the bar shocks simulated by Athanassoula (1992), who carried out a detailed theoretical study of the properties of periodic orbits in barred galaxy model potentials, in order to reproduce and explain the morphologies and kinematics of gas observed in barred galaxies.

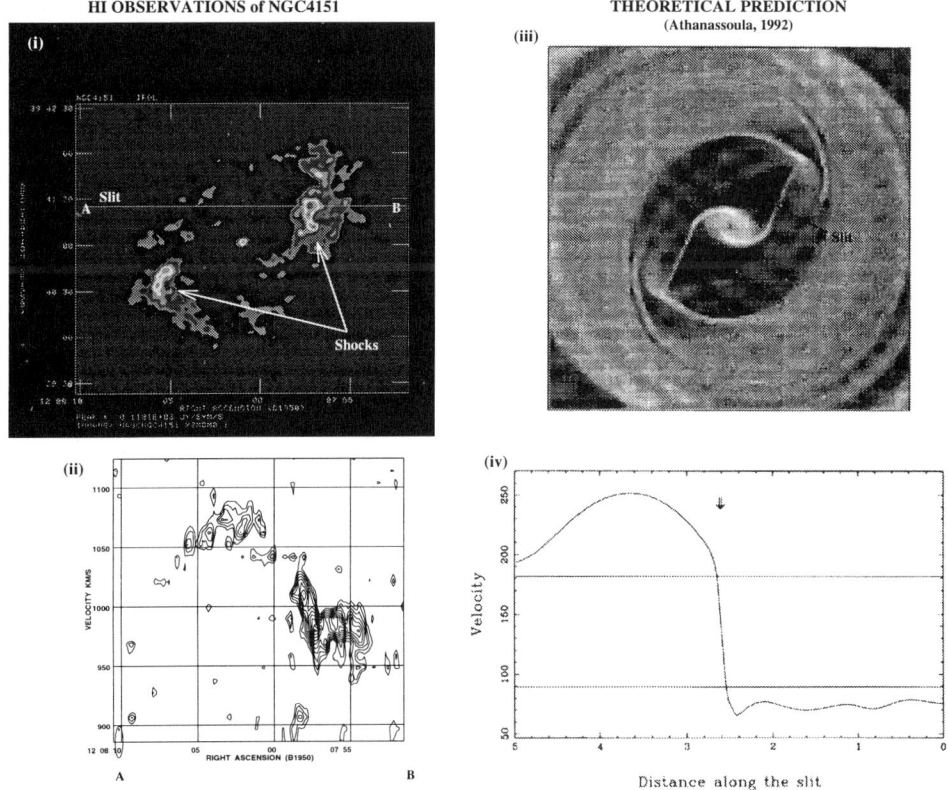

FIGURE 1. Comparison of observations with numerical simulations of gas flows in a weak barred potential. (i) Shocks in the neutral hydrogen in the bar of NGC4151. The long-slit position across the NW shock is marked A–B (ii) The velocity jump across the northern shock; the slit width was 6″, comparable to the beam (iii) Numerical simulations of gas in a weak barred potential (iv) predicted velocity jump across gas shocks in bar when a long-slit is placed across the shock (Athanassoula, 1992)

The most straightforward way to determine whether the arcs are shocks is to measure the velocities across each one and look for the predicted velocity jumps that are characteristic of shocks in neutral gas. Until now, direct evidence of such shocks has been difficult to obtain (Sellwood & Wilkinson, 1988); optical studies are restricted by the small amount of ionised gas in bar shocks (Lindblad & Jörsater, 1988) and HI observations, although ideal for this type of measurement, have proved useless due to limitations in angular resolution and sensitivity. However, the resolution of the HI observations presented here (6″ × 5″) permit detailed investigation of the shocks in NGC4151.

I extracted emission from the HI cube along the equivalent of a 6″-wide slit, across each emission arc in NGC4151. The observed change in velocity along the slit, placed across the NW arc, is shown in Fig.1. This slit position, perpendicular to the emission arc and across its brightness peak, provides the maximum velocity jump, but similarly oriented slits, centred at different positions along the arc also showed this type of velocity structure. The SE shock shows a similar velocity jump.

It is interesting to compare this observed velocity profile with the theoretical predictions of Athanassoula (1992). Fig.1 shows the predicted velocity jump across the enhanced density region in her simulations. The maximum jump is achieved when the

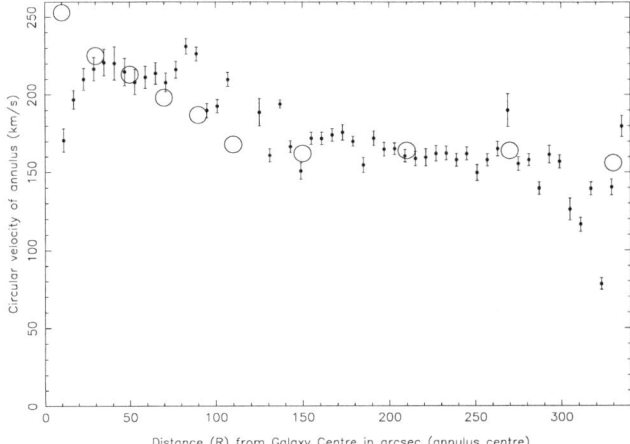

FIGURE 2. Rotation curve from the new high resolution data (small dots). The turnover at a radius of ∼35″ is not evident in the data of Pedlar et al.(1992) (open circles), due to the lower resolution of their observations and the contamination from the central absorption feature.

slit is perpendicular to the shock, as in the observations of NGC4151, and it can be seen that the observations and simulations are remarkably similar.

2.1. Bar Pattern Speed and Resonances

The evidence that the bright regions in the bar are streaming shocks seems quite compelling, so one may use the detection of the shocks to constrain some of the bar properties such as the pattern speed and the possible location of the ILRs. Athanassoula (1992) has shown that, in order to produce shocks similar to those in NGC4151, the bar pattern speed should be such that the co-rotation radius lies in the range 1.2±0.1a (a is the semi-major axis of the bar). Therefore, using the new rotation curve shown in Fig.2 to derive resonance curves for NGC4151 and, assuming a co-rotation radius of 1.2a, the bar pattern speed is found to be ∼29 km s^{-1}, giving one inner Lindblad resonance at a radius of 1.9 kpc. If a second ILR does exist it must lie much closer to the nucleus and therefore would not be detected in the H I data.

2.2. Bars, Inflow and Fuelling

Bars are thought to be an efficient mechanism for transporting galactic material into the central regions of the galaxy (Larson, 1994). In a barred potential, gas flow has a complicated structure, with regions of both outflow and inflow (Athanassoula, 1995). If there are no shocks in the bar, the gas follows quasi-elliptical orbits and there is only a small net inflow due to viscosity. However, as in NGC4151, the presence of ILR, central mass concentration, and slow pattern speed leads to the formation of leading-edge shocks which in turn cause inflow, albeit at a smaller rate than the initial bar-formation stage.

This rather undramatic type of fuelling mechanism may in fact be ideal for Seyferts. Statistical evidence shows that strongly interacting galaxies and highly disturbed systems do not show an excess of Seyfert activity (Bushouse, 1986). This suggests that Seyfert activity is strongly related to the host galaxy properties and that nuclear activity may be the result of the galaxy responding coherently with well-structured features to non-axisymmetric perturbations of the potential rather than due to major disruption (Moles et al., 1995).

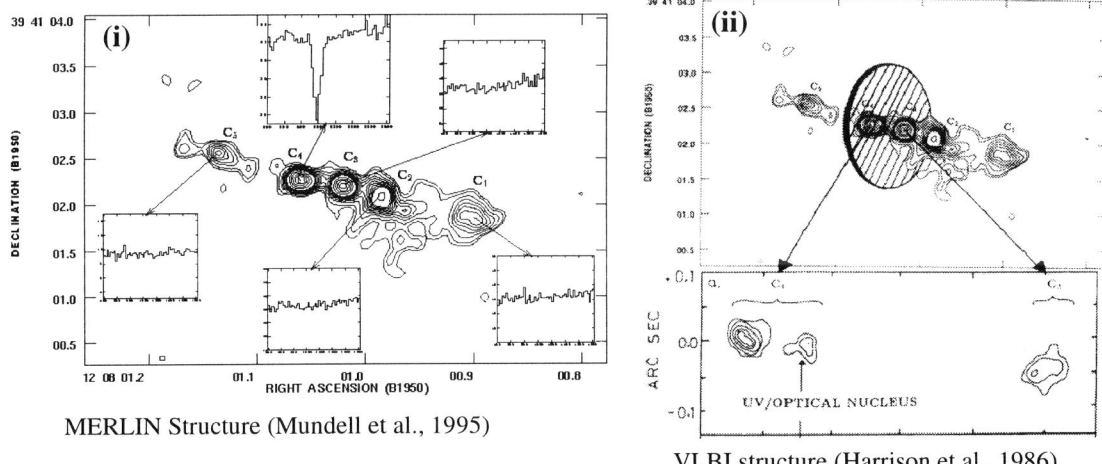

MERLIN Structure (Mundell et al., 1995)

VLBI structure (Harrison et al., 1986)

FIGURE 3. (i) The jet in NGC4151 with H I absorption seen against C4 only (ii) model to account for the difference in UV and radio column densities.

3. Subarcsecond Absorption in NGC4151

MERLIN H I absorption observations of the central few hundred parsecs of NGC4151 reveal absorption against only one of the 5 components of the nuclear jet, that component thought to contain the active nucleus, C4 (Mundell *et al.*, 1995a). The lack of absorption against components C1, C2 and C3 is not due to sensitivity limitations and is indeed consistent with them being part of the jet that lies in front of any H I disk containing the nucleus. Component C5 may be behind the neutral disk but is too weak, at $10\,\mathrm{mJy\,beam^{-1}}$, to show absorption in the present data.

Harrison *et al.*(1986) have shown that most of the λ18-cm flux density of C4 is from a region only 22×25 mas in extent. Hence the column density of 4.8×10^{21} atoms cm^{-1} over this area could be provided by an H I cloud of only 90 $M_\odot$, 1.5 pc in extent.

Lyman absorption measurements (Kriss *et al.*, 1992) give column densities in the range 6×10^{17} to 6×10^{20} which is significantly lower than the MERLIN measurements. Given the inaccuracy of the UV positions, it is unclear where the UV continuum is located relative to the radio structure, but it is clear from the high blue-shifted velocities and large intrinsic widths ($\sim$1000 km s^{-1}) of the UV lines that H I absorption (90 km s^{-1} linewidth centred on systemic) does not originate from the same gas. VLBI observations have shown that C4 consists of two components separated by 0.1″ ($\sim$7 pc) and which of these two contains the radio nucleus is the crucial issue. Pedlar et al (1993) assumed that the stronger, eastern component contained the nucleus, but this need not necessarily be the case. If the weaker, western component were the radio nucleus, most of the H I absorption would then be taking place against the stronger eastern component (now the first knot in the eastern jet). If the western radio component and the optical/UV continuum nucleus were coincident, this would also explain the lower column density in the line of sight to the UV continuum. Fig. 3(ii) shows a sketch of the possible geometry. If the gas is in the form of a torus and the radio jet makes an angle of 40° with the line of sight the torus can be no thicker than $\sim$40 pc and a radial extent of $\sim$70 pc.

REFERENCES

ATHANASSOULA, E. 1992. *MNRAS* **259**, 345

ATHANASSOULA, E. 1995. ed. I. Appenzeller, *Highlights of Astronomy*, **IAU Vol. 10**, 547
BINNEY, J. & TREMAINE, S. 1987. in *Galactic Dynamics*, Princeton University Press
BOSMA, A., EKERS, R.D. & LEQUEUX, J. 1977. *A&A* **57**, 97
BOSMA, A. 1981. *AJ* **86**, 1791
BRINKS, E., SKILLMAN, E.D., TERLEVICH, R.J. & TERLEVICH, E. 1993. *in "Multi-Wavelength Continuum Emission of AGN"* **IAU Symposium 159**, p42
BUSHOUSE, H.A. 1986. *AJ* **91**, 255
HARRISON, B., PEDLAR, A., UNGER, S.W., BURGESS, P., GRAHAM, D.A., PREUSS, E. 1986. *MNRAS* **218**, 775
HILL, J.K. & HOLLENBACH, D.J. 1978. *ApJ* **225**, 390
IRWIN, J.A. & SEAQUIST, E.R. 1991. *ApJ* **371**, 11
KRISS, G. A., ET AL. 1992. *ApJ* **392**, 485
LARSON, R.B. 1994. In "Mass-Transfer Induced Activity in Galaxies", ed. I. Shlosman, CUP, p489
LISZT, H.S. & DICKEY, J.M. 1995. *AJ* **110**, 998
MEIXNER, M., PUCHALSKY, R., BLITZ, L., WRIGHT, M. & HECKMAN, T. 1990. *ApJ* **354**, 158
MOLES, M., MARQUEZ, I. & PEREZ, E. 1995. *ApJ* **438**, 604
MUNDELL, C.G., ET AL. 1995a. *MNRAS* **272**, 355
MUNDELL, C.G., ET AL. 1995b. *MNRAS* **277**, 641
MUNDELL, C.G. 1995T. PhD Thesis, Victoria University of Manchester
OSTERBROCK, D.E. & KOSKI, A.T. 1976. *MNRAS* **176**, 61p
PEDLAR, A., HOWLEY, P., AXON, D.J., UNGER, S.W. 1992. *MNRAS* **259**, 369
PEDLAR, A., KUKULA, M., LONGLEY, D.P.T., MUXLOW, T.W.B., AXON, D.J., BAUM, S., O'DEA, C., UNGER, S.W. 1993. *MNRAS* **263**, 471
PEDLAR, A., MUNDELL, C.G., GALLIMORE, J.F., BAUM, S.A., O'DEA, C.P. 1994. In "Proceedings of the Oxford Torus Workshop", ed. M.J. Ward, p85
PEREZ-FOURNON, I. & WILSON, A.S. 1990. *ApJ* **356**, 456
PRIETO, M.A. & FREUDLING, W. 1993. *ApJ* **418**, 668
SELLWOOD, J.A. & WILKINSON, A. 1993. *Rep. Prog. Phys.* **56**, 173
SHLOSMAN, I, BEGELMAN, M.C. AND FRANK, J. 1990. *Nature* **345**, 679
SIMKIN, S.M., SU, H.J. & SCHWARZ, M.P. 1980. *ApJ* **237**, 404
TACCONI, L.J., ET AL., 1994. In *"Proceedings of the Oxford Torus Workshop"* ed. M.J. Ward, p85
TEUBEN, P.J., SANDERS, R.H., ATHERTON, P.D. & VAN ALBADA, G.D. 1986. *MNRAS* **221**, 1
THEAN, A. 1995. MSc Thesis, Victoria University of Manchester
DE VAUCOULEURS, G., DE VAUCOULEURS, A., CORWIN, H.G., BUTA, R.J., PATUREL, G & FOUQUE, P. 1991. *Third Reference Catalogue of Bright Galaxies*, Springer-Verlag

SNR and ionised gas in M82

By T. W. B. MUXLOW, A. PEDLAR, K. A. WILLS, P. N. WILKINSON, AND D. J. AXON

NRAL, University of Manchester, Jodrell Bank, Macclesfield, UK

We present the latest MERLIN and VLA results which are part of a continuing study of the starburst phenomenon in M82. The compact radio features are now established to be recent supernova remnants with ages of a few hundreds to around a thousand years. A new remnant is thought to appear about every 20-30 years. A previous study at 5 GHz showed the M82 remnants are similar to, but younger and smaller than the equivalent remnants in the LMC and our own galaxy. These new results extend this study to include larger older remnants. Many of the remnants show significant low–frequency spectral turnovers which are probably due to free–free absorption with an emission measure of $\sim 10^6$ pc cm^{-6}. The central 12 arcsec (180 pc) around the inner part of M82 is seen to possess a steeper radio spectral index than regions further out.

A study of the most compact 30 SNR with diameters less than 5 pc show that there is evidence that the rate of expansion is slowing as $\mathbf{D} \propto \mathbf{T}^{0.6}$ which is slower than for pure Sedov expansion implying that even the largest and oldest of the 30 SNR studied cannot have swept up more ISM material than their original ejected mass. This sets an upper limit to the density of the ISM in M82. The observed free-free absorption sets a lower limit to this density. Within these constraints we suggest that the ISM in M82 consists of material with a number density of 30 cm^{-3}, and a filling factor of 0.1. SNR with diameters larger than about 5 pc are lost from our study. At this size their average surface brightness matches that of the background extended emission and thus they become blended with this background.

1. Introduction

Previous MERLIN observations at 5 GHz (Muxlow et al, 1994) have shown that the compact radio features in M82 are recent supernova remnants (SNR) with ages of a few hundreds to around a thousand years with a new remnant expected to appear about every 20-30 years. The remnants are similar to, but younger and smaller than the equivalent remnants in the LMC and our own galaxy. Because of a lack of short-spacing coverage, this study was limited to the most compact remnants with diameters smaller than about 3 pc. This latest work extends this study to include larger older remnants and the extended background emission by incorporating the VLA within MERLIN at both L- and C-band.

2. New MERLIN+VLA Combination Images

We have combined data from MERLIN and the VLA to produce images with full spatial frequency coverage allowing us to study radio emission on all scale sizes within the central 900 pc (1 arcmin) of M82. In each case, Multi Frequency Synthesis (MFS) MERLIN data have been added to A-array VLA data. MFS observations at a number of frequencies spread by $\approx 15\%$ around the centre of the band fill the outer spatial frequency coverage. Figure 1 shows the L-band image at 300-mas resolution with a field of view of 1.5 arcmin. Figure 2 is an interim C-band image at 50-mas resolution of the central 1 arcmin; additional VLA B-array data are still to be added to this image.

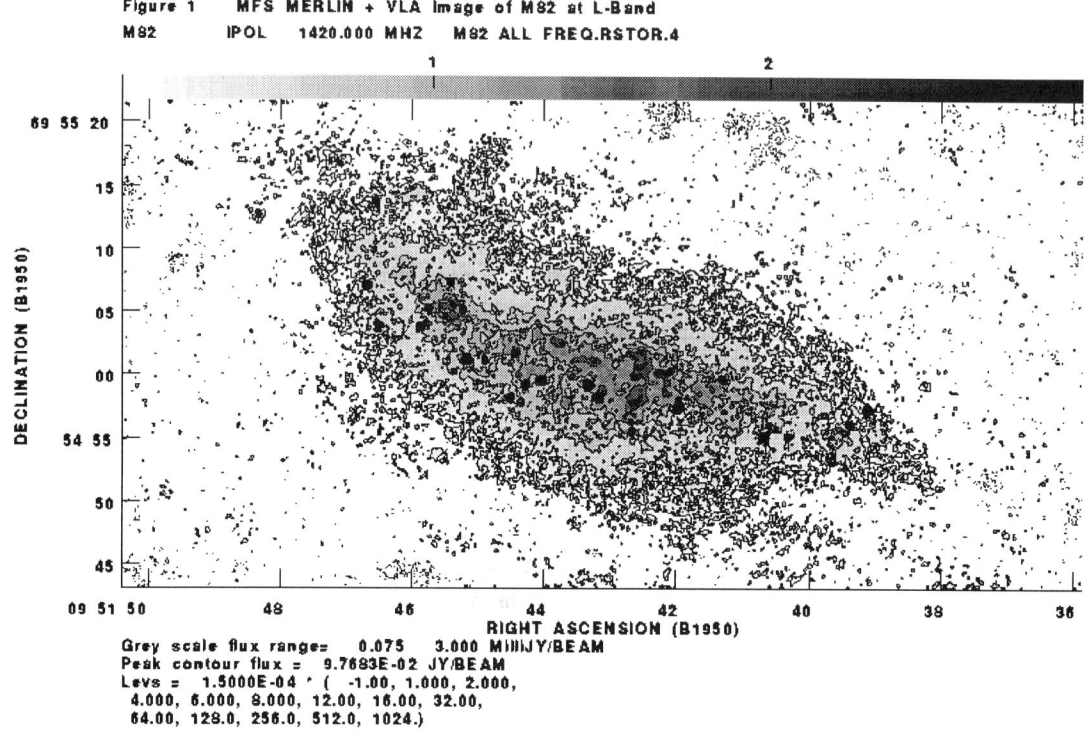

FIGURE 1. MERLIN+VLA L-band Image of M82

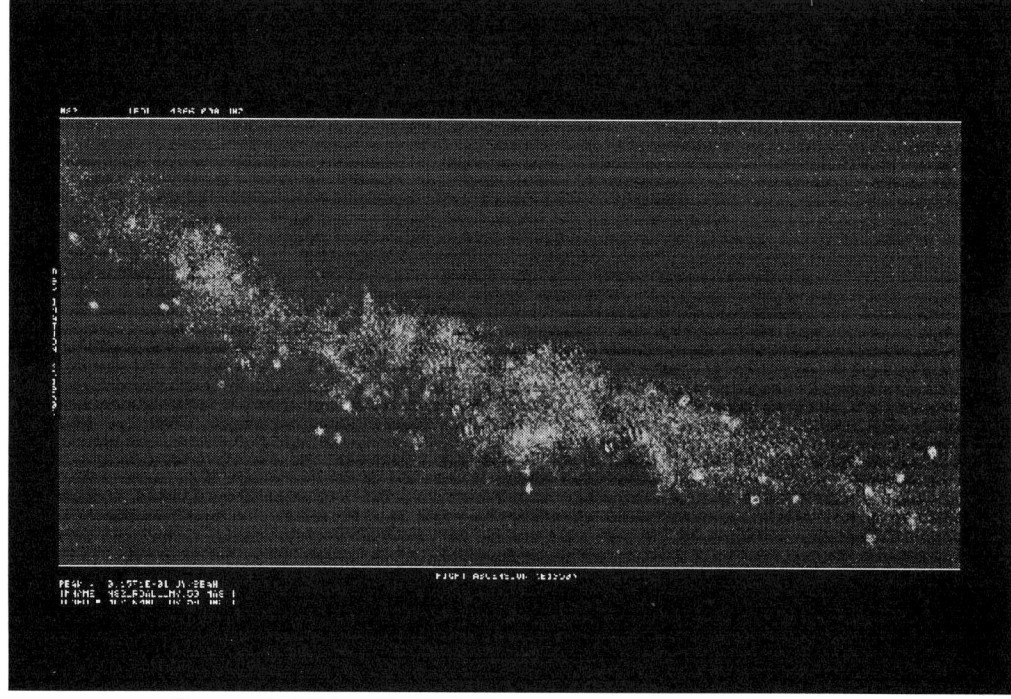

FIGURE 2. MERLIN+VLA C-band Image of M82

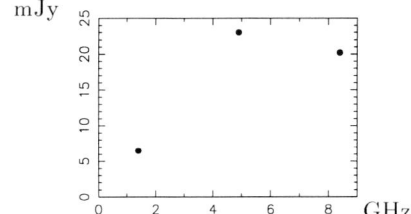

FIGURE 3. Low Frequency Spectral Turnover in the SNR 44.01+596

3. C/L-band Spectral Index Comparison at 350-mas Resolution

We have derived a spectral-index image from the VLA A-array C-band image and the VLA+MERLIN L-band image smoothed to the C-band resolution of 350 mas. Much of the extended background emission has a spectral index of $-0.2 \leq \alpha \leq 0.2$ where $F_\nu = S^\alpha$. However, the central 12 arcsec (180 pc) around the inner core of M82 is found to possess a steeper radio spectral index ($\alpha \leq -0.8$) than regions further from the core. This steeper region corresponds to that lying within the established ring of molecular gas which surrounds the nucleus of M82 (e.g. Nakai et al., 1987).

Many remnants show significant low-frequency spectral turnover which is probably due to free-free absorption in the gas surrounding the remnant (Synchrotron self-absorption can be discounted since all the SNR are resolved with MERLIN at 50-mas resolution). The required emission measure of 10^6 pc cm^{-6} is similar to that of giant Galactic HII regions. The spatial distribution of those SNR showing strong spectral turnover appears random implying that the gas distribution is clumpy. The spectrum for an SNR (44.01+596) showing strong absorption is shown in Figure 3.

4. C-band Imaging at 50 mas Resolution

The image shown in Figure 2 is an interim result awaiting additional VLA B-array data. All the SNR are resolved at 50 mas resolution. We have measured the sizes of 35 identifiable SNR within M82 and present the histogram of their diameters in Figure 4. A comparison of this figure with the equivalent distribution in Muxlow et al. (1994) shows that the earlier investigation was indeed insensitive to older remnants with sizes larger than about 3 pc as noted at that time. The present sample appears complete to diameters around 4.5 to 5 pc. Muxlow et al (1994) found no statistical evidence to suggest that the SNR sizes expand other than linearly. The latest work, however, shows that the inclusion of older larger SNR provides strong evidence of a slowdown in the expansion rate. The sizes measured in the latest image for the smaller SNR agree with the earlier image but there are many more detected remnants in the diameter range 3 to 5 pc. A plot of the cumulative number diameter distribution is also shown in Figure 4.

It is clear that we are failing to detect SNR with sizes greater than about 5 pc, but for the most compact 30 SNR with diameters less than 5 pc there is evidence that the rate of expansion is slowing as $\mathbf{D} \propto \mathbf{T}^{0.6}$. This is slower than for pure Sedov expansion, $\mathbf{D} \propto \mathbf{T}^{0.4}$, implying that even the largest and oldest of the 30 SNR studied cannot have swept up more ISM material than their original ejected mass. The measured value is, however, lower than the predicted relation for very young SNR, $\mathbf{D} \propto \mathbf{T}^{0.8}$ (Chevalier, 1982). The smallest, most luminous and youngest SNR (41.95+575) has an expansion velocity of ~ 6000 km s^{-1} which implies a zero age date of circa. 1955 (Bartel et al., 1987). If we assume that this remnant is typical of the population in M82, this implies a supernova rate of ~ 0.03 per year.

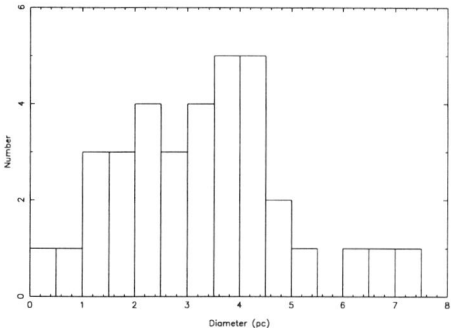 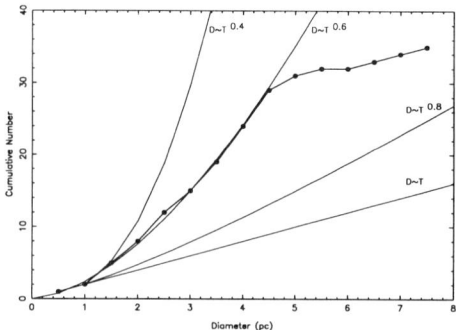

FIGURE 4. Histogram and Cumulative Number Distribution of the SNR in M82

5. The Nature of the ISM in M82

The detection of substantial free-free absorption at L-band implies high densities (at least in clumps) within the M82 ISM. Typical required emission measures of $10^6 \mathrm{pc\,cm^{-6}}$ are similar to that of giant Galactic HII regions which have number densities of around $100 \mathrm{\,cm^{-3}}$ and length scales of about 100 pc. Conversely, the observed slowdown rate in SNR expansion places an upper limit on the density since none of the 30 most compact SNR are yet in the Sedov phase and may not therefore have swept up more mass from the ISM than was originally expelled in the supernova explosion. Roelfsema and Goss (1992) have suggested from H166α observations that the ISM has a number density of $30 \mathrm{\,cm^{-3}}$ and a filling factor of 0.1. With this type of medium, emission measures of $10^6 \mathrm{\,pc\,cm^{-6}}$ are achievable over length scales of 2 kpc which is not too unreasonable. Furthermore, an SNR with a diameter of 5 pc will have swept up a mass of $\approx 5 \mathrm{\,M_\odot}$ which is again acceptable.

SNR with diameters greater than 5 pc are not in general visible on our latest MERLIN+VLA C-band image. Furthermore, the distribution of SNR in M82 is very spatially asymmetric with a zone of avoidance to the NW of the nuclear region. The first effect can be readily explained. At a diameter of around 5 pc the average surface brightness of SNR matches that of the extended background in M82. SNR larger than this thus become blended with this background and cannot be separated from it. The second effect is difficult to explain. If the starburst phenomenon in M82 is in the form of an expanding ring and M82 is viewed close to edge-on, then the SNR to the SE of the nuclear region may be in the foreground and those to the NW in the background. Increased free-free absorption could thus hide the SNR in the NW. This is feasible at L-band, but the effect is still seen at C-band where emission measures of $10^8 \mathrm{\,pc\,cm^{-6}}$ would be required. The distribution of SNR may be intrinsically asymmetric.

REFERENCES

CHEVALIER 1982. *ApJ* **259**, 302
MUXLOW ET AL 1994. *MNRAS* **266**, 455
NAKAI ET AL 1987. *PASJ* **39**, 685
ROELFSEMA & GOSS 1992. *A&A Rev* **4**, 161
BARTEL ET AL 1987. *ApJ* **323**, 505

Atomic hydrogen absorption in radio-loud sources

By JOHN E. CONWAY[1]

[1] Onsala Space Observatory, Onsala, Sweden

In the last few years HST imaging and ground-based spectral-line radio and millimetre observations have provided direct evidence for circumnuclear disks or tori of gas in the centres of AGN. Such observations are consistent with the expectations of AGN unified schemes. While much of this circumnuclear material is probably in a molecular state increasingly both observations and theory suggest that this gas may also have an atomic phase. In light of this we investigate again the incidence of H I absorption toward radio loud sources in ellipticals. We find a strong preponderance of absorption in the FRI, CSO and SSC classes which can be best be explained assuming a circumnuclear origin. Recent VLBI H I absorption observations (particularly of the CSO 4C31.04) provide even stronger evidence for such H I disks.

1. Introduction

Recently it has become evident that the centres of Active Galactic Nuclei contain large amounts of neutral gas in disks or tori. Such structures have been revealed on a variety of scales by HST imaging of gas and dust (e.g. in M87, Ford et al 1994 and NGC4261 Jaffe et al 1993), by observations of water megamasers (Miyoshi, et. al 1995, Gallimore et al 1996) and by mapping in nearby AGN of thermal molecular emission in CO (in Centaurus A, Rydbeck et al 1993) or the high density tracer HCN (e.g. in M51 (Kohno et al 1996), or in NGC1068 (Tacconi et al 1994)). Such tori or disk structures are also inferred indirectly from the ionization cones seen in Seyfert galaxies (Wilson & Tsvetanov 1994) and even more indirectly from AGN unification models. While the inner edge of such obscuring tori must be of order a parsec, so they can hide the BLR but not the NLR, the outer radii can be much larger. HST observations of dust in ellipticals (Dokkum and Franx 1996) indicate that gas out to several hundred parsecs preferentially lies in a plane perpendicular to the radio axis and so can be considered circumnuclear.

It is often supposed that the obscuring matter in the circumnuclear disks or tori is purely molecular. While this may be true in many objects in fact a wide range of physical conditions for the circumnuclear gas are possible depending on its density, the presence or absence of heating by the AGN and the distance of the gas from the central engine (see Maloney et al 1994, Neufeld & Maloney 1995). Given the strong hard X-ray sources in the centres of some radio-loud AGNs an abundant atomic phase is quite possible (see Maloney et al 1994). Recent work with MERLIN has shown that the H I absorption in the Seyfert galaxy NGC4151 (Mundell et al 1995) has a distribution consistent with a disk like geometry. Furthermore in NGC4261 there appears to be H I absorption associated with the 100 pc-radius HST disk (Jaffe and McNamara 1994). Most recently it has been shown (see Gallimore et al 1996) that the OH and H_2O masers in NGC1068 have the same velocity width as the nuclear H I absorption and that this H I absorption can be modelled by a Keplerian annulus with similar radius to the H_2O disk. Given the new observational developments it is interesting to consider again the class of radio loud objects which show broad (> 100 km s^{-1}) H I absorption. Perhaps the majority of these broad absorption systems are due to gas associated with the central engine rather than, as has been previously assumed, atomic ISM gas along the line of sight.

2. Surveys of H I Absorption

There has been surprisingly little published on systematic surveys for H I absorption in powerful radio loud AGN (from which we exclude Seyfert galaxies). The most systematic search is that published by van Gorkom et al. (1989) of 28 radio-bright sources (> 0.35Jy) with $z < 0.13$ and elliptical hosts. Only three of the objects observed from this well defined sample showed H I absorption. Searching the literature we find a total of 16 radio loud objects which have broad (FWHM > 100km s^{-1}) H I absorption at the same redshift as that of a confirmed normal elliptical or cD host which has $z < 0.13$ and hence is easily accessible to H I absorption observations.

The four cD sources detected in broad H I absorption (3C84, 2A0335+096, NGC5920, PKS2322-123) exist in clusters with high X-ray luminosity and it is thought that in these objects the absorption might be associated with cooling-flow gas. Amongst the remaining sources it is interesting that none of the H I detections are against strong, presumably beamed, core-dominated radio sources. This result is consistent with the H I absorption in the majority of these sources being preferentially in a plane perpendicular to the radio jet. Five of these sources are FRI radio galaxies (namely in Cen A, NGC315, NGC1052, NGC4261 and Hydra A). One of these systems (NGC4261) shows a 100 pc-radius HST disk (Jaffe et al 1993). Recent VLBA observations of Hydra A shows H I absorption against the core, the jet and counterjet (Taylor 1996). The absorption is claimed to be narrower in velocity width against the jets than against the core suggesting that the absorbing H I is in a flattened distribution of approximate thickness 20pc. Only one of the H I absorption detections is in an FRII galaxy, in the hidden-quasar candidate Cygnus A (Conway & Blanco 1995). The lack of FRIIs can be ascribed to their higher redshift and relatively weak cores making observations difficult. Considering their rarity there is are a surprisingly large fraction of H I absorbers which belong to either the Compact Symmetric Object (CSO) class (at least 3 objects, 4C31.04, 1146+596, 1946+708 and possibly also 1353+055) or the related Steep Spectrum Core (SSC) class (2 objects, 3C236 and 3C293). The CSOs show the same double or triple morphology as classical sources but are factors of thousands of times smaller (see Wilkinson et al 1994). The SSC objects, of which the best example is 3C236 (see Barthel et al 1985), are sources with radio lobes hundreds of kpc away from a central steep spectrum core, which when observed at high resolution have similar morphology to the CSOs. CSOs are probably a young bright phase in the evolution of classical radio sources, and SSCs are perhaps objects in which a new epoch of activity is occurring within a previously dormant AGN.

Since CSOs and SSC sources comprise less than 10% of the sources selected on total flux density at 21-cm wavelength the large fraction of H I absorbers that are of this type is significant. In the well-defined sample of van Gorkom et al (1989) out of 28 objects searched (excluding 1353+055 which as far as we can tell did not fulfil the selection criteria but was included because it was a known H I absorber) two are CSOs (4C31.04 and 1146+59) and the third (3C236) a SSC source. The vast majority of the sources sampled were FRI cores and these did not yield any detections. In contrast amongst CSOs 3 out of the 5 CSOs searched for H I absorption at $z < 0.15$ have been detected at the 5% opacity level over 100 km s^{-1}. The high incidence of H I absorption in CSOs or SSC sources might be taken as evidence that these radio morphologies are preferentially associated with the presence of dense gas (the so called 'frustration' hypothesis). An alternative explanation consistent with the presence of circumnuclear disks is that these objects just form ideal background objects in which to see absorption. Unlike other sources the continuum emission from these objects is not beamed and the more distant minilobe contains about half the flux density. Since the sizes of CSOs are comparable to that of 100pc-scale circumnuclear disks we can expect that in a large fraction of orientations the more distant minilobe will be covered by the disk. To test this model we have begun a programme of spectral-line VLBI observations of CSOs/SSCs in order to spatially resolve the HI absorption; preliminary results are described below.

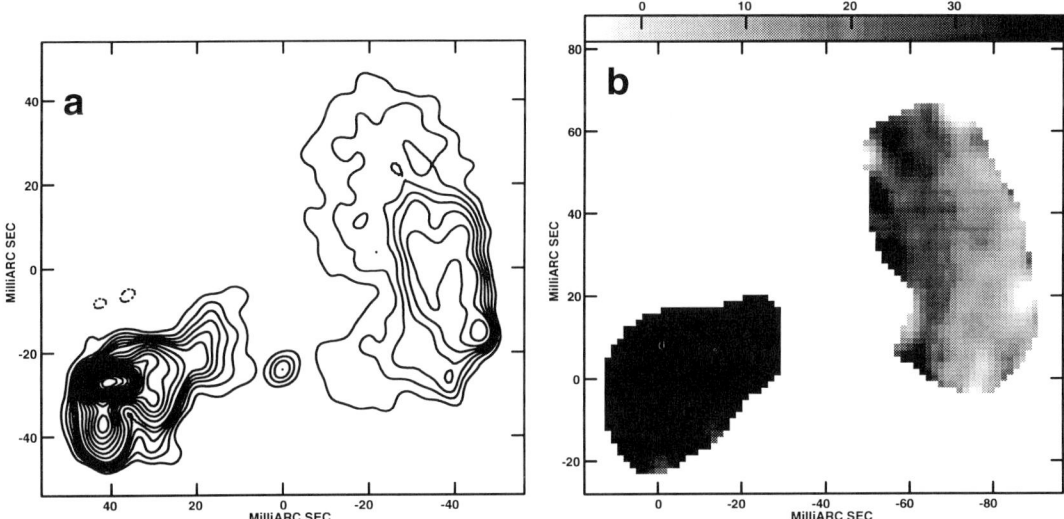

FIGURE 1. (a) Continuum (1340MHz) MEM image of the the CSO 4C31.04, 3 mas-FWHM resolution, first 6 contours are at 1.5-mJy/beam intervals, rest at 3.0-mJy/beam intervals. (b) HI opacity averaged over central part of broad absorption line. Greyscale image from -0.005 to 0.040, resolution 12 by 9mas, PA=5 degrees.

3. VLBI Observations of HI absorption in CSO/SSCs

4C31.04: This CSO object which is at z=0.06 (1mas = 1pc for $H_o = 75 \text{km s}^{-1} \text{ Mpc}^{-1}$) was discovered (Mirabel 1990) to have strong HI absorption with both broad (FWHM 133 km s^{-1}) and narrow (< 20km s^{-1}) absorption systems. VLBA observations of this source were made in July 1995. Figure 1a shows the resulting continuum map confirming the CSO classification. Figure 1b shows the HI opacity integrated over the deepest part of the broad absorption line. The opacity is large ($\tau \approx 0.07$) and fairly uniform over the E lobe but in the W lobe there is a sharp 'edge' to the opacity; on the core side of the edge the HI opacity is approximately 0.02 while on the other side it is consistent with zero.

The fact that the HI edge is roughly perpendicular to the vector from core to W hotspot strongly suggests that the obscuring gas 'knows' about the radio axis and is associated with the AGN rather than being random foreground gas in the host galaxy. The observations can be explained by a model with a > 100pc-radius HI disk whose axis is that of the radio jet, and in which this radio-jet axis is close to the sky plane (as is likely for an isotropically emitting CSO selected on total flux density). In this model the disk completely covers the more distant, Eastern lobe, generating a large opacity, while the HI edge in the Western lobe is caused by the finite thickness of the disk or torus.

1946+708: This z=0.101 source has one of the most unusual VLBI continuum structures known (Taylor et al 1995) comprising a symmetric string of almost equally bright components arranged in a line. This object fulfils the definition of the CSO class and it is probable that the outer components comprise hotspots while the inner components are knots in a two sided jet. Broad ≈150-km s^{-1} HI absorption was detected in a recent VLA survey of CSOs. Short follow up VLBA observations (made in collaboration with G.Taylor) have shown that while HI opacity is significant against the outer components it is consistent with being zero in the centre of the source. However this result must be confirmed in higher-SNR observations. In some models of disks (Neufeld & Maloney 1995) the midplane of the circumnuclear disk contains mainly cold molecular material while the top and bottom surfaces are heated by the central engine, either directly or by reflection, and contain atomic gas. We can speculate that given the very symmetrical continuum structure in 1946+708 the jet lies close to the sky plane so that we view the central part of the source through the disk/torus midplane.

3C236: Finally we describe a preliminary global VLBI observation of the SSC source 3C236 (at z=0.099) which has HI absorption of FWHM 140 km s^{-1} (van Gorkom et al 1989). So far only the shorter baselines have been analysed but these already show that the absorption is confined to the compact components B and C (using the nomenclature of Barthel et al 1985) with no significant absorption against the 1.5 kpc-distant diffuse component A. This observation and the possible sign of position-dependent velocity centroids are consistent with a disk-like structure rotating around B and C.

REFERENCES

BARTHEL, P.D., SCHILIZZI, R.T., MILEY, G.K., JAGERS, W.J., & STROM. R.J. 1985 *A and A* **148**, 243.

CONWAY J.E., & BLANCO, P.R. 1995 *ApJL* **449**, L131

DOKKUM, P.G., & FRANX, M. 1996 *A.J. in press*

FORD. H.C., ET AL 1994 *ApJ* **435**, L27.

GALLIMORE, J.F., BAUM, S.A., O'DEA, C.P., BRINKS, E., & PEDLAR, A. 1996 *Ap J, in press*

VAN GORKOM, J.H., KNAPP, G.R., EKERS, R.D., EKERS, D.D., LAING, R.A., & POLK, K.S. 1989 *AJ* **97**, 708.

JAFFE, W., FORD, H.C., FERRARESE, L., VAN DEN BOSCH, & F., O'CONNELL, R.W. 1993 *Nature* **364**, 213.

JAFFE, W., & MCNAMARA, B.R. 1994 *Ap J* **434**, 110.

KOHNO, K., KAWABE, TOSAKIM T., & OKUMURA, S.K. 1996 *ApJ* **461**, L29.

MALONEY, P.R., BEGELMAN, M.C., & REES, M.J., 1994 *ApJ* **432**, 606.

MIRABEL, I.F. 1990 *ApJ* **352**, L37.

MIYOSHI,M., MORAN, J., HERRNSTEIN,J., NAKAI., N. DIAMOND, P., & INOUE, M. 1995, *Nature* **373**, 127.

MUNDELL, C.G., PEDLAR, A., BAUM, S.A., O'DEA, C.P., GALLIMORE, G.F., & BRINKS, E. 1995 *MNRAS*, **272**, 355.

NEUFELD, D.A., & MALONEY, P.R., 1995 *ApJL* **447**, L17.

RYDBECK, G., WIKLIND, T., CAMERON, M., WILD, W., ECKART, A., GENZEL, R, & ROTHERMEL, H., 1993, *A and A* **270**, L13.

TACCONI, L.J. GENZEL, R., BLIETZ, M., CAMERON, M., HARRIS, & A.I., MADDEN, S. 1994 *Ap J* **426**, L77.

TAYLOR, G.B., VERMEULEN, R.C., & PEARSON, T.J., 1995 *'Quasars and Active Galactic Nuclei: High Resolution Radio Imaging' Proc Natl Acad Sci USA, Vol 92,11381*

TAYLOR, G.B. 1996 *'Extragalactic Radio Sources' IAU Symposium 175, Ed C.Fanti*

WILKINSON P.N., POLATIDIS, A.G., READHEAD, A.C.S., XU, W., & PEARSON, T.J. 1994 *ApJ* **432**, L87.

WILSON, A.S., & TSVETANOV, Z. 1994 *AJ* **107**, 1227.

VLBI observations of low-power radio galaxies

By G. GIOVANNINI[1,2], W.D. COTTON[3], L. FERETTI[1,2], L. LARA[2,4] AND T. VENTURI[2]

[1]Dipartimento di Astronomia, via Zamboni 33, 40126 Bologna, Italy

[2]Istituto di Radioastronomia, via Gobetti 101, 40129 Bologna, Italy

[3]NRAO, 520 Edgemont Rd, Charlottesville, VA 22903-2475, USA

[4]IAA, CSIC, Apdo 3004, 18080 Granada, Spain

We present the parsec-scale properties of 8 low-power radio galaxies. One of them shows a symmetric parsec-scale morphology that we briefly discuss. All the other radio galaxies show an asymmetric one-sided jet emission. We use observational data to constrain the parsec-scale jet velocity and orientation with respect to the line of sight and compare our results with available data of BL-Lac type objects to test unified scheme predictions.

1. Introduction

VLBI observations of low-power radio galaxies are necessary to test unified scheme models, which predict that low-power radio galaxies should be at angles $\theta \geq 30°$, with respect to the line of sight. Radio galaxies at a smaller angle with respect to the line of sight will appear as a BL Lac (see e.g. Urry and Padovani, 1995). According to this model, low-power radio galaxies should also have parsec-scale jets moving at a velocity close to the speed of light. In this paper we will use radio data on 8 low-power radio galaxies (FR I, see Fanaroff and Riley, 1974) from the sample currently under study by us (Giovannini et al., 1990), to test the low-power unified scheme. These radio galaxies have a total radio power at 408 MHz $\leq 10^{25}$ W/Hz using a Hubble constant $H_0 = 100$ km sec^{-1} Mpc^{-1} (see Table 1). A morphological analysis based on available VLBI maps indicates that an asymmetric morphology, i.e. core and one-sided jet, is the most frequent radio structure. It is shared by 7 out of the 8 radio FR I radio galaxies mapped by us. A clear symmetric structure is found only in 3C338. The parsec-scale jet is always oriented on the same side as the main kpc-scale jet. This correlation implies either that jets are intrinsically asymmetric, or that parsec and kpc-scale jets are both relativistic. A detailed study of the inner kpc-scale properties of low-power radio galaxies and the evidence of proper motion at high velocities in some galaxies (see Giovannini et al., 1995 and references therein) confirm that radio jets in FR I radio galaxies are initially relativistic. For these reasons, we interpret the radio structures presented here as affected by Doppler favouritism and will use the available data to constrain the possible values of the jet velocity ($\beta = v/c$) and of the orientation of the radio source with respect to the line of sight (θ).

2. Jet Velocity and Orientation

We can constrain the jet velocities and orientation in three different ways: a) from the jet to counter-jet brightness ratio, b) from the prominence of the core radio power with respect to the total radio power, c) from comparing the observed X-ray nuclear emission with that expected by the self-Compton model. A detailed discussion on these methods can be found in Giovannini et al., 1994. The allowed values for the jet velocity β and

Name	z	$Log P_{408}$ W/Hz	θ °	β v/c	γ
NGC315	0.0167	23.95	$30\leq \theta \leq 41$	$0.78\leq \beta \leq 0.962$	$1.60\leq \gamma \leq 4.03$
3C31	0.0169	24.50	$34\leq \theta \leq 60$	$0.62\leq \beta \leq 0.997$	$1.27\leq \gamma \leq 12.10$
4C35.03	0.0375	24.85	$0\leq \theta \leq 54$	$0.37\leq \beta \leq 0.996$	$1.08\leq \gamma \leq 11.32$
NGC2484	0.0413	25.04	$0\leq \theta \leq 46$	$0.55\leq \beta \leq 0.999$	$1.20\leq \gamma \leq 26.54$
0836+29	0.0790	25.08	$0\leq \theta \leq 37$	$0.75\leq \beta \leq 0.999$	$1.51\leq \gamma \leq 24.06$
3C264	0.0206	24.85	$0\leq \theta \leq 55$	$0.56\leq \beta \leq 0.992$	$1.21\leq \gamma \leq 8.15$
3C338	0.0303	25.25	$0\leq \theta \leq 90$	$0.00\leq \beta \leq 0.711$	$1.00\leq \gamma \leq 1.42$
3C465	0.0301	25.30	$0\leq \theta \leq 53$	$0.60\leq \beta \leq 1.00$	$1.25\leq \gamma$

TABLE 1. Source details and orientation - velocity constraints

its orientation with respect to the line of sight θ are given in Table 1. The range of θ and β is fully consistent with the unified schemes which predict FR I radio galaxies to be at angles $\geq 30°$ (see e.g. Ghisellini et al., 1993). We note that for a low value of θ the whole β range is not allowed (see e.g. Giovannini et al., 1994), but that a high jet velocity is allowed only if $\theta \geq 30°$.

Urry and Padovani (1995), comparing the FR I and BL Lac luminosity functions, derive values of gamma in the range of 2 to 20 for radio BL Lacs, with $<\gamma>=3$. Our estimates of γ in Table 1 are fully consistent with this result. We note however that we have very few sources with large angles allowed with respect to the line of sight. This result is expected since the radio galaxies present in our sample have been selected from the B2 and 3CR catalogues also on the basis of their arcsecond core flux density ($\geq$ 100 mJy, Giovannini et al., 1990) for obvious observational reasons. This selection effect means that most radio galaxies in our sample have a relatively small angle with respect to the line of sight.

The parsec-scale radio properties of the FR I sources presented here are very similar with the possible exception of 3C338. However their kpc-scale morphologies are quite different. These differences in the kpc-scale morphology and in the linear sizes of these sources, despite their having similar parsec-scale morphology and the similar total radio power at 408 MHz, suggest that the interaction with the external medium plays an important role in the extended structure of these FR I sources. It is important to note that FR I radio galaxies are usually found in environments rich in galaxies. In particular, all sources discussed here belong to clusters or groups of galaxies, in which the existence of an intra-cluster medium is supported by the detection of diffuse thermal X-ray emission associated with these sources. It seems clear that as soon as the jet decelerates to a nonrelativistic velocity the external medium strongly affects the source properties to produce the very different kpc-scale morphologies independent of the pc-scale properties.

3. The symmetric source 3C338

The radio source 3C338 has a very steep global radio spectrum and is associated with the multiple nuclei cD galaxy NGC6166. The arcsecond-resolution radio maps reveal the core radio emission and a peculiar bridge south of the core embedded in low-brightness extended lobes (Fig. 1). Burns et al. (1983) interpreted this structure as evidence of an intermittent radio activity connected with the motion of the radio nucleus within the cD envelope. New VLA high resolution maps (A configuration) taken by us at 20, 6 and 3.6 cm reveal the presence of a faint symmetric radio jet connecting the core radio emission

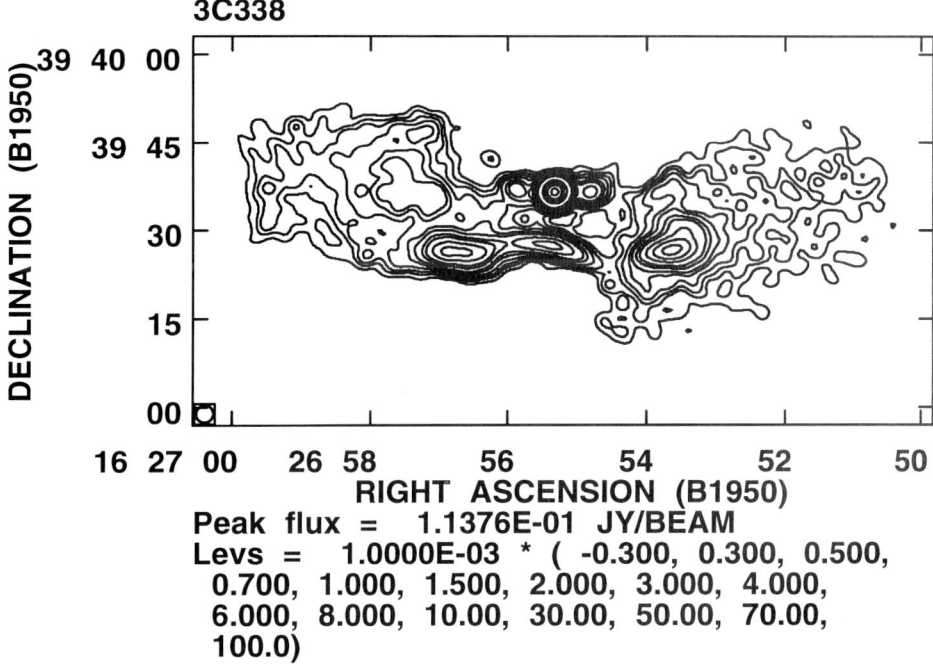

FIGURE 1. VLA map at 6cm of the large-scale radio emission of 3C338

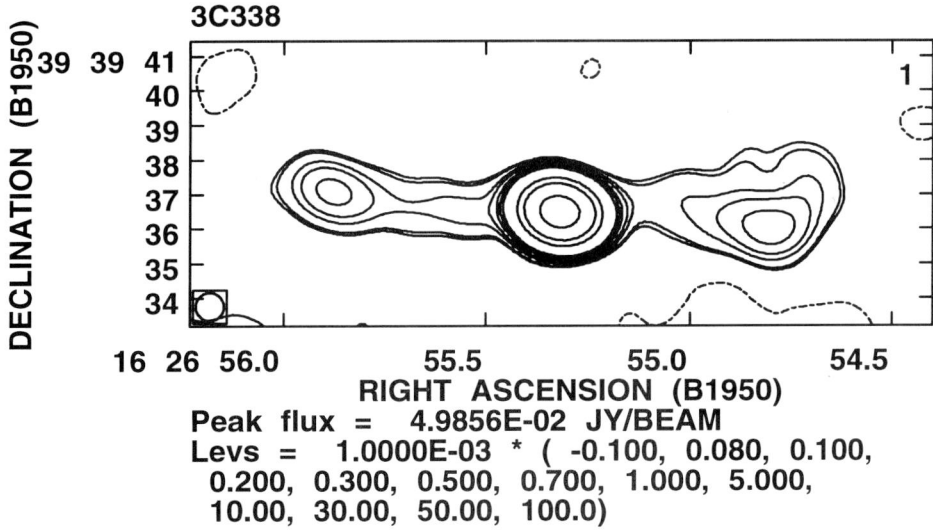

FIGURE 2. High resolution VLA map at 6cm of the central radio emission of 3C338

with two faint hot spots (Fig. 2). The spectral index of these two hot spots is 0.3 – 0.4 ($S(\nu) \propto \nu^{-\alpha}$) and the emission appears distinct from the steep spectrum extended emission, in agreement with the Burns et al. model. The core radio emission is strongly variable in time (Giovannini et al., in preparation) and has a moderately steep spectrum (0.4 – 0.5). In agreement with the moderately steep-spectrum core, the parsec-scale

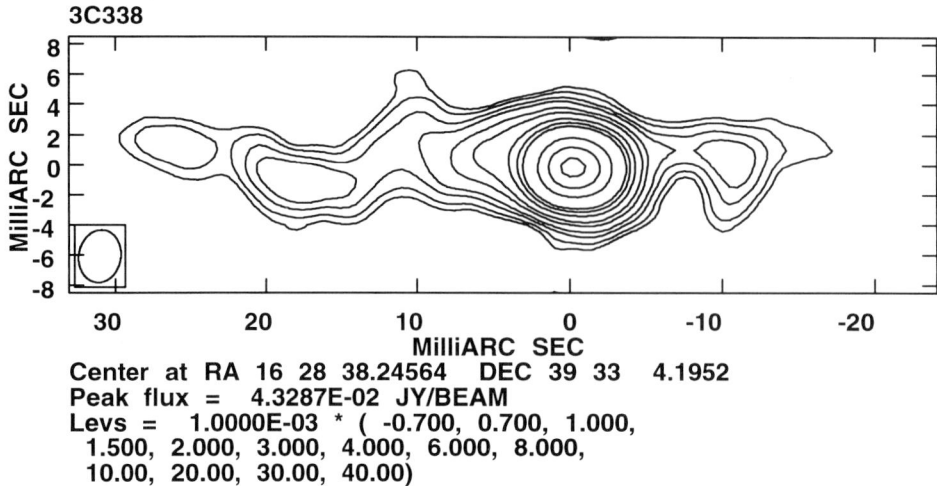

FIGURE 3. VLBA map of 3C338 at 6 cm

structure shows a flat-spectrum core with two symmetric extended jets oriented in E-W direction (Feretti et al., 1993; Fig. 3). This is in agreement with the very symmetric large-scale structure and the arcsecond VLA map (Fig. 2). The observational data indicate that a high jet velocity is not allowed even if the source is near the plane of the sky. The highest possible velocity for the two symmetric jets has been estimated to be 0.7c assuming an angle θ near to $90°$. New observations of this source at different epochs show clear morphological changes in the parsec-scale emission in agreement with the large core flux density variability found at arcsecond resolution. However, an unambiguous measure of a proper motion is not trivial due to the low brightness of the parsec-scale jets and the absence of prominent substructures inside them. A preliminary comparison of different epoch data suggests a possible motion corresponding to an apparent velocity ≤ 0.5 c.

REFERENCES

BURNS, J.O., SCHWENDEMAN, E., & WHITE, R.A. 1983. *ApJ* **271**, 575
FANAROFF, B.L. & RILEY, J.M. 1974. *M.N.R.A.S.* **167**, 31
FERETTI, L., COMORETTO, G., GIOVANNINI, G. & AL. 1993. *ApJ* **408** 446
GHISELLINI, G., PADOVANI, P., CELOTTI, A. & MARASCHI L. 1993. *ApJ* **407**, 65
GIOVANNINI, G., FERETTI, L. & COMORETTO, G. 1990. *ApJ* **358**, 159
GIOVANNINI, G., FERETTI, L., VENTURI, T. & AL. 1994. *ApJ* **435**, 116
GIOVANNINI, G., COTTON, W.D., FERETTI, L. & AL. 1995. *Proc. Natl. Acad. S.* **92**, in press
URRY, C.M. & PADOVANI, P. 1995. *P.A.S.P.* **107**, 803.

5-GHz EVN polarization of 3C286

By H. S. SANGHERA[1], D. R. JIANG[2],
D. DALLACASA,[3] R. T. SCHILIZZI[1,4],
E. LUDKE[5], AND W. D. COTTON[6]

[1]JIVE, Postbus 2, NL-7990 AA Dwingeloo, The Netherlands.

[2]Shanghai Observatory, Chinese Academy of Sciences, Shanghai 200030, China.

[3]NFRA, Postbus 2, NL-7990 AA Dwingeloo, The Netherlands.

[4]Sterrenwacht Leiden, Postbus 9513, NL–2300 RA Leiden, The Netherlands.

[5]Universidade Federal de Santa Maria, 97119-900 Santa Maria RS, Brazil.

[6]NRAO, 520 Edgemont Road, Charlottesville, VA 22903-2475, USA.

EVN polarization images at 5 GHz of the CSS radio quasar 3C286 are presented. The brightest region of this source is strongly polarized with the magnetic field dominated by the component perpendicular to the source axis. The electric vector orientation follows the stream of the emitting plasma along most of the structure revealed by the EVN data. We consider two possible interpretations for the observational data on this source: core–jet or lobe with hotspots, although the latter scenario is favoured by the EVN results.

1. Introduction

3C286 (1328+307) is a powerful radio source identified with a quasar at z=0.849, and has a steep spectrum ($\alpha = -0.61$, $S_\nu \propto \nu^\alpha$ between 1.4 and 15 GHz). Subarcsecond-resolution radio images show a misaligned triple structure, dominated by the central component which accounts for at least 95% of the total flux density at all frequencies. At mas resolution it shows a knotty jet-like structure immersed in a broad, low-surface-brightness cocoon (Zhang et al. 1994). At the north-eastern end there are two bright compact components, both with a flat spectrum between 1.4–5 GHz (Zhang et al. 1994).

The second brightest region of radio emission is located 2."6 to the southwest and is well resolved by the VLA at 5 GHz (Hines et al. 1989). The third radio component, 0."8 to the east, is visible at 4.6 GHz with the VLA (Hines et al. 1989), and at 1.7 GHz in the MERLIN image by Spencer et al. (1989). The projected linear size of the radio emission is about 14 kpc (H_o=100 km/s/Mpc and q_o=0.5), and therefore it is classified as a compact steep-spectrum (CSS) radio source (Fanti et al. 1985). 3C286 is one of the strongest extragalactic sources in polarized emission, with fractional polarization of $\sim$10% at 1.6 and 5 GHz (Perley 1982), and the position angle of the electric vector of the central component is 33° at both frequencies (Perley 1982).

2. Observation and data reduction

Polarization VLBI data at 5 GHz were recorded on 17/06/93, in MKIII mode A with an effective bandwidth of 28 MHz for both left and right hands of circular polarization at six EVN antennas: Effelsberg, Medicina and WSRT recorded in both hands, while Jodrell Bank, Noto and Onsala recorded in LCP only. The data were correlated at the MPIfR MK3 correlator in Bonn. The correlated data were read into AIPS for processing, calibration and imaging. The method used to calibrate VLBI polarization data is described in Cotton (1993). In order to improve the calibration and the sampling of the uv-plane we combined the EVN data with MERLIN data (taken in 1983) and obtained

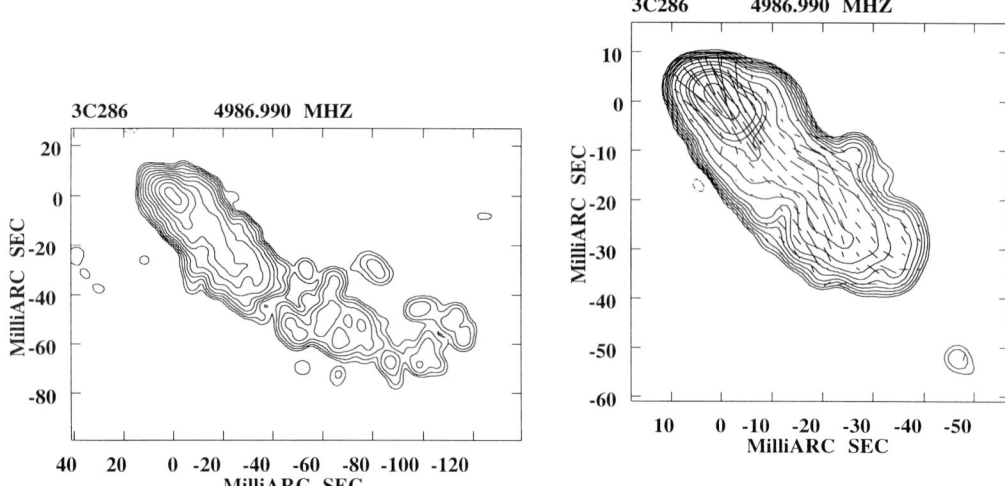

FIGURE 1. Left - Combined EVN + MERLIN image of 3C286 restored by 5-mas Gaussian beam: Right - EVN-only image, the vectors represent the projected orientation of the electric vector; their length is proportional to polarized flux density

a combined Stokes I image which was used as the model for the phase self-calibration of the EVN-only data.

3. Results

3.1. Total intensity

3C286 is heavily resolved on all the EVN baselines. Our EVN+MERLIN image (Fig. 1, left) accounts for 6.95 Jy, about 97% of the total flux density of the central component of 3C286 at 5 GHz. The two arcsecond-scale components to the east and west are resolved out by MERLIN. The image in Fig. 1 has been restored with a 5×5-mas Gaussian beam, and it is super-resolved by a factor of 1.36. We did not find any significant emission (at the mJy/beam level) on the northeast side, in contradiction to the Spencer et al. (1991) result. Our image shows radio emission extending up to $\sim$ 120 mas from the position of peak brightness. As the surface brightness decreases, the source axis progressively changes in orientation from $\approx 45°$ to $\approx 90°$. The EVN-only data accounted for only 5.1 Jy which is $\sim$ 70% of the total flux density. The eastern part of the structure is very similar to that in Fig. 1, suggesting that the missing flux density in the EVN image is mostly in the southwestern, more extended region.

3.2. The polarization results

The integrated polarized flux density of 3C286 is 0.84 Jy. To calibrate the absolute orientation of the electric vector, we applied a rotation to our polarization vectors to obtain an orientation of 33°, in accordance with the VLA result. In polarized intensity the brightest region contains two peaks separated by about 8 mas, and the maximum polarized emission is located $\sim$6 mas southwest of the peak in total intensity. In the EVN data, the scalar mean polarization is about 12% and the fractional polarization of the north-eastern peak of the total intensity is about 8%. At ~15–20 mas from the brightest peak, where the jet changes direction, there is a second region of low fractional polarization. The orientation of the electric vector is well aligned with the source axis, and diverges only at the northeastern edge of the radio emission. The integrated polarized flux density (840 mJy) in our EVN polarization image is in good agreement with the VLA results (see van Breugel et al. 1984).

4. Discussion

4.1. Scenario 1: The brightest region of 3C286 harbours the core

The standard interpretation is that the central bright radio emitting region contains the source nucleus so the overall arcsecond morphology is that of a misaligned, asymmetric and core-dominated triple. The following features support this scenario:

1) The presence of two bright but slightly resolved components, both with a *flat spectrum* between 1.7 and 5 GHz can be interpreted as being a core plus a bright shock in the jet or the nuclei of a binary system (Zhang et al. 1994).

2) The presence of an elongated structure to the north-east of the bright component, detected at 2.3 GHz by Phillips and Shaffer (1983) and in the MERLIN+EVN image at 5 GHz (Spencer et al. 1991), which can be interpreted as a counterjet.

4.2. Problems with scenario 1

In this scenario the quasar would be unusual, as the core would be strongly polarized and the magnetic field in the jet dominated by the component perpendicular to the jet axis. This configuration of the magnetic field can be found in arcsecond scale jets in some low-luminosity radio galaxies (Birkinshaw et al. 1981), but also in these cases the core region is weakly polarized, if at all.

The angular sizes of the brightest compact components at 5 GHz are significantly larger than are typical for cores in AGNs (Ghisellini et al. 1993). Lack of significant variability in the total flux density of 3C286 suggest that relativistic beaming may not be relevant and the source orientation is likely to be near the plane of the sky.

Finally, a counterjet to the east of the brightest region has not been detected at 327 MHz above a surface-brightness level of 0.02 mJy/mas^2 (Dallacasa et al. in preparation), nor in our 5-GHz combined image, although in both cases the uv-plane sampling was not comprehensive. This is not consistent with the interpretation that the source orientation is near the plane of the sky, and calls into question the earlier detections of the counterjet.

4.3. Scenario 2: The bright region of 3C286 is a lobe with hot-spots

A second possible interpretation of 3C286 is that the bright structure we imaged with the EVN is the hot-spot region of the main lobe in an asymmetric compact FR II radio source. The other lobe is 2.6 arcsec to the southwest (Spencer et al. 1989) and the core has not yet been identified, and should be located somewhere between them.

The strong asymmetry between the sizes and flux densities of the two lobes may be due to different local environments, so that the main lobe appears brighter and smaller due to a stronger confinement of the radio-emitting plasma. Conversely the asymmetry may arise from orientation effects, the brighter lobe being the approaching one; in this second case however we would require a misalignment between the pc-scale and kpc-scale axes of the radio sources, in order to have a very weak core, with initial jet direction in the plane of the sky. Evidence in favour of this "lobe – hot-spots" interpretation can be found in the objections to *Scenario* 1. In addition we make the following remarks:

1) The bright knots of emission are immersed in a large cocoon, as can be seen in the

high-resolution images at 1.7 and 5 GHz (Zhang et al. 1994). They resemble therefore the morphology of the hot-spots at the jet termination where the radio-emitting plasma interacts with the ambient medium.

2) The steep radio spectrum, high fractional polarization and perpendicular field orientation seen in 3C286 are consistent with the observed properties of the hot-spot regions of FR II sources (Laing 1988), where the high polarization and transverse fields are thought to originate in a compression shock (Meisenheimer et al. 1989).

4.4. Problems with scenario 2

The lack of a clear and unambiguous core identification is a point against "*Scenario* 2". In fact, in steep-spectrum radio quasars the cores account for a few percent of the total flux density at 5 GHz. In 3C286 the core has not been detected at a level of 10–20 mJy in any currently available image at any observing frequency. Considering an upper limit of 20 mJy, the maximum fractional contribution to the total flux density at 5 GHz would be 0.3%, which is similar to that found in radio galaxies.

The asymmetry between the lobes and the lack of a detected core require a very special combination of misalignment between the pc-scale jet and the larger kpc-scale structure, if the asymmetry is to be ascribed to orientation effects. We could have highly beamed hot-spots with high contrast with the lobes if the jet is oriented away from us in its inner region and then bends towards us at its termination. Moreover, if the brightest lobe is confined by a much denser medium, we would expect ionization of the ambient medium in the region where interaction with the radio jet takes place. This in turn would cause Faraday rotation and possibly depolarization even for a weak ambient magnetic field. Since significant Faraday rotation is not observed, we would require a very peculiar ambient magnetic field geometry.

Although a straightforward interpretation of the data is not possible we consider the "hot-spot" scenario is the more likely, but polarization images at higher resolution, and a more thorough search for the radio core, should help to resolve this dilemma.

REFERENCES

Birkinshaw, M., Laing, R.A. & Peacock, J.A. 1981 *MNRAS* **197**, 253.

Cotton, W.D. 1993 *A.J.* **106**, 1241.

Fanti, C., Fanti, R., Parma, P., Schilizzi, R.T. & van Breugel 1985 *A&A* **143**, 329.

Ghisellini, G., Padovani, P., Celotti, A. & Maraschi, L. 1993 *ApJ* **407**, 65.

Hines, D.C., Owen, F.N. & Eilek, J.A. 1989 *ApJ* **347**, 713.

Laing, R.A. 1988. In *Hot Spots in Extragalactic Radio Sources* (ed. K. Meisenheimer & H.J. Roser,), pp. 27.

Meisenheimer, K., Roser, H.J., Hiltner, P.R., Yates, M.G., Longair, M.S., Chini, R. & Perley, R.A. 1989 *A&A* **219**, 63.

Perley, R.A. 1982 *AJ* **87**, 859.

Phillips & Shaffer, D.B. 1983 *ApJ* **271**, 32.

Spencer, R.E., McDowell, J.C., Charlesworth, M., Fanti, C., Parma, P. & Peacock, J.A. 1989 *MNRAS* **240**, 657.

Spencer, R.E., Schilizzi, R.T.,Fanti, C., Fanti, R., Parma, P., van Breugel, W.J.M., Venturi, T., Muxlow, T.W.B. & Nan, R. 1991 *MNRAS* **250**, 225.

van Breugel, W.J.M., Miley, G.K. & Heckman, T.M. 1984 *AJ* **84**, 5.

Zhang, F.J., Spencer, R.E., Schilizzi, R.T., Fanti, C., Bååth, L.B. & Su, B.M. 1994 *A&A* **287**, 32.

Polarization imaging with MERLIN at K-band to resolve Faraday effects in CSS jets

By E. LÜDKE[1], H. S. SANGHERA[2], AND W. D. COTTON[3]

[1] UFSM/CCNE-Departamento de Física, 97119-900, Santa Maria/RS, Brazil
[2] Joint Institute for VLBI in Europe, P.O. Box 2, 7990 AA Dwingeloo, The Netherlands
[3] NRAO Headquarters, 520 Edgemont Road, 22903-2475 Charlottesville-VA, USA

In this paper we report preliminary results concerning the polarization mapping of luminous CSS jets with MERLIN. At 22 GHz and 12-mas resolution, the Faraday effects on the jet of 3C147 appear to be unresolved, suggesting a scale size of ~ 28 pc for the thermal irregularities which depolarize the source radiation leading to extremely high depolarization for CSS sources with the largest Faraday effects.

1. Introduction

Luminous quasars associated with prototypical compact steep-spectrum (CSS) sources at $z > 0.6$ often exhibit exceedingly strong Faraday rotation and depolarization of the radio radiation with Faraday dispersion in the range 10–900 cm$^{-3}\mu$G pc (Lüdke & Bellincanta, 1996) and average rotation measures up to $\sim$3000 rad m^{-2} (Lüdke, 1994) respectively, which are about two orders of magnitude larger than most extragalactic radio sources.

Such values support the idea that these sources are lying within a magnetoionic medium with much higher densities and/or magnetic fields than the environment of typical extended radiogalaxies and quasars (Kato et al, 1987; Lüdke et al. 1994). Further evidence for a gas-rich environment is given by the detection of strong broad forbidden optical emission lines with unusual velocity structure (Gelderman and Whittle, 1994). In this paper, we present a preliminary quantitative analysis of the physical properties of the magnetoionic medium which surrounds the CSS sources.

2. Discussion and Conclusions

Observations of 3C147 ($z = 0.54$) were made at 22 GHz with the MERLIN network in 1995. Since the instrumental polarization may reach $\sim 8\%$ in a few telescopes and the source brightness is high, the data have been calibrated and imaged with the NRAO/AIPS package using the polarization self-calibration algorithm devised by Lüdke (1994) where we could obtain a calibration accuracy of $\sim 1\%$, which is due to nonclosing phase errors due to bad weather.

Figure 1 shows the first 22-GHz image of 3C147 at 12-mas resolution with an r.m.s. noise level of about 720 μJy/beam. No polarized emission is seen along the jet of 3C147 although the integrated fractional polarization is similar to that obtained in VLA observations.

This map is useful to provide an upper limit for the sizes of the thermal irregularities which depolarize the radio radiation, which can be taken from the fact that no polarized flux is detected along the jet at VLBI resolutions and at wavelengths shorter than the zero-wavelength. This can be explained in terms of abnormal Faraday effects which can be produced if magnetic fields and densities are high enough in thermal irregularities which constitute the depolarizing medium, following the standard theory (Burn 1966;

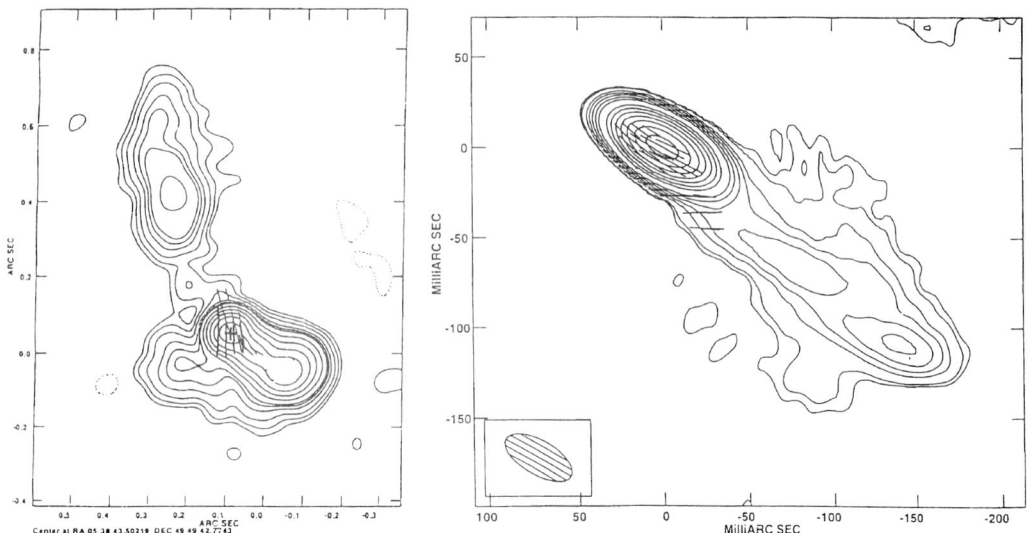

FIGURE 1. MERLIN polarization maps at 5 and 22 GHz.

Tribble 1991). Using the results of 'random walk' distribution functions for the Faraday depths along the line-of-sight, the size limit is about a third of a beamwidth (Garrington & Conway, 1991). If H_0=50 km s^{-1} Mpc^{-1} and q_0=0.5, we take $d \sim 28$ pc as an upper limit for the cell sizes in the depolarizing medium. In order to produce the observed depolarization of 680 cm$^{-3}\mu$G pc, we obtain an overall figure of $n_e B_z$ which is much larger than central values of $n_e B_z$ seen in the central regions of extended FRII radio sources ($\sim (1 - 40) \times 10^{-3}$ cm$^{-3}\mu$G; Garrington & Conway 1991), characterizing extremely abnormal depolarization for CSS sources with the largest Faraday effects. The main polarized emission arises in the central region, of a size of less than a beamwidth, where exceedingly high Faraday rotation with source rotation measure ~ -2900 rad m^{-2} is detected. The presence of such a region is consistent with the idea that a very narrow nozzle in the VLBI jet is seen behind a single irregularity, where less depolarization and enhanced Faraday rotation would take place, as our limit on the cell sizes would suggest.

Acknowledgments: EL would like to thank CNPq/Brazilian Government for a research fellowship and the MERLIN National Facility for hospitality and support.

REFERENCES

BURN B.J. 1966. *MNRAS* **133**, 67
GARRINGTON S.T., CONWAY R.G. 1991. *MNRAS* **250**, 198
GELDERMAN R. AND WHITTLE M. 1994. *ApJS* **91**, 491
KATO T., TABARA H., IONUE M. AND AIZU K. 1987. *Nature* **329**, 223
LÜDKE E. 1994. *PhD thesis, University of Manchester*, UK
LÜDKE E., BELLINCANTA A. 1996. *Publ. Astron. Soc. Pacific*, in press
TRIBBLE P.C. 1991. *MNRAS* **250**, 276

The cosmological evolution of radio sources

By JAMES S. DUNLOP

Institute for Astronomy, Department of Physics & Astronomy, University of Edinburgh, Edinburgh, EH9 3HJ, UK

Our current knowledge of the cosmological evolution of powerful radio sources is described, with particular emphasis on the high-redshift evolution of steep-spectrum radio galaxies. The predictions of the alternative high-redshift 'cut-off' models of Dunlop & Peacock (1990) are compared with the actual redshift distributions of two new complete samples; a 'bright' (2 Jy) sample with complete redshift coverage (the 6C/B2 sub-sample), and the first 'faint' sample (1 mJy) with spectroscopic redshifts or 'reliable' estimates for all sources (the LBDS sub-sample). The implications of the recent discovery of the most distant known galaxy – 6C 0140+326 at $z = 4.41$ – are also considered. It is concluded that the available evidence strongly supports the existence of a high-redshift decline in the steep-spectrum radio-source population similar to that established for flat-spectrum sources. In addition these new samples provide a sufficiently large base-line in radio-luminosity to justify a first attempt at distinguishing between alternative forms of the redshift cut-off; the present data favour a luminosity-dependent cut-off, with the more powerful radio sources suffering a less dramatic although still significant decline in space density at $z > 2.5$.

1. Introduction

It was the study of the radio source number counts which, over 30 years ago, first revealed the existence of cosmological evolution in our Universe (see Scheuer (1990) for an entertaining account of the events of the late 1950s), and even today radio sources remain the only class of object whose evolving luminosity function can be tracked with real confidence over virtually a Hubble time. The cosmological evolution of radio sources is thus a subject which is (or certainly should be) of wide-spread interest both to those interested in the physics of active galactic nuclei, and to cosmologists interested in the formation and evolution of structure in the Universe.

In this brief article I attempt to summarize, in broad terms, our current knowledge of the evolution of powerful radio sources between $z = 5$ and the present day, and how this compares with the evolution of active objects determined in other wavebands. I have taken the liberty of deliberately distorting the balance of this review to re-consider in some detail the crucial issue of the high-redshift cut-off in the light of recently published data.

2. 'Low-Redshift' Evolution

Out to redshifts $z \simeq 2$ the evolution of powerful ($P_{2.7GHz} > 10^{26} \mathrm{WHz}^{-1}\mathrm{sr}^{-1}$) radio sources is reasonably well-constrained, and in 1990 John Peacock and I showed that, given the existing complete-sample database, the evolution of both the flat-spectrum ($\alpha < 0.5$ where $f_\nu \propto \nu^{-\alpha}$) and steep-spectrum ($\alpha > 0.5$) radio-source populations is at least consistent with pure luminosity evolution (PLE) (Dunlop & Peacock 1990). The fact that both the form and rate of this evolution ($P(z) \propto (1+z)^3$) is so similar to that which is seen in both the optical and X-ray luminosity functions of QSOs (Hewett *et al.* 1993, Boyle *et al.* 1993) suggests, perhaps not surprisingly, that the cosmological evolution of the radio luminosity function (RLF) has little to do with the evolution of individual radio sources, but rather traces an overall cosmological decline in the frequency and/or strength of activity in galaxies (Dunlop 1994). Moreover, the fact that very similar evolution is apparently displayed by the IRAS starburst population (Rowan-Robinson et al. 1993) suggests that what we are in fact tracing is a cosmological climate change in the strength/frequency of galaxy interactions capable of triggering both starburst activity and nuclear activity, probably in the majority of bright ($L \geq L^\star$) galaxies (Taylor et al. 1996).

Interesting though this result is, it obviously raises the question of whether/how the study of the evolving RLF can be of any help in illuminating the evolution of individual radio sources. In fact it has already provided some important and useful constraints – for example the fact that flat-spectrum and steep-spectrum sources display very similar evolution is consistent with unification through orientation (Peacock 1987, Barthel 1989), as is the demonstration that the flat-spectrum RLF can be derived from the steep-spectrum RLF using an appropriate range of beaming parameters (Padovani & Urry 1992, Urry & Padovani 1996). Such results indicate that the traditional division between flat-spectrum and steep-spectrum sources may be somewhat artificial and have motivated recent attempts to interpret the overall evolution of the radio-source population in the context of unified models (Jackson & Wall 1996).

However, significant further progress requires extension of the complete-sample database to the point where the simple PLE model can no longer provide an adequate description of the data (as is beginning to happen for the QSO OLF and XLF; Miller et al. 1993, Reid (private communication)), alongside the development of physical models of individual radio sources to assist the interpretation of the evolving RLF. An example of recent progress in this latter area is provided by the multi-frequency study of compact symmetric objects by Readhead et al. (1996), from which they have inferred a primary evolutionary path for powerful radio sources in which sources advance at constant speed, but decline in luminosity by a factor $\simeq 10$ between a size of 500 pc and 200 kpc. In this picture, compact steep-spectrum sources are simply bright juvenile extended sources (Cygnus A would have been much more luminous in its youth) rather than a separate population, and the brightest steep-spectrum radio sources at any epoch should be very compact. However, there are problems in applying such a simple and elegant picture over cosmological time-scales since, in essence, it relies on the assumption of an external medium of smoothly declining density ($\rho_{ext} \propto R^{-1.3}$). In particular, recent 3-dimensional hydrodynamic simulations (Dunlop 1996) indicate that the most luminous steep-spectrum radio sources are likely to be produced during or just after a jet-cloud interaction. Since such interactions may have been more frequent in the high-redshift universe, this effect may at least in part be responsible for the apparent luminosity evolution seen in the RLF, as well as explaining the optical-radio alignment effect (Dunlop & Peacock 1993).

3. High-Redshift Evolution: the Redshift Cut-Off Revisited

I turn now to an area which is perhaps of more wide-spread interest, namely the high-redshift evolution of the radio-source population. Amid the current excitement over the detection of significant numbers of radio-quiet galaxies at $z > 3$ (Steidel et al. 1996) it is worth re-emphasizing that due to the great virtue of completeness offered by radio selection, and the immunity of radio waves to dust obscuration, radio galaxies remain at present the only class of galaxy whose comoving space-density can be determined with any real reliability at $z > 2$. However this advantage of completeness can of course only be realised with reliable redshift information.

3.1. The high-redshift evolution of radio-loud quasars

Dunlop & Peacock (1990) deduced that the comoving density of flat-spectrum radio sources ($\simeq 85\%$ of which are quasars), after reaching a peak at $z \simeq 2 \to 2.5$ must decline at higher redshifts (the redshift 'cut-off'). This result was really rather robust, because of the ease with which quasar redshifts can be determined, and the resultant lack of much need to resort to methods of redshift estimation for distant flat-spectrum sources. Moreover, despite the discovery of radio-loud quasars out to $z \simeq 4.46$, the recent studies by Hook et al. (1995) and Shaver et al. (1996) both support the existence of a high-redshift decline in the radio-quasar population.

3.2. The high-redshift evolution of radio galaxies

In the remainder of this article I will therefore concentrate on the more controversial issue of the high-redshift decline in the steep-spectrum ($\equiv$ radio-galaxy) population, the first evidence for which was also presented by Dunlop & Peacock (1990). In particular, given that at least two radio galaxies are now known with $z > 4$ it is of interest to consider the extent to which new data support the existence of the redshift cut-off and can constrain its form.

The key component of the complete-sample database used by Dunlop & Peacock (1990) to constrain the high-redshift evolution of the RLF was the Parkes Selected Regions, a sample of 178 sources with $S_{2.7GHz} > 100$ mJy over an area of 0.075 sr. Much of the evidence for a high-redshift decline in the steep-spectrum population then rested on the use of K-band photometry to estimate the redshifts of the faintest galaxies in this sample. The reliability of this method has since been called into question (Eales et al. 1993) although it is worth noting i) that a renewed spectroscopic campaign on the Parkes Selected Regions is currently revealing that virtually all the galaxies have true redshifts which are *smaller* than their $K - z$ estimates, and ii) that Dunlop & Peacock (1990) explored the effect of varying the redshift-estimating relation and found it that it was virtually impossible to remove the cut-off at all powers.

The revised redshift distribution for the Parkes Selected Regions, and its implications for the high-redshift evolution of the RLF will be presented elsewhere. However, a complete redshift distribution has recently been obtained for a somewhat smaller and brighter complete sub-sample of 65 radio galaxies from the 6C/B2 survey (Eales 1985). Below I compare this redshift distribution with those which are predicted by the redshift cut-off models of Dunlop & Peacock (1990), and also consider the implications of the very recent discovery of a radio galaxy at $z = 4.41$ in a slightly deeper sub-sample of the 6C survey (Rawlings et al. 1996). Finally I consider the redshift distribution that John Peacock and I have recently derived for a much deeper sample from the Leiden Berkeley Deep Survey, a sample which is of sufficient depth to resolve the issue of the redshift cut-off beyond any doubt.

3.2.1. Alternative models for the redshift cut-off

Although Dunlop & Peacock (1990) concluded that the evidence for a high-redshift cut-off in the steep-spectrum luminosity function was strong, uncertainties in redshift estimation, combined with lack of coverage of the luminosity baseline, meant that they were unable to distinguish between universal negative pure luminosity evolution at $z > 2.5$ (the PLE model) and an alternative model involving continuing positive evolution combined with negative density evolution at high z (the LDE model). These two alternative models are illustrated in Figure 1, the principal difference between them being that in the LDE model the strength of the decline in comoving number density is a function of radio power.

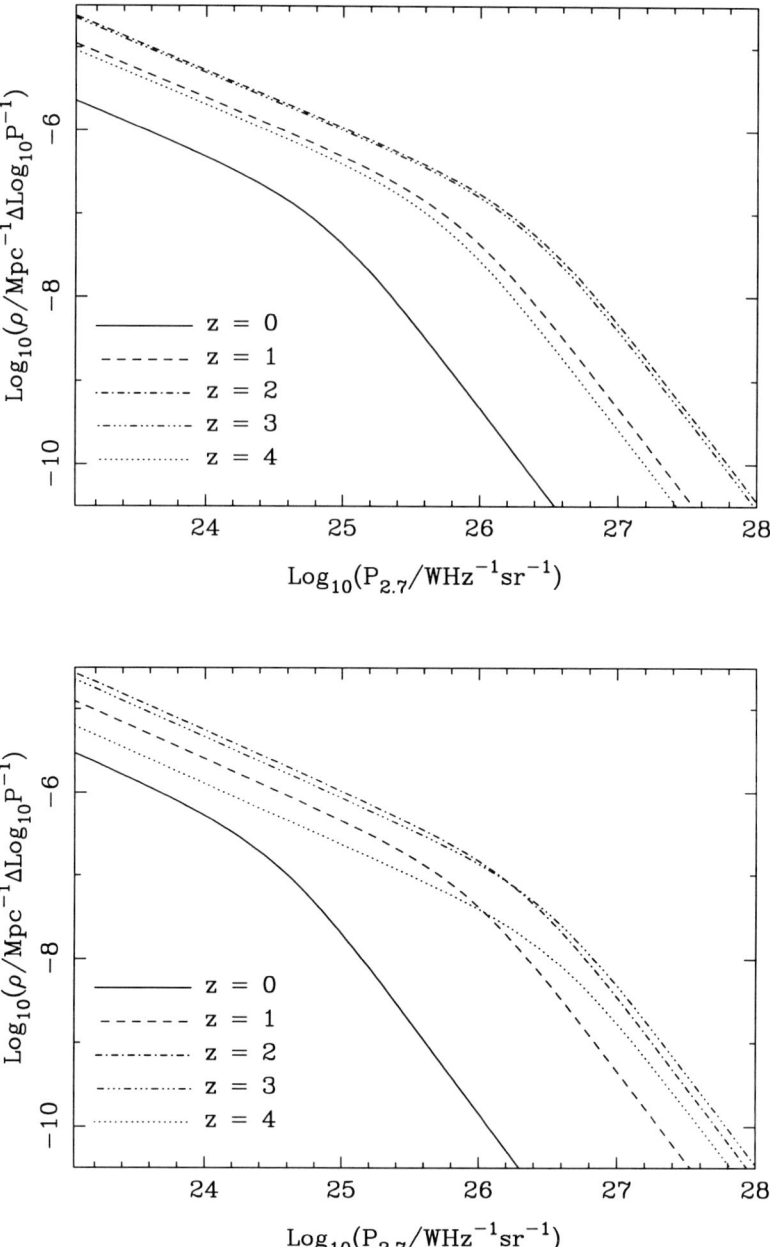

FIGURE 1. Two simple alternative models produced by Dunlop & Peacock (1990) to describe the high-redshift decline in the comoving number density of steep-spectrum radio sources. The upper panel shows the PLE model in which the redshift cut-off is parameterized in terms of universal negative luminosity evolution at high z. The lower panel shows the LDE model in which continuing positive luminosity evolution is overcome by negative density evolution at all powers for $z > 3$. In the latter model the apparent severity of the redshift cut-off becomes a function of radio luminosity, with the most luminous sources suffering a less dramatic, but still significant decline in comoving number density.

3.2.2. *Comparison with the complete 6C sub-sample*

The complete 6C/B2 sub-sample consists of 65 sources detected at 151 MHz with $2.2\text{Jy} < S_{151MHz} < 4\text{Jy}$ in 0.1 sr of sky. Eales, Rawlings and collaborators have recently completed the measurement of spectroscopic redshifts for all the sources in this sample; 11 galaxies lie at $z > 2$ (Eales & Rawlings 1996) and two of these lie at $z > 3$ (6C 1232+39 with $z = 3.221$ (Eales *et al.* 1993) and B2 0902+34 with $z = 3.395$ (Lilly 1988). The cumulative redshift distribution for this sample for $z > 2$ is shown in Figure 2, where it is compared with the distributions predicted by the PLE and LDE redshift cut-off models shown in Figure 1, and also with the prediction of a 'no-cutoff' model in which the form and normalization of the RLF is frozen for $z > 2$. In the upper panel the predicted number counts at 151 MHz have been produced from the models (which are defined at 2.7 GHz) by using the average spectral index displayed between 151 MHz and 2.7 GHz by the 11 $z > 2$ sources in the sample ($\langle \alpha_{151MHz}^{2.7GHz} \rangle = 0.87$) to convert the 151 MHz flux-density boundaries to their equivalent values at 2.7 GHz. ($0.164\text{mJy} < S_{2.7GHz} < 0.298\text{mJy}$). On this basis it is clear that the redshift distribution of this sample strongly supports the existence of the redshift cut-off, but that the form of this cut-off seems better described by the LDE model than the PLE model.

The lower panel shows the results of taking a different approach to comparing the observed redshift distribution of the 6C/B2 sub-sample with the predictions of the models. In this case the weakest 2.7-GHz flux density actually displayed by any of the 6C sources at 2.7 GHz ($S_{2.7GHz} \simeq 0.1$ Jy) has been taken as the equivalent lower flux-density limit for the purpose of predicting the redshift distribution. When this is done the PLE cut-off model correctly reproduces the required 2 sources with $z > 3$, but of course the total number of sources in the sample is grossly over-predicted because in effect it has been presumed that all sources detected at $S_{2.7GHz} > 0.1$ Jy are also guaranteed to be detected at $S_{151MHz} > 2$ Jy. This is equivalent to assuming that all sources at these flux-density levels have $\alpha_{151MHz}^{2.7GHz} > 1$, which is clearly incorrect. However the PLE model can in fact be reconciled both with the total number of sources in the sample and with the high-redshift distribution of the 6C/B2 sample by introducing an average spectral index which becomes a function of redshift for $z > 2$; the lower panel in Figure 2 illustrates the effect on the model predictions of assuming that the fraction of sources with $\alpha_{151MHz}^{2.7GHz} > 1$ changes smoothly from 40% for $z < 2$ to effectively 100% at $z > 3.5$. As yet there exists no clear evidence for such a redshift dependence of spectral index, although its existence may be hinted at by the spectacular success of the method of steep-spectrum selection in locating high-redshift radio galaxies (Chambers & Miley 1989). Certainly, if the updated redshift distribution for the Parkes Selected Regions favours the PLE model over the LDE model, this figure indicates that a redshift-dependent spectral index may offer a means of reconciling the high- and low-frequency data.

3.2.3. *Implications of the discovery of 6C 0140+326 at $z = 4.41$*

As Figure 2 clearly demonstrates, the absence of any galaxies with $z > 3.5$ in the 2.2-Jy 6C/B2 sub-sample is perfectly consistent with the predictions of the high-redshift cut-off models. However the most distant known galaxy – 6C 0140+326 with $z = 4.41$ – has recently been discovered in a separate sub-sample of the 6C survey, covering a very similar area of sky with a flux density limit only a factor of two lower than that of the complete 2-Jy sample considered above (Rawlings *et al.* 1996). While it is obviously dangerous to attempt to draw any major conclusions from a single object, it is nevertheless of interest to calculate whether, taken at face value, its existence is consistent or inconsistent with the redshift cut-off. In Figure 3 I have therefore plotted the redshift distributions predicted for this deeper 6C sub-sample using the same models and alternative spectral

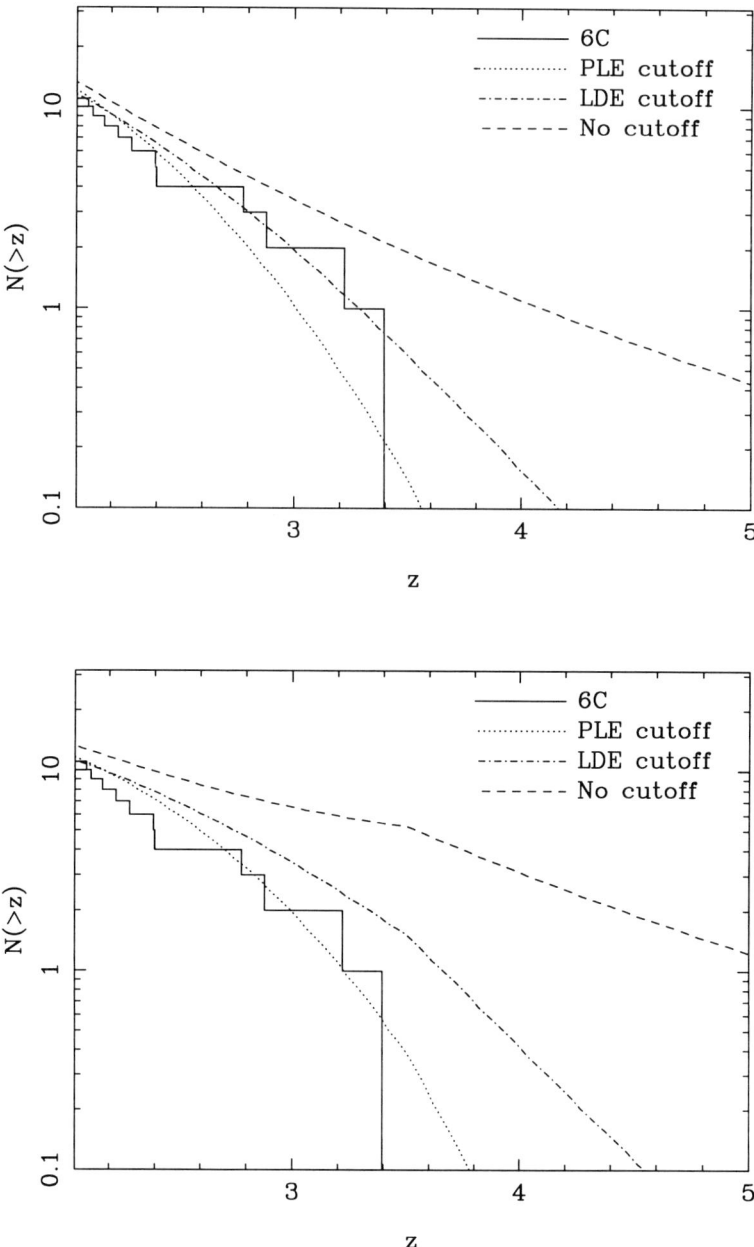

FIGURE 2. The observed high-redshift cumulative redshift distribution for the complete 6C/B2 sub-sample, compared with the predictions of the PLE and LDE cut-off models presented in Figure 1, and with the predicted redshift distribution for a model in which the radio luminosity function is fixed at its $z \simeq 2$ form for all higher redshifts. The upper panel presents a straight-forward comparison, in which the average 151 MHz $\rightarrow$ 2.7 GHz spectral index displayed by the high-redshift 6C/B2 sources has been used to predict the low-frequency counts from the models (which are defined at 2.7 GHz). The model predictions in the lower panel were produced by adopting a sufficiently deep 2.7 GHz flux limit to detect all 11 of the 6C/B2 high-redshift sources, and then introducing a redshift-dependent spectral index to reconcile the 6C/B2 data with the PLE model (see text for further details).

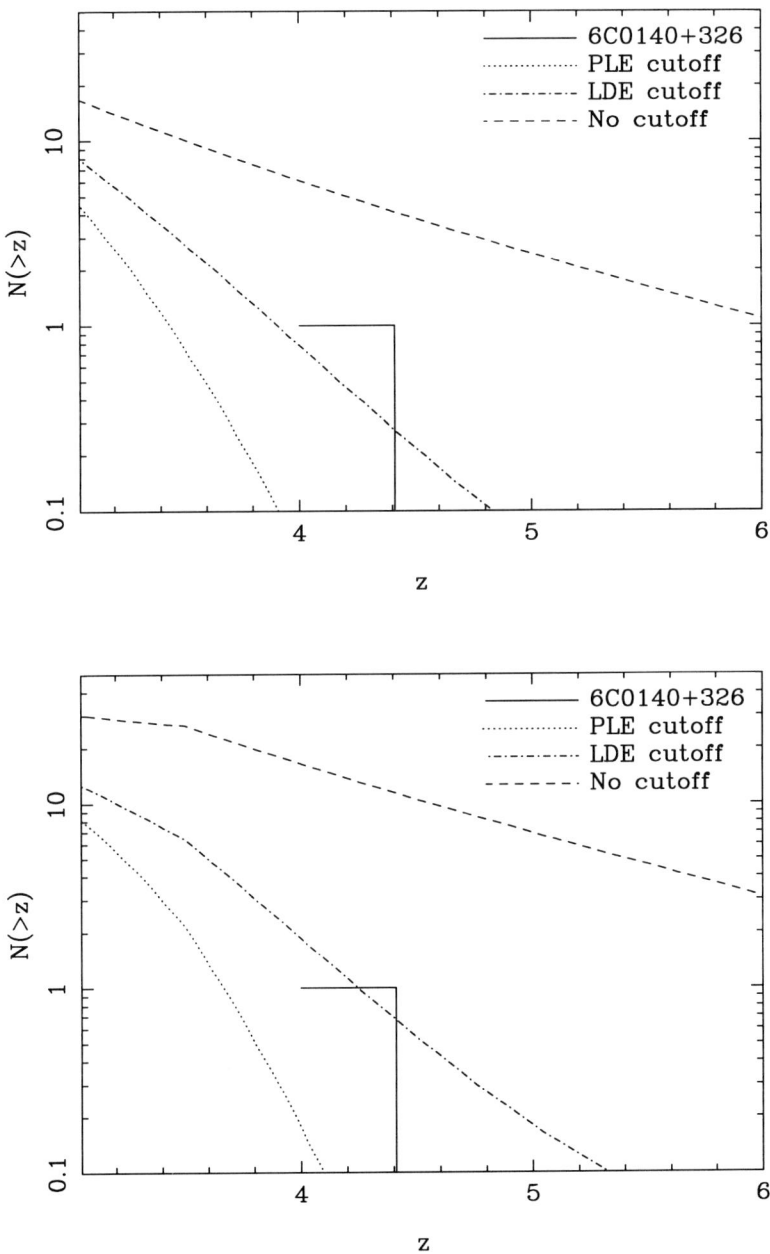

FIGURE 3. The 'single-object' high-redshift cumulative redshift distribution defined by 6C 0140+326 for the deeper 6C ($S_{151MHz} = 1 \to 2$ Jy) 0.1 sr sample, compared with the appropriate PLE, LDE and no-cutoff model predictions. The upper and lower panels show the effect of varying the spectral index assumptions as in Figure 2.

index assumptions as in Figure 2. It can be seen that the existence of 6C 0140+326 is perfectly consistent with the LDE model of the high-redshift cut-off.

3.2.4. *Comparison with the new LBDS sample*

The 6C/B2 sub-sample considered in section 3.2.2 is of interest because of its complete redshift information rather than because of its power to unambiguously confirm or refute the existence of the redshift cut-off. Indeed it is a factor $\simeq 2$ *less* deep than the Parkes Selected Regions, and thus at $z > 3$ can only be used to estimate the comoving space density of objects brighter than $P_{2.7GHz} \simeq 10^{27} \mathrm{WHz}^{-1}\mathrm{sr}^{-1}$ (which are intrinsically very rare). To unambiguously confirm or refute the existence of the redshift cut-off really requires the study of a sample with a flux density which is $\simeq 100$ times fainter, and is thus capable of sampling the radio luminosity function down to $P_{2.7} \simeq 10^{24} \mathrm{WHz}^{-1}\mathrm{sr}^{-1}$ out to $z \simeq 4$ (*i.e.* below the break luminosity at all redshifts – see Figure 1).

Accordingly, over the last few years we have been attempting to determine the redshift distribution of a statistically-complete sample of 77 galaxies with $S_{1.4GHz} > 1$ mJy selected from the Leiden Berkeley Deep Survey (LBDS) (Windhorst et al. 1984a, 1984b, Kron et al. 1985). We now possess g, r, i, K photometry for all the galaxies in this sample (plus J and H for a subset of sources) enabling us to estimate redshifts both from spectral fitting and from a modified version of the infrared Hubble diagram (Dunlop, Peacock & Windhorst 1995). Optical spectroscopy of a subset of sources with the Keck telescope indicates that this dual-pronged approach to redshift estimation appears to be reliable, certainly out to $z \simeq 2.5$, principally because the starlight from these more moderate-luminosity radio galaxies is less contaminated by strong emission lines or scattered AGN light than is the case in the more extreme objects found at high-redshift in the brighter radio samples (Dunlop et al. 1996).

The resulting redshift distribution of this 1-mJy sample is compared with that predicted by the PLE-cutoff, LDE-cutoff and no-cutoff models in Figure 4, where the predicted redshift distributions have been produced assuming $\alpha_{1.4GHz}^{2.7GHz} \simeq 0.8$. Comparison of the number count predictions in this figure with those presented in Figures 2 and 3 makes clear the enormous power of this much deeper sample, despite the need to resort to redshift estimation. In fact, to remove the cut-off, 10 of the 77 sources in this sample need to lie at $z > 4$, whereas in fact our best estimate of the redshift distribution follows almost exactly the predictions of the cut-off models.

3.3. *The form of the high-redshift cut-off*

The depth of the LBDS sample means that while its redshift distribution provides unambiguous confirmation of the existence of the redshift cut-off, it cannot be used to differentiate between the PLE and LDE models (as is clear from the alternative predicted redshift distributions in Figure 4). However the combination of the LBDS sample and the 6C data offer sufficient baseline in radio power to attempt to do so, and it is clear that, taken together, Figures 2, 3 and 4 favour a luminosity dependent cut-off of a form very similar to that which is produced by the LDE model.

In other words it seems that for radio galaxies the cut-off is least drastic for the most luminous sources. Large redshift surveys of bright radio quasars are beginning to indicate that a a very similar luminosity-dependent redshift cut-off is also displayed by the quasar RLF (Hook & McMahon 1996). Furthermore, the high-redshift evolution of the QSO OLF, also appears to have an at least qualitatively similar luminosity dependence (Warren, Hewett & Osmer 1994). The implication is that the similarity between the evolving flat-spectrum RLF, steep-spectrum RLF and QSO OLF seen at $z < 2$ extends right out to $z \simeq 4$, and that all powerful AGN suffer a similar luminosity-dependent

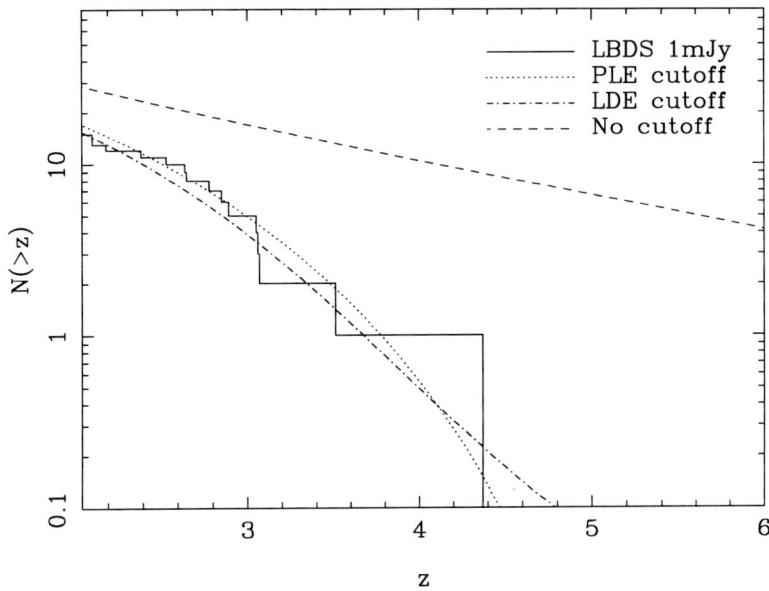

FIGURE 4. The high-redshift cumulative redshift distribution of the LBDS ($S_{1.4GHz} > 1$ mJy) sample, compared with the PLE, LDE and no-cutoff model predictions assuming $\alpha_{1.4GHz}^{2.7GHz} \simeq 0.8$.

redshift cut-off at $z > 2.5$. This apparent bias towards the most luminous sources at $z > 3$ may in part be due to distortion of the luminosity function by lensing (Rawlings et al. 1996) (although for extended radio sources the magnification factors are expected to be relatively modest). Alternatively it may simply be telling us that the most powerful active nuclei formed first.

REFERENCES

BARTHEL, P.D., 1989, *ApJ* **336**, 606.

BOYLE, B.J., GRIFFITHS, R.E., SHANKS, T., STEWART, G.C. & GEORGANTOPOULOS, I., 1993. *MNRAS* **260**, 49.

CHAMBERS, K.C. & MILEY, G., 1989. In: *'The Evolution of the Universe of Galaxies: The Edwin Hubble Centennial Symposium'*, p.373, ed. Kron, R.G., ASP

DUNLOP, J.S., 1994. In: *'Frontiers of Space and Ground-based Astronomy: The Astrophysics of the 21st Century'*, ESLAB Symp. No. 27, p.395. eds. Wamsteker, W., Longair, M.S. & Kondo, Y., Kluwer.

DUNLOP, J.S., 1996. In: *'Examining the Big Bang and Diffuse Background Radiations.* p.79. ed. Kafatos, M., Kluwer

DUNLOP, J.S. & PEACOCK, J.A., 1990. *MNRAS* **247**, 19.

DUNLOP, J.S. & PEACOCK, J.A., 1993. *MNRAS* **263**, 936.

DUNLOP, J.S., PEACOCK, J.A. & WINDHORST, R.A., 1995. In: *'Galaxies in the Young Universe'*, p.84. eds. Hippelein, H., Meisenheimer, K. & Roser, H., Springer-Verlag.

DUNLOP, J.S., PEACOCK, J.A., SPINRAD, H., DEY, A., JIMEMEZ, R.. STERN, D. & WINDHORST, R.A., 1996. *Nature* , in press.

EALES, S.A., 1985. *MNRAS* **217**, 149.

EALES, S.A. & RAWLINGS, S., 1996. *ApJ* **460**, 68.

EALES, S.A., RAWLINGS, S., DICKINSON, M., SPINRAD, H., LACY, M. & HILL, G., 1993. *ApJ* **409**, 578.

HEWETT, P.C., FOLTZ, C.B. & CHAFFEE, F.H., 1993. *ApJ* **406**, L43.

HOOK, I.M. & MCMAHON, R.G., 1996. *MNRAS*, in preparation.

HOOK, I.M., MCMAHON, R.G., PATNAIK, A.R., BROWNE, I.W.A., WILKINSON, P.N., IRWIN, M.J. & HAZARD, C., 1995. *MNRAS* **273**, L63.

JACKSON, C.A. & WALL, J.V., 1996. In: *'Extragalactic Radio Sources'*, Proc. IAU Symp. No. **175**. eds. Ekers, R., Fanti, C. & Padrielli, L., Kluwer, in press

KRON, R.G., KOO, D.C. & WINDHORST, R.A., 1985. *A&A* **146**, 38.

LILLY, S.J. 1988. *ApJ* **333**, 161.

MILLER, L., GOLDSCHMIDT, P., LA FRANCA, F. & CHRISTIANI, S., 1993. In: *'Observational Cosmology'*, ASP Conf. Ser. **21**, p.614. eds. Chincarini, G. et al., ASP

PADOVANI, P. & URRY, C.M., 1992. *ApJ* **387**, 449.

PEACOCK, J.A., 1987. In: *'Astrophysical Jets and their Engines'*, NATO ASI Series C. Vol. **208**, p.171, ed. Kundt, W., Reidel

RAWLINGS, S., LACY, M., BLUNDELL, K.M., EALES, S.A., BUNKER. A.J. & GARRINGTON, S., 1996. *Nature*, submitted

READHEAD, A.C.S., TAYLOR, G.B., PEARSON, T.J. & WILKINSON. P.N., 1996. *ApJ*, in press

ROWAN-ROBINSON, M., BENN, C.R., LAWRENCE, A., MCMAHON, R.G. & BROADHURST, T.J., 1993. *MNRAS* **263**, 123

SCHEUER, P.A.G., 1990. In: *'Modern Cosmology in Retrospect'*, p.331. eds. Bertotti, B. et al., CUP

SHAVER, P.A., WALL, J.V. & KELLERMAN, K.I., 1996. *MNRAS* **278**, L11

STEIDEL, C.C., GIAVALISCO, M., PETTINI, M., DICKINSON, M. & ADELBERGER, K.L., 1996. *ApJ*, in press.

TAYLOR, G.L., DUNLOP, J.S., HUGHES, D.H. & ROBSON. E.I., 1996. *MNRAS*, in press.

URRY, C.M. & PADOVANI, P., 1995. *PASP* **107**, 803.

WARREN, S.J., HEWETT, P.C. & OSMER, P.S., 1994. *ApJ* **421**, 412.

WINDHORST, R.A., VAN HEERDE, G.M. & KATGERT, P., 1984. *A&AS* **58**, 1.

WINDHORST, R.A., KRON, R.G. & KOO, D.C., 1984. *A&AS* **58**, 39.

Determining H_o

By MICHAEL ROWAN-ROBINSON

Blackett Laboratory, Imperial College, Prince Consort Rd, London SW7 2BZ

Recent developments on the distance scale are reviewed, including HST measurements of Cepheids in galaxies in Virgo and other clusters and groups, new studies of Type I and II supernovae, and work on the gravitational lens time delay and Sunyaev-Zel'dovich methods. There is a good consensus on distances to the Local and M81 Groups, but still controversy about the distance to the Virgo cluster. Combining the best-quality estimates of the Hubble constant gives a formal value $H_o = 65 \pm 4 \text{ km s}^{-1} \text{ Mpc}^{-1}$.

Estimates of the age of our Galaxy are briefly reviewed and a best value of $t_o = 13 \pm 1$ Gyr is adopted. Dynamical estimates of the value of the density parameter by a variety of methods using IRAS galaxies are consistent with a value for Ω_o close to 1. There is then a problem for an Einstein-de Sitter model ($\Omega_o=1$, $\Lambda=0$), for which $H_o=65$ implies an age for the universe of 10 Gyr. For this model to survive either the Hubble constant or the age of the Galaxy must be overestimated.

1. Reviews of the distance scale and H_o

I have not attempted here a major, complete review of the distance scale, along the lines of my earlier reviews (Rowan-Robinson 1985, 1988). In The Cosmological Distance Ladder (Rowan-Robinson 1985) I gave a reasonably complete review of work on the distance scale up to the end of 1982. My 1988 review covered the period 1983-7 in comparable detail. The reader is referred to these reviews for further details.

Several other reviews on the distance scale and the value of H_o have appeared in the period 1988-93. Dressler (1988) reviewed the values of H_o, q_o, Λ_o, Ω_o, and t_o. Tully (1989) reviewed the distance to Virgo and the value of H_o. van den Bergh (1989, 1993) reviewed distances out to Virgo and the value of H_o. Sandage and Tammann (1990) reviewed the distance to Virgo, and of more distant clusters relative to Virgo, arriving at a value for H_o "freed from all local velocity anomalies". A brief discussion of the distance to Virgo, H_o, and the implications for inflation, is given in Rowan-Robinson (1991). Jacoby et al. (1992) have reviewed 8 distance methods in detail and summarised estimates of the distance to Virgo and the value of H_o. Kennicutt et al. (1995) have reviewed recent distance estimates and the plans of the HST Key Project on the distance scale. Table 1 summarises the values of the distances to Virgo and the values of H_o adopted in these reviews. This illustrates that there is still deep controversy about the value of H_o, with values in the range 50–80 $\text{km s}^{-1} \text{ Mpc}^{-1}$ being advocated.

It is convenient to define the distance modulus $\mu_o = 5 \log_{10} (d/10\text{pc})$. A 10% error in d therefore corresponds to an error of ± 0.2 in μ_o.

2. The local distance scale

Population I indicators (Cepheids) are based on Hyades ($\mu_o=3.30$, Hanson 1980; 3.45, Vandenbergh and Bridges 1984) and Pleiades ($\mu_o=5.60$) clusters. **Population II indicators** (RR Lyrae stars, globular clusters, novae, planetary nebulae) are based on a few local subdwarfs with known trigonometric parallaxes. Both calibrations will be enormously strengthened by results from the Hipparcos satellite, expected soon.

authors	dVirgo (Mpc)	H_o (km s^{-1} Mpc^{-1})
RR 85	18.4±1.6	67±15
RR 88	19.0±1.2	66±10
Dressler 88		60-80
Tully 89	15.2±1	87±10
vdB 89	20±2	67±8
ST 90	21.9±0.9	52±2
Jacoby et al. 92	16.0±1.7	80±11
vdB 93	16.0±1.3	76±9

TABLE 1. Distances to the Virgo cluster

3. Primary extragalactic distance indicators

These are methods which can be calibrated from observations in our Galaxy or from theoretical considerations.

3.1. Cepheid variable stars

These are massive stars crossing the instability strip in the HR diagram and becoming pulsationally unstable. As a result their visual luminosity undergoes regular, large-amplitude variations with period 2-150 days. Levitt (1907) showed that there is a correlation between mean luminosity and period. A simple derivation of the physical basis for this was given by Sandage (1958) – we have:

$$p = Q\rho^{-1/2} \text{(natural oscillation period)}$$

$$\rho = 3M/4\pi R^3$$

$$L = 4\pi R^2 \sigma T_{eff}^4$$

Over the mass-range of Cepheids, we can also expect that:

$$\log M = hM_{bol} + \text{constant}$$

$$M_{bol} = M_V + d(B - V) + \text{constant}$$

$$\log T_{eff} = a(B - V) + \text{constant}.$$

These give

$$\log p + (0.5h + 0.3)M_V + (0.5hd + 0.3d + 3a)(B - V) = \text{const},$$

which is the period-luminosity-colour relation for Cepheids. Thus the spread in the $p - L$ diagram at B or V is correlated with colour. We therefore need multi-wavelength observations (and these are also needed to correct for the effects of extinction). Alternatively we can shift to infrared wavelengths, where the intrinsic spread in the $p - L$ diagram is much lower and the effects of extinction are also much reduced.

Cepheids have been studied in the galaxies of the Local Group, Sculptor group, M81 group and M101 (2 Cepheids). Altogether 50 galaxies have been or are being studied (Jacoby 1992, Kennicutt et al. 1995). Recent distance measurements using Cepheids include those in table 2, all but the first two being results from the Hubble Space Telescope. The distance to M81 is important for calibration of the Tully-Fisher method (see below). IC4182 is important as the nearest galaxy in which a Type Ia supernova has occurred (SN1937c). There is good agreement of Cepheid distances with other methods (Freedman and Madore 1993). A key step for the Hubble constant is the detection and

M33	$\mu_o = 24.64\pm0.09$	(Freedman et al. 1991), d = 0.85 Mpc
NGC300	$\mu_o = 26.66\pm0.10$	(Freedman et al. 1992), d = 2.15 Mpc
M81	$\mu_o = 27.80\pm0.20$	(Freedman et al. 1993), d = 3.6 Mpc
IC4182	$\mu_o = 28.36\pm0.09$	(Sandage et al. 1992, Saha et al. 1994), d = 4.7 Mpc
NGC5253	$\mu_o = 28.06\pm0.06$	(Sandage et al. 1994, Saha et al. 1995), d = 4.1 Mpc
M100	$\mu_o = 31.16\pm0.22$	(Freedman et al. 1994), d = 17.1 Mpc
M96 (Leo)	$\mu_o = 30.52\pm0.15$	(Tanvir et al. 1995), d = 11.6 Mpc
N4496(Vir)	$\mu_o = 31.10\pm0.15$	(Saha et al. 1996), d = 16.6 Mpc
N4536(Vir)	$\mu_o = 31.05\pm0.15$	(Saha et al. 1996), d = 16.2 Mpc

TABLE 2. Recent Cepheid distance measurements. All but the first two are results from the Hubble Space Telescope.

study of Cepheids in M100 in Virgo, which has been carried out as part of the HST Key Project led by J.Mould (Freedman et al. 1994). Many other Cepheid distances will be determined during this programme (Kennicutt et al. 1995).

3.2. Supernovae

The main types are:
- Type Ia (no H lines): deflagration of CO white dwarf following accretion from a companion
- Type Ib (6115 absorption absent): explosion of 1–20 $M_\odot$ star which has previously lost its envelope
- Type II (strong H lines): explosion of star of mass >8 $M_\odot$, following formation of an iron core
- Type IIb (no H lines, SN1987A): explosion of 20 $M_\odot$ star of low heavy element abundance which has lost its envelope.

In all cases the declining portion of the light curve is due to the radioactive decay of 0.5–1 solar mass of ^{56}Ni through ^{56}Co to ^{56}Fe, with a half-life of 60 days.

The 'Baade-Wesselink' method (Baade 1926) for determining the distance to pulsating or exploding stars assumes, in its simplest form, that the star has a blackbody spectrum, so that the colour of the star gives the surface temperature. The surface brightness in, say, the V band is then given by the Planck law. Measurement of the flux in this band then allows the angular radius of the star to be calculated. By integrating the observed radial velocity, measured from spectroscopic observations, with respect to time, the linear radius of the star can be calculated and hence the distance of the star determined. This method was first applied to supernovae by Arp (1961). It has been applied to both Type Ia and Type II supernovae (and SN1987A), with detailed model atmospheres in place of the blackbody assumption. It is worth emphasizing that this is a purely geometrical distance.

Recent developments include the use of circumstellar ring of gas around SN1987A to derive a distance of 50.1±3.1 kpc for the LMC (Panagia et al. 1991), in good agreement with the distance of 55±5 kpc derived by the Baade method (Branch 1987). Branch (1992) determined the mass of ^{56}Ni from spectroscopic detection of elements of intermediate mass and by the explosion kinetic energy as inferred from the spectroscopic velocities in Type Ia supernovae, and derived $H_o = 61\pm10$ km/s. Shigeyama et al. (1992) have discussed detailed models for the Type Ia supernovae 1990N and conclude that a white dwarf detonation model following a merger may be needed to explain the observed high velocity Si and Ca. Branch and Miller (1993) have discussed whether Type Ia supernovae are good standard candles, using 41 supernovae whose parent galaxies have relative distances from the Tully-Fisher or $D_n - \sigma$ methods. Apart from supernovae which are spectroscopically peculiar or have high extinction, Type Ia supernovae show a small dispersion in their absolute magnitudes, only 0.36m. They derive

$$M_B = -19.72 \pm 0.06 + 5\log(H_o/50).$$

Combined with the Sandage *et al.* (1994a,b) Cepheid distances to IC4182, NGC5253, this gives $H_o = 55\pm8$ km s^{-1} Mpc^{-1}. Schmidt *et al.* (1992, 1994) have applied the Baade method to 16 Type II supernovae and derive a distance to Virgo of 22±3 Mpc, and a Hubble constant $H_o = 73\pm6$ km s^{-1} Mpc^{-1}. Riess *et al.* (1995) have studied 13 Type Ia supernovae and deduce that $H_o = 67\pm7$ km s^{-1} Mpc^{-1}, after correction for the correlation of luminosity at maximum light with rate of decline (see also Hamuy *et al.* 1995, Tammann and Sandage 1995).

3.3. *Novae*

These are outbursts due to accretion by a white dwarf from a companion. They are less violent than Type Ia supernovae and are recurrent. The absolute magnitude at maximum light is correlated with the rate of decline of the light curve and there is a theoretical explanation of this in terms of models for the expanding nova atmosphere. They have been studied in the Galaxy, M31 and in Virgo cluster galaxies by Pritchet and van den Bergh (1985, 1987), who have a distance for Virgo of 19.5±4 Mpc.

The method has been reviewed by Jacoby *et al.* (1992), who draw attention to inconsistencies between the calibrations in M31 and the Galaxy. Della Valle and Livio (1995) have reexamined the calibration of novae and derive a distance for Virgo of 18.6±3.2 Mpc.

3.4. *Tip of the red giant branch*

The I-magnitude of the tip of the first-ascent red giant branch of low mass stars has been found to be a good distance indicator for resolved galaxies with metal-poor old populations (Lee 1993, Lee and Freedman 1993). It has been applied only to Local Group galaxies and NGC3109 to date.

3.5. *Gravitational lens time delay*

This method has been applied to only one system to date, the gravitationally lensed quasar 0957+561. The system has been modelled by Young *et al.* (1981), Greenfield *et al.* (1985), Falco *et al.* (1985, 1991). The optical image consists of two quasar images, A, B, and a galaxy, G1, whose image is merged with B. Radio maps show a further 3 features, C, D, E. Falco *et al.* (1991) model G1 with a King profile (tidally truncated isothermal sphere) with parameters: bulge velocity dispersion, $\sigma_v = 390$ km s^{-1}, core radius, $\theta_c = 2.9''$, with an additional point mass, $M_c = 11 \times 10^{10} h^{-1} M_\odot$. They include the contribution of about 100 other galaxies in the field which affect the image.

Refsdal (1966) pointed out that if time variations could be detected in 2 images of a gravitational lens, H_o could in principle be determined. For 0957+561, the time delay between the two quasar images has been determined as $t_{AB} = 540 \pm 12$ days (Lehar et al 1992), by combining optical and VLA data. This results in

$$H_o = 60 \pm 10 (\sigma_v/390\text{km s}^{-1})^2 (t_{AB}/540\text{days})^{-1}.$$

The main uncertainty is in σ_v, which could be significantly lower if there is a smooth dark matter component in the cluster. Thus the method essentially gives $H_o \sim 60$ km s^{-1} Mpc^{-1}.

A second gravitational lens system with a measurable time-delay, 0218+357, has been found by Wilkinson *et al.* (1996).

3.6. Sunyaev-Zel'dovich effect

We see a distortion of the microwave background spectrum in the direction of rich clusters of galaxies due to inverse-Compton scattering of CBR photons by hot gas. On the low-frequency side of the CBR peak, the cluster appears as a dark patch, on the high-frequency side as a bright patch. By mapping this S-Z defect or excess, and by mapping the same cluster at X-rays to determine the temperature structure of the hot gas, we can determine H_o.

The amplitude of the S-Z effect $\propto$ (inverse-Compton scattering optical depth through cluster) $\times$ (mean energy change of scattered photons), i.e. $\Delta T_{RJ} \propto <n_e T_e> L$.

The X-ray flux $S_X \propto (<n_e n_p T_e^{1/2}>/D_L^2)\times$ volume, where D_L is the luminosity distance. If the thermal and density structure of the cluster gas are known, we can eliminate n_e (and $n_p \propto n_e$).

Now $V = L\theta^2 D_A^2$, where D_A is the diameter distance, so $L \propto \Delta T_{RJ}^2 S_X^{-1} T_e^{-3/2}$, and this determines D_A from L/θ if the shape of the cluster is known (taken as spherical). Hence H_o can be determined (with a weak dependence on Ω_o). For the method to work, we need a good X-ray spectrum (T_e), the 3-dimensional density and thermal structure (from S-Z and X-ray maps), and no contamination by sources.

For the cluster A665 at $z = 0.182$, there is a good resolved detection of the S-Z effect (Birkinshaw et al. 1991). They assume $n_e = n_{e,o}(1 + r^2/r_c^2)^{-3\beta/2}$, which is a good approximation to an isothermal sphere, and find a good fit with $\beta = 0.66, \theta_c = 1.6'$. This yields $H_o = 40\pm 9$ km s^{-1} Mpc^{-1}. A worst case combination of systematic errors gives $H_o = 26\pm 8$ to 65 ± 10.

The main uncertainties are:
- H_o could be underestimated by a factor of 2 if A665 is prolate with the long axis close to the line of sight
- clumpiness of the gas - could only reduce H_o
- thermal structure could be non-isothermal - gives an uncertainty of a factor of 2 (need resolved X-ray spectra).

These uncertainties can be reduced with improved observational data and studies of several different clusters.

McHardy et al. (1990) have used this method on A2218 and found $H_o = 24^{+13}_{-10}$, which seems on the low side compared with other distance methods. Birkinshaw et al. (1993) comment that the S-Z effect is unresolved in this cluster and that the simple isothermal model may not be a good fit in this cluster. Birkinshaw et al. (1994) have analyzed the S-Z effect in A665 and A2218 and conclude that $H_o = 55\pm 17$ km s^{-1} Mpc^{-1}.

4. Secondary extragalactic distance indicators

These are methods whose calibration is via galaxies of the Local Group and other nearby groups of galaxies, whose distances are known from primary methods.

4.1. Tully-Fisher method for spirals

This depends on a correlation between the absolute magnitude of spiral galaxies and the maximum rotation speed. The method is extensively reviewed by Rowan-Robinson (1988) and Jacoby et al. (1992). It has been applied to many hundreds of galaxies, reaching out to the Coma cluster and beyond, and has been successfully used as a relative distance indicator to map large-scale velocity flows.

4.2. $D_n - \sigma$ method for ellipticals

This method depends on a correlation between the angular radius within which there is a fixed mean surface brightness and the velocity dispersion in the central regions of the galaxy. It has also been applied to many hundreds of galaxies and successfully used as a relative distance indicator to map large-scale velocity flows. However as there are no nearby normal ellipticals (only dwarfs), there is no real calibration of the method with galaxies for which primary distances are known. Dressler (1987) used the bulges of M31 and M81 to calibrate the method, assuming that spiral bulges are identical to elliptical galaxies. Pierce (1989) used the planetary nebula distance to the Leo group, and Tonry (1991) used the surface brightness fluctuation distance to the same group, but these are both secondary methods. For this and the subsequent three methods, see the review by Jacoby *et al.* (1992).

4.3. Globular cluster luminosity function

This is based on the assumption that the shape of the luminosity function in globular clusters is the same in all galaxies. It has been applied to the Virgo cluster and to a few other clusters.

4.4. Planetary nebula luminosity function

This is based on the assumption that the shape of the luminosity function for planetary nebulae is the same in all galaxies. It agrees well with other methods out to M81. It has been applied to the Virgo cluster.

4.5. Surface brightness fluctuations

This is based on the fact that rms variations in pixel brightness reflect an incipient resolution into stars. The calibration is via M31 and M32, but could in principle be via the Galaxy. It agrees well with other methods out to M81. It has been applied to the Virgo cluster.

4.6. Brightest red stars in galaxies

See Rowan-Robinson (1985, 1988) for some of the past history of this method. There have been recent claims by Shanks *et al.* (1992) and Pierce *et al.* (1992) to have measured the distance to Virgo by this method. Rozanski and Rowan-Robinson (1993) show that the method is much less accurate than has been claimed and that it is incapable of distinguishing between the long and short distance scales.

5. Summary of distances to Local Group galaxies, M81 and Virgo

Table 3 shows a comparison of Local Group distance moduli. This illustrates that there is a good consensus on the distances of galaxies in the Local Group, and good agreement between different methods. Table 4 shows a comparison of distance moduli beyond the Local Group; this illustrates that there is good agreement now about the distance to M81, but still major disagreement about the distance to Virgo.

6. The Hubble constant, H_o

We now have a route to H_o entirely via primary distance indicators, using Cepheids, supernovae, the gravitational lens time delay and the Sunyaev-Zel'dovich effect. Since these methods give broadly consistent results, I do not see that secondary methods now add anything to our knowledge of H_o, though of course the latter have proved very

	LMC	SMC	M31	M33	IC1613	N6822
CDL (RR85)	18.70	19.12	24.25	24.65	24.29	23.77
RR88	18.64	19.09	24.25	24.43	24.21	23.55
Panagia et al. (1991)	18.51					
Freedman et al. (1991)				24.64		
van den Bergh (1992)			24.3	24.5	24.42	23.66
Lee et al. (1993):Ceph	18.50		24.44	24.63	24.42	23.62
RR Lyr	18.28		24.36	24.71	24.27	
rg branch	18.42		24.36	24.70	24.27	23.46
unweighted average	18.51	19.10	24.33	24.61	24.31	23.61

TABLE 3. Summary of distance moduli to Local Group galaxies

	NGC300	M81	IC4182	Virgo
CDL (RR85)	26.44	27.31	28.15	31.32
RR88	26.70	27.44	28.13	31.40
van den Bergh (1992)	26.0	27.6		31.0
Freedman et al. (1993):				
br stars,HII		27.70±0.3		
IR T-F		27.86±0.3		
(B-band Ceph		28.73)		
B-band T-F		27.60±0.16		
I-band Ceph		27.59±0.31		
plan neb LF		27.72±0.25		
surf br fluctns		27.72±0.18		
HST Ceph		27.80±0.20		
Freedman et al. (1992)	26.66			
Sandage et al. (1992)			28.47	
Jacoby et al. (1992):				
glob cl LF				31.34
novae				31.59
SN Ia				31.40
T-F				30.97
plan neb LF				30.90
surf br fluctns				30.84
Dn-s				31.10
Freedman et al. (1994)				31.16

TABLE 4. Summary of distance moduli beyond the Local Group

valuable in mapping the peculiar velocities of galaxies. Table 5 summarises the estimates of H_o by primary methods.

Alternatively we can follow the approach of CDL and RR88 and combine all distance estimates to clusters and groups, with a weighting depending on the accuracy of the method. Including all the recent distance estimates reviewed here, I find, once again $H_o = 65$ km s^{-1} Mpc^{-1}. If only galaxies with distances greater than 100 Mpc are used, a slightly lower value of $H_o = 61$ km s^{-1} Mpc^{-1} is found.

It will clearly be valuable to have Cepheid distances to other galaxies in Virgo, to check the Freedman et al. value. All distance indicators are in good agreement out to

Cepheids in Virgo (M100)	$H_o = 80\pm17$ (Freedman et al. 1994)
Cepheids + SN Ia	$H_o = 55\pm8$ (Sandage et al. 1994)
	(supported by Baade-Wesselink method
	and by CO white dwarf deflagration models)
	$H_o = 67\pm7$ (Riess et al. 1995)
SN II, Baade-Wesselink	$H_o = 73\pm13$ (Schmidt et al. 1994)
gravitational lens time delay	$H_o = 60\pm10$ (Lehar et al. 1992)
Sunyaev-Zel'dovich, A665,2218	$H_o = 55\pm17$ (Birkinshaw et al.1994)
average of these 6	$H_o = 65\pm4$ km s^{-1} Mpc^{-1}

TABLE 5. Summary of estimates of H_o by primary methods

Symbalisty and Schramm (1981)	8.7-15.8 $\times 10^9$ years
Thieleman et al. (1983)	$20.8^{+2}_{-4} \times 10^9$ years
Meyer and Schramm (1986)	8.7-18.8 $\times 10^9$ years
Fowler (1987)	$11.0\pm1.6 \times 10^9$ years
Cowan et al. (1987)	12.4-14.7 $\times 10^9$ years
summary	10-16 $\times 10^9$ years

TABLE 6. Age of Galaxy by nucleocosmochronology

M81 and there is good agreement between different distance estimators about the relative distances of clusters beyond Virgo (Rowan-Robinson 1988).

7. Age of the universe, t_o

7.1. Nucleocosmochronology

Dating of terrestrial and lunar rocks, and of meteoritic material, by the abundance of radioactive elements and their daughter elements gives an age for the solar system of 4.55×10^9 years (see Rowan-Robinson 1985, 1988). The useful isotopes for determining the age of the Galaxy are

	^{232}Th	^{235}U	^{238}U	^{187}Re
daughter:	^{208}Pb	^{207}Pb	^{206}Pb	^{187}Os (β-decay)
half-life (yr):	13.9	0.7	4.5	43×10^9

These elements are made by rapid neutron capture (r-process) during supernova explosions and theory predicts the initial ratios of the number of atoms of different isotopes. For example the initial ratio of ^{232}Th to ^{238}U is predicted to be ^{232}Th/^{238}U = 1.6. To get the ratio observed today in terrestrial rocks (^{232}Th/^{238}U = 4.0), the radioactive isotopes must have been decaying for 9.8 $\times 10^9$ years if they were all made at the same time. If, on the other hand, these elements were made at a uniform rate right up to the formation of the solar system, then an age of 15 $\times 10^9$ years is found. Table 6 summarises age estimates from nucleocosmochronology.

7.2. Globular cluster ages

The ages of globular clusters are determined by fitting isochrones from stellar evolution models to observed HR diagrams. The parameters for the models are (Demarque et al. 1991):

He abundance, Y:	use primordial value 0.23 or 0.24
heavy element abundance, Z:	[Fe/H] = −1 to −2.3 (±0.15 dex)
oxygen abundance:	investigators adopt [O/Fe] = 0.0 or 0.5–0.7 (this is hard to measure for halo dwarfs)
globular cluster distances:	use RR Lyrae or horiz. branch stars: there is controversy on whether $M_V(RR)$ depends on [Fe/H]
helium diffusion:	neglect leads to age overestimate by 20-30% (not included in most calculations)
lithium diffusion:	a smaller effect
reddening:	but there are several globular clusters with E(B−V)<0.1
rotation:	does not affect ages

There is some evidence for an age spread of 3–4 Gyr, reflecting the time to form the halo (but in contradiction to the Eggen et al. 10^8 year time-scale). VandenBergh (1988) argues that all globular clusters have the same age, ~14 Gyr. For M92, one of the metal-poorest (and oldest ?) globular clusters, Demarque (1991) adopts the following age:

standard models	17 Gyr
include [O/Fe] = 0.4	16 Gyr
include He diffusion	14 Gyr.

Globular cluster ages have an uncertainty of at least 10% due to uncertainty in distances alone (Hipparcos should improve the RR Lyrae calibration).

Profitt and Michaud (1991) argue that gravitational settling of He and Li decreases ages by 20% (slightly more than allowed for by Demarque). Canuto and Mazzitelli (1991) propose a new convective theory which results in younger ages. Wood (1992) suggests that these effects reduce globular cluster ages to 10-13 Gyr.

7.3. White dwarf luminosity function and age of the Galaxy

The white dwarf luminosity function shows a turndown at the faint end due to the finite age of the Galactic disc. To determine this age we need a luminosity-age relation for white dwarfs. Winget et al. (1987) derived an age for the disc of 10.3±2.2 Gyr. Wood (1992) has explored a wide range of models and concludes that the disc age lies in the range 6–13.5 Gyr, with 40% of this uncertainty being observational. His best estimate for the age of the Galactic disc is 7.5–11 Gyr. He suggests that star formation began in the halo and bulge 10–13 Gyr ago and began at our Galactocentric radius 7.5–11 Gyr ago, the difference between these supporting pressure-supported collapse models of the formation of the Galaxy.

In summary, I adopt as the age of the Galaxy (and of the universe)

$$t_o = 13 \pm 1 \, \text{Gyr}.$$

	QDOT	2 Jy	1.2 Jy	$<\beta>$
dipole	0.94±0.2 R90,L95		$0.55^{+0.20}_{-0.12}$ S92	0.75
vel. field vs. dens. field z-space anisotr.	0.86±0.14(K91) 0.83±0.10(T94) 0.54±0.3 (C95)	1.28±0.34(D93) 0.6(R94) 0.69±0.27(H93) 0.84±0.45(FG93)	0.6 (ND 94) 0.55±0.13 (W95) 0.45±0.22(F94b) 0.52±0.3 (C95)	0.80 0.60
power spectrum spherical harmonics			1.0±0.2 (PD94) 0.94±0.17(F94c) 1.1±0.3 (HT94)	1.0 1.0
β	0.80	0.85	0.69	0.8±0.15

TABLE 7. IRAS estimates of $\beta = \Omega_o^{0.6}/b$ (Rowan-Robinson 1996)

8. The value of Ω_o

Large-scale flow studies based on the IRAS survey now consistently give $\Omega_o \sim 0.8\pm0.1$, by a variety of independent methods (Rowan-Robinson 1993). This lends support to an $\Omega_o = 1$ universe in which IRAS galaxies trace the matter with a rather low degree of bias (b = 1.2±0.1). Low Ω_o universes, for example, the vacuum-energy dominated model proposed by Efstathiou et al. (1990), would appear to be inconsistent with these data unless an unphysical level of anti-bias (b < 0.5) is invoked.

Quite apart from the inherent limitations of working at optical wavelengths due to interstellar extinction, there are no deep all-sky optical galaxy surveys of sufficient quality to provide a credible value of Ω_o for the dipole method of section 2.1. The low values of Ω_o found in most optical studies to date may indicate a higher level of bias towards regions of high total matter-density for elliptical galaxies. However from the IRAS analyses it is quite hard to see how inclusion of elliptical galaxies alone can reduce Ω_o to the low values found in optical studies.

An obvious implication of the high values of Ω_o from dynamical studies is that 95% of the universe is non-baryonic.

9. Implications of observed values of H_o, t_o, Ω_o

My best estimates of cosmological parameters in sections 6-8 above were:
- $H_o = 65\pm10$ km s^{-1} Mpc^{-1}, corresponding to a Hubble time $\tau_o = 15.1$ Gyr
- $t_o = 13\pm1$ Gyr
- $\Omega_o \simeq 1$.

There is therefore a problem for an Einstein-de Sitter model ($\Lambda = 0$, $\Omega_o = 1$) for which

$$H_o = 50 \quad \text{implies} \quad t_o = 13.0\,\text{Gyr}$$
$$65 \quad\quad\quad\quad\quad 10.0\,\text{Gyr}$$
$$80 \quad\quad\quad\quad\quad 8.1\,\text{Gyr}$$

unless H_o is < 55 km s^{-1} Mpc^{-1} (or the age estimates are seriously out).

A model with $k = 0$, $\Omega_o = \Omega_b = 0.05$, $\lambda = 0.95$ gives $t_o = 14.6$ Gyr if $H_o = 100$, so this model is credible only if H_o is high (≥ 80). For $k = 0$, $\Omega_o = 0.3$, $\lambda = 0.7$, then t_o

$= 14.4$ Gyr if $H_o = 65$, so this model is consistent with all observations except the high value of Ω_o from IRAS galaxy redshift surveys.

Two other possible scenarios are:

(1) $\Lambda = 0$, $\Omega_o = \Omega_b = 0.05$, $H_o = 65$, $t_o = 14.3$ Gyr.

In this case, the age of the universe at z = 5 is 12 Gyr (corresponding to bulge and elliptical formation?) and at z = 2 is 9.5 Gyr (corresponding to spiral disc formation?). However this scenario is inconsistent with Ω_o from IRAS galaxy redshift surveys, is inconsistent with standard inflation, and contains no non-baryonic dark matter so would have difficulty accounting for the origin of structure.

(2) $\Lambda = 0$, $\Omega_o = 1$, $\Omega_b = 0.05$, $H_o = 60$, $t_o = 11$ Gyr.

In this case, the age of the universe at z = 5 is 10.2 Gyr (bulge and elliptical formation?) and at z = 2 is 8.9 Gyr (spiral disc formation?). This scenario would require the ages of globular clusters to have been overestimated.

REFERENCES

Arp, H.L. 1961. *ApJ* **133**, 874

Birkinshaw, M., Hughes, J.P., & Arnaud, K.A. 1991. *ApJ* **379**, 466

Birkinshaw M., et al. 1994. *ApJ* **420**, 33

Branch, D. 1979. *MNRAS* **186**, 609

Branch, D. 1987. *ApJ* **320**, L23

Branch, D. 1992. *ApJ* **392**, 35

Branch, D., & Miller, D.L. 1993. *ApJ* **405**, L5

Canuto, V.M., & Mazzitelli, I. 1991. *ApJ* **370**, 295

Cowan, J.J., Thielemann, F.-K., & Truran, J.W. 1987. *ApJ* **323**, 543

Della Valle M., & Livio M. 1995. *ApJ* **452**, 704

Demarque, P., Deliyannis, C.P., & Sarajedini, A. 1991, in Observational Tests ofInflation, eds T.Shanks et al. (Kluwer), p.111

Dressler, A. 1987. *ApJ* **371**, 1

Dressler, A. 1988, 3rd ESO-CERN Symposium

Efstathiou, G., Sutherland, W.J., & Maddox, S.J. 1990. *Nature* **348**, 705

Falco, E.E., Gorenstein, M.V., & Shapiro, I.I 1985. *ApJ* **289**, L1

Falco, E.E., Gorenstein, M.V., & Shapiro, I.I 1991. *ApJ* **372**, 364

Ford, H.C., Harms, R.J., Ciardullo, R., & Bartok F. 1981. *ApJ* **245**, L55

Fowler, W.A. 1987. *QJRAS* **28**, 87

Freedman, W.L., Wilson, CD., & Madore, B.F. 1991. *ApJ* **372**, 455

Freedman, W.L., et al. 1992. *ApJ* **396**, 80

Freedman, W.L., et al. 1994. *ApJ* **427**, 628

Freedman, W.L., et al. 1994. *Nature* **371**, 757

Freedman, W.L., & Madore,B.F. 1993, in New Persp. on Stellar Pulsation & Pulsating Stars

Greenfield, P.E., Roberts, D.H., & Burke, B.F. 1985. *ApJ* **293**, 370

Hamuy M., Phillips M.M., Maza J., Suntzeff N.B., Schommer R.A., Ariles R. 1995. *AJ* **109**, 1

Hanson, R.B. 1980, in Star Clusters, ed. J.E.Hessler, IAU Symp. 85,71

Jacoby, G.H., et al. 1992, *PASP* **104**, 599

Kennicutt R.G.Jr, Freedman W.L., & Mould J.R. 1995. *AJ* **110**, 1476

Lee, M.G. 1993. *ApJ* **408**, 409

Lee, M.G., Freedman, W.L., & Madore B.F. 1993. *ApJ* **417**, 553

Lehar, J., Hewitt, J.N., Roberts, D.H., & Burke, B.F. 1992. *ApJ* **384**, 453
McHardy, I., et al. 1990. *MNRAS* **242**, 215
Meyer, B.S., & Schramm, D.N. 1986. *ApJ* **311**, 406
Mould, J., et al. 1995. *ApJ* **449**, 413
Panagia, N., Gilmozzi, R., Machetto, F., Adorf, H.-M., & Kirschner, R.P. 1991. *ApJ* **380**, L23
Pierce, M.J. 1989. *ApJ* **344**, L57
Pierce, M.J., McClure, R.D., & Racine, R. 1992. *ApJ* **393**, 523
Pritchett, C.J., & van den Bergh, S. 1985. *ApJ* **288**, L41
Pritchett, C.J., & van den Bergh, S. 1987. *ApJ* **318**, 507
Proffit, C., & Michaud, G. 1991. *ApJ* **371**, 584
Refsdal, S. 1966. *MNRAS* **132**, 101
Riess, A.G., Press W.H., & Kirschner R.P. 1995. *ApJ* **438**, L17 & 445, L91
Rowan-Robinson, M. 1985, The Cosmological Distance Ladder (W.H.Freeman)
Rowan-Robinson, M. 1988, Advances in Space Science, **48**, 1
Rowan-Robinson, M. 1991, in Observational Tests of Inflation, eds T.Shanks et al. (Kluwer) p.161
Rowan-Robinson, M. 1993. *Proc.Nat.Acad.Sci.* **90**, 4822
Rozanski, R., & Rowan-Robinson, M. 1993. *MNRAS* **271**, 530
Sandage,A.R. 1958. *ApJ* **127**, 513
Saha, A., et al. 1994. *ApJ* **425**, 14
Saha, A., et al. 1995. *ApJ* **438**, 8
Sandage, A., & Tammann, G.A. 1990. *ApJ* **365**, 1
Sandage, A., Saha, A., Tammann, G.A., Panagia, N., & Macchetto, D. 1992. *ApJ* **401**, L7
Sandage, A. 1993. *ApJ* **402**, 3
Sandage, A., et al. 1994. *ApJ* **423**, L13
Schmidt, B.P., Kirschner, R.P., & Eastman, R.G. 1992. *ApJ* **395**, 366
Schmidt, B.P., et al. 1994. *ApJ* **432**, 42
Shara, M.M. 1981. *ApJ* **243**, 268 & 926
Shigeyama, T., Nomoto, K., Yamaoka, H., & Thielemann, F.-K. 1992. *ApJ* **386**, L13
Symbalisty, E.M.D., & Schramm, D.N. 1981. *Rep.Progr.Phys.* **44**, 293
Tammann, G.A. 1991, in Observational Tests of Inflation, eds T.Shanks et al. (Kluwer) p.179
Tammann, G.A., & Sandage, A. 1995. *ApJ* **425**, 16
Tanvir, N., et al. 1995. *Nature* **377**, 27
Thielemann, F.-K., et al. 1983. *A&A* **123**, 162
Tonry, J.L. 1991. *ApJ* **373**, 21
Tully, R.B. 1989, 5th IAP Meeting Astrophysical Ages and Dating Methods
van den Bergh, S. 1989. *A&ARev* **1**,111
van den Bergh, S. 1993. *PASP* (in press)
VandenBergh, D.A., & Bridges, J.J. 1984. *ApJ* **278**, 679
VandenBergh, D.A. 1988, in Globular Cluster Systems in Galaxies, ed. J.E.Grindlay & A.G.D.Philip (Kluwer) p.107
Walker, T.P., et al. 1991. *ApJ* **376**, 51
Winget, D.E., et al. 1987. *ApJ* **315**, L77
Wood, M.A. 1992. *ApJ* **386**, 539
Young, P., et al. 1981. *ApJ* **244**, 736

Radio objects at redshifts 300>z>10.

By V. K. DUBROVICH

Special Astrophysical Observatory, Russia.

The spectrum and intensity of radiation from primordial objects is considered. It is shown that only some of the simplest molecular lines could be observed. Such molecules as LiH, HD^+, H_2D^+, HeH^+ should be taken into account. Peculiar motion of objects relative to the cosmic background radiation produced values for these lines' intensity at the level of about 1 mK for molecular abundances of 10^{-12} relative to neutral hydrogen. Investigation of these objects will give information about primordial small-scale fluctuations of matter density at redshifts $z \geq 100$. The reionization stage (at $z \leq 30$) of the Universe could be investigated by observations of the HeH^+ lines in the mm-wavelength band. The crucial test for nonstandard nucleosynthesis could be done by searching for H_2O lines at high redshifts. This gives unique information about models of phase transitions in the early Universe which have been predicted by modern theory.

The theory of evolution of the Universe is now well enough developed to consider in detail the whole picture of the distribution of matter at high redshifts. The effect of the formation of Spectral Spatial Fluctuations (SSF) of the cosmic background radiation (CBR), suggested by Dubrovich (1994), can be used for observational verification of the theoretical prediction. In this paper the manifestations of small-scale fluctuations of density at high redshifts, that could be observed, will be considered. The main interest is in small objects with average mass of $M \leq 10^{10} M_\odot$, due to their role in formation of the modern matter distribution and to the nature of their evolution. In standard cosmology the primary fluctuations of matter density can be divided into so-called adiabatic and entropic fluctuations. In adiabatic fluctuations, both baryon and photon components are perturbed. During the recombination period of primary hydrogen at $z=1400$ their evolution is strongly dependent on the scale. The entropic fluctuations, by contrast, correspond to the perturbation of the baryon component only. Their evolution is almost unrelated to the moment of recombination and they are only apparent on small scales. The mass spectrum of those objects is very interesting because it persists from the period of the early Universe and includes information about fundamental parameters of overdense matter.

After recombination the Jeans mass becomes about $10^5 M_\odot$ and this leads to the possibility that the collapse of proto-objects with small masses can begin at high redshifts. The $\Delta \rho$ spectrum of one of the types of primordial fluctuations at small scales is as follows:

$$\Delta\rho/\rho = (M/M_0)^{(-1/2+n/6)}/(1+z) \qquad (0.1)$$

where n is a model parameter. Gott and Rees (1975) considered the conditions which make $n=1$. Puy and Signore (1995) used this model for detailed calculations of the objects' parameters. Angular sizes of these objects are small, but we can see a cluster of those objects – a protocluster of galaxies – whose quantity is determined by the spectrum of a large scale inhomogeneity. In this case quite a large density of "point" objects is possible at scales of $0.5'-2'$. If the proportion of matter which is contained in such clusters is about 10^{-3}, the angular interval between them will be about $10'-20'$. Some parameters of these objects at the moment when their expansion stopped – z_0 – are given in Table 1.

Let us consider the spectrum of radiation of a single object as a function of z. As was shown by Dubrovich (1994,1996) the main mechanisms of forming this radiation

M	z_0	R_0	n_0	T_0	θ_0
10^6	564	1.00	281	4496	0.3
10^7	261	4.66	28.1	1536	0.67
10^8	121	21.65	2.81	447	1.44
10^9	56	100.5	0.28	119	3.1
10^{10}	26	466.5	0.03	25	6.7

TABLE 1. Parameters of protoclusters at the moment when their expansion stopped. Here M is the mass of object in $M_\odot$, $R_\odot$ is the size in 10^{20} cm, n_o is the concentration of hydrogen in cm^{-3}, T_o is the temperature of matter in K, and θ_o is the angular size in arcseconds.

are based on two different interactions of primordial molecules with the CBR at narrow resonant lines. The efficiency is determined either by the degree of non-equilibrium of CBR or by the value of the peculiar velocity of the object and, of course, by the quantity of molecules. The continuum will be practically absent, because the density is very small and the density of electrons will be 3–4 orders of magnitude smaller. For the same reason, the intensity of lines formed by collisional excitation will be small too. The phenomena which are caused by the presence of the peculiar velocity of proto-objects are pure scattering and fluorescence of CBR. The latter is, for most molecules, the main effect. The final spectrum depends on the size of the object, because for different objects there are different moments at which the collapse begins and further evolution leads to different concentrations of molecules. However this is a second-order effect and we will not consider it at present. The basic factors which determine the intensity of lines are:

a) correlation between the potential of dissociation of the molecule and the temperature of the CBR at the relevant z,

b) increase of the relative density at first (at the moment of slow-down in the extension of the proto-object) and the absolute density (at the moment of collapse) of the substance,

c) position of the frequency of the vibrational transition relative to the maximum of the spectrum of the CBR.

We will consider LiH, H_2D^+, HD^+, HeH^+ as the basic molecules at high z. Unfortunately, their abundance can be predicted only inaccurately because of some uncertainties in the process of radiative association. Besides, the ionization degree of the hydrogen at that moment is very important and it can be determined by the presence of reionization radiation from non-stable particles. So we consider only a lower limit for the molecules' abundance, which can be constrained by observations. Some individual spectra of a single object at different z are shown in Fig.2. The dependence of amplitudes of some lines on z is shown on Fig 1. All quantities correspond to a peculiar velocity of about 30 km s^{-1} and molecular abundance of about 10^{-10} (Dubrovich, 1996).

We comment briefly on line widths. In case of free Hubble expansion of the matter there is a correlation between size of the object and line width. However, taking into account selfgravitation, this dependence must be corrected. Thus, by the end of the expansion the width of the line does not depend on the size of the object and is determined only by the thermal velocity of the molecules. In this case the line becomes very narrow – 10^{-5} – and the optical depth increases.

Let us consider molecules based on heavier atoms – C, N, O. There are many theo-

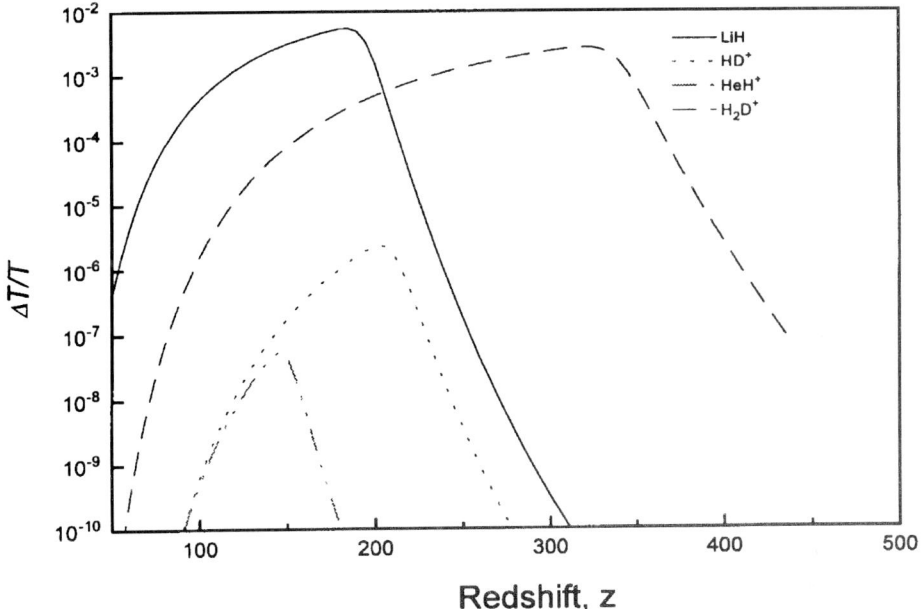

FIGURE 1. Evolution of $\Delta T/T$ with z for different molecules at high redshift

retical investigations of nonstandard primordial nucleosynthesis which give quite a large abundance of these elements (see Rauscher et al., 1994). The physical basis of these declines could be the presence of a microinhomogeneous medium at very high redshifts. The most probable molecule to find there is H_2O. It has energy of dissociation H-OH of 5.12 eV, which is 8000K more than the binding energy of H-H. When the temperature of the matter is $T_m \leq 500K$ and the concentration of atomic hydrogen is about $1\,cm^{-3}$, we can expect that almost all primordial oxygen in the proto-object will be bound in H_2O due to the very fast reactions: $H_2+O=OH+H$ and $H_2+OH=H_2O+H$. In the first reaction, the energies on the right and on the left are almost the same, but in the second one there is a difference of energies which was mentioned previously. Other elements, such as those formed between C and N with H, have binding energy smaller than that in H_2 and so their abundance will be small. This result holds for the objects at $z \geq 100$. For later stages of evolution of the Universe ($z \leq 30$), the physical parameters of the matter may differ considerably from quasi-equilibrium, perhaps because of the formation of primordial stars. Here enrichment of the surrounding matter by heavy elements from supernovae may be weak because of the relatively low velocity of the material expelled. Much more important may be the reionization of the primordial H. In this case the concentration of such molecules as H_2^+, HeH^+ increases and the temperature of the matter (not the radiation!) may become $\geq 1000K$. The HeH^+ molecule has a large dipole moment and the first rotational transitions will give SSF at the wavelength $\lambda = 149.13\mu(1+z)$. These SSF will show large-scale fluctuations during the nonlinear stage of their evolution.

The author is grateful to Yu.N. Parijskij and R. Braun for stimulating interest and support of this work. This work was financially supported in part by the Russian fund of fundamental research and by SEC "KOSMION" (Russia).

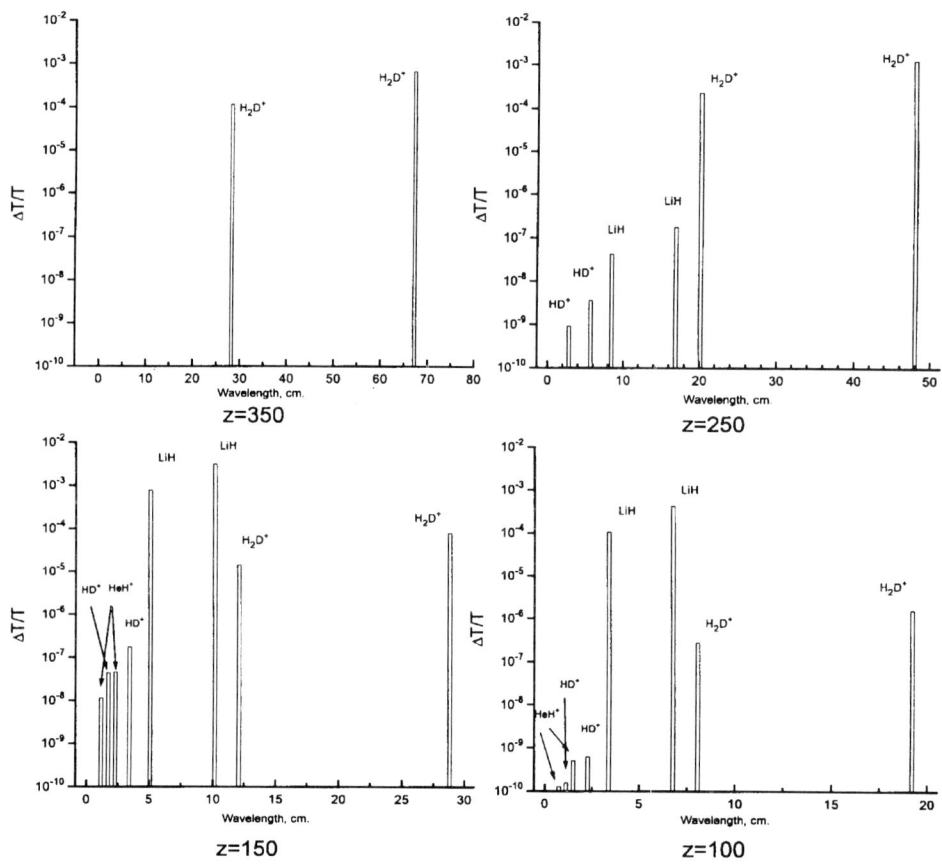

FIGURE 2. $\Delta T/T$ spectra at different redshifts

REFERENCES

DUBROVICH V.K. 1994. A&A Transactions **5**, 57
DUBROVICH V.K. 1996. *A&A* in press.
GOTT III R.J., REES M.J. 1975. *A&A* **45**, 265
PUY D., SIGNORE M. 1996. *A&A***305**, 371
RAUSCHESTER, T., APPLEGATE, J.H., COWAN, J.J., & THIELEMANN F.-K.
WIESCHER, H. 1994. *ApJ* **429**, 499

New-generation telescopes

The Millimetre Array projects

By ROY S. BOOTH[1]

[1]Onsala Space Observatory, Onsala, Sweden

Among the highest-priority projects in radio astronomy today are the large millimetre arrays. Three projects for the 21st century are currently under discussion by groups in Europe, Japan and the USA. These projects are presented together with some of the fascinating research programmes for which they will be used. The scientific goals are to study all types of millimetre radiation – molecular line emission and absorption from cool gas, continuum emission from dust, line and continuum emission from ionized gas and synchrotron emission from relativistic particles. Scientific research ranges from studies of stellar and planetary formation to observations of dust and molecules in galaxies in the early Universe.

1. Introduction

At this meeting on high-sensitivity radio astronomy it is important to remind ourselves that advances in radio astronomy come not only from increased sensitivity but also from improved resolution (spatial, frequency and temporal) and through extending our spectral range. The extension of radio astronomy into the millimetre wavelength band has proven to be very rewarding scientifically and a whole new branch of astronomy, cosmochemistry, has opened up through the observations of spectral lines from interstellar molecules. However, equally important are continuum observations of cool dust and the cosmologically-significant observations, both line and continuum, of distant galaxies and of the microwave background radiation, which peaks in the millimetre band. Thanks to a new generation of precision radio telescopes and interferometers which have become operational during the past twenty years, millimetre-wave astronomy is now a mature branch of astronomy. One might liken this stage of development in millimetre astronomy to that of decimetre/centimetre radio astronomy before the advent of the 100-m class of antennae, or the VLA. It is poised to yield more and more exciting results but now requires a significant leap in sensitivity for their final realisation. The best way to realise an increase in sensitivity in millimetre astronomy today is through the construction of large arrays of telescopes. Receiver performance is very close to the theoretical quantum limit and large, efficient, single-aperture telescopes, bigger than the IRAM 30 m-diameter telescope on Pico Veleta, or the Nobeyama 45-m antenna will be difficult to build for the highest frequencies. In this report, I will discuss three proposed millimetre arrays, - the Millimetre Array (MMA), an array of 40×8-m antennae, being studied by the US National Radio Astronomy Observatory, NRAO, the Japanese Large Millimetre-Submillimetre Array (LMSA) which will consist of 50×10-m telescopes, and the Large Southern Array (LSA), a European project being studied by a team from the consortium of European Observatories: the Institute de Radio Astronomie Millimetrique (IRAM), Onsala Space Observatory, the Netherlands Foundation for Research in Astronomy (NFRA) and the European Southern Observatory (ESO). The LSA will comprise 50×16-m or 100×11-m telescopes and will have a collecting area 5 times as great as the MMA. A fourth, pioneering array for sub-millimetre wavelengths is under construction on Mauna Kea, in Hawaii. This project, initiated by the Smithsonian Astrophysical Observatory, is now paving the way for future developments at frequencies up to 890 GHz, and testing technology and atmospheric transmission to their ultimate limits. The proposed array studies will all hope to benefit from the experience of the Smithsonian group, and the current specifications of the LMSA and even the MMA include 890-GHz capabilities.

2. Millimetre wave astronomy

All three large array projects are driven scientifically by the richness and astronomical potential of the millimetre band. (see *e.g.* the MMA proposal; NRAO, 1990) and that for the LSA (LSA 1995). It contains the peak of the microwave-background emission, the spectral tails of the free-free and synchrotron radiation from Galactic and extragalactic radio sources and especially the thermal emission from dust. In addition, the band is populated by rotational transitions of interstellar molecules, and their association with dense dust clouds make it of fundamental importance for studies of the physics and chemistry of regions of star formation in the Milky Way and other nearby galaxies, and of mass loss from evolved stars as a function of their position on the asymptotic giant branch of the H-R diagram. While this has been the essence of millimetre astronomy, recent work has shown the importance of the genre for studies of central tori in Seyfert galaxies and the detection of CO and dust at $z = 2.5$ demonstrates the significance of millimetre wave observations in studies of the early Universe.

2.1. Stellar evolution

Because millimetre waves can penetrate the dusty envelopes associated with stellar birth and death, observations at these wavelengths provide astronomers with their most intimate view of these important evolutionary stages of stellar evolution. The significance of millimetre interferometry in the study of stellar evolution is shown by the remarkable detection of Keplerian rotation in the disc surrounding the binary star GG Tau (Fig. 1). These ^{13}CO-line IRAM interferometer images of a probable protoplanetary disc are the work of Dutrey *et al.* (1994). Future high-resolution, high-sensitivity observations will not only enhance this work but enable astronomers to extend such observations to many more systems. For example, an interferometer with a resolution of 0.05 arc seconds at 230 GHz would resolve a 5-AU disc at a distance of 100 pc. Studies of the chemical properties of circumstellar material surrounding evolved stars are highlighted by another IRAM interferometer study of the carbon star IRC 10216 (Lucas 1994).

More than 50 different molecules have been detected in the dusty envelope of this star and Fig. 2 shows the distribution of several species with different chemical characteristics. For example, sodium chloride, a highly refractory molecule, seems to be produced exclusively in the stellar atmosphere while CCH, a product of photo/ion molecule chemistry, is seen only in the outer envelope.

2.2. Extragalactic molecules

CO and other molecules are detected in galaxies out to rather large distances. Interferometers have been used to measure the association of molecular gas with spiral structure and this is well illustrated in the observations of M51 by *e.g.* Rand & Kulkani 1990. Density, excitation and ionisation vary across galactic discs and affect, and are affected by, the onset of star formation, through shocks, tidal effects and cloud-cloud collisions. Millimetre lines can trace these parameters in galaxies; sufficient resolution and a combination of *e.g.* CO, CS, HCN and HCO^+ will trace lower density gas (CO), high density gas (CS & HCN) and ionisation (HCO^+). Of striking interest are the recent observations of central molecular tori in Centaurus A by Rydbeck using SEST at 230 GHz (see LSA 1995), and in NGC 1808 by Tacconi et al 1994, using the IRAM interferometer. The Cen A molecular ring has a diameter of 200 pc and lies orthogonal to the continuum jet, while

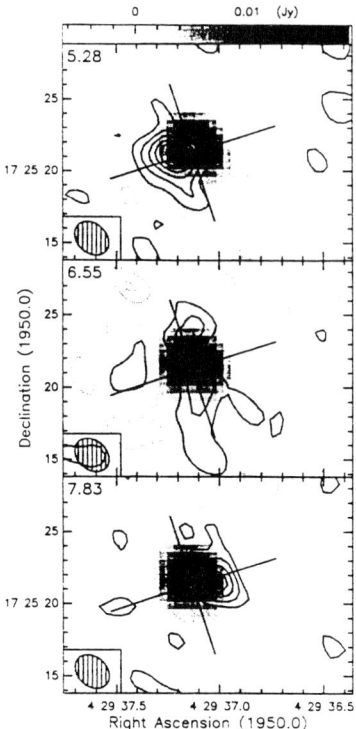

FIGURE 1. IRAM array data on the disk around the young double star GG Tau. The ^{13}CO(1-0) line (contours) is superimposed on the 3mm dust emission (grey scale), at three velocities, showing the existence of a rotating ring (Dutrey et al. 1994).

the central torus or disc seen in NGC 1808 has a diameter of 400 pc. In the latter case the authors suggest that it is entirely plausible that the thick, turbulent molecular cloud layer is responsible for obscuring the Seyfert 1 nucleus in the visible, UV and even in X-rays! The projected millimetre arrays, especially the LSA, are essential tools to study the central molecular gas and to resolve it into individual clumps, which are presumably being eroded as they fall into the central black hole. The recent VLBI observations of the water masers in NGC 4258 by Myoshi et al. 1995 show that dense molecular gas still exists within 0.1 pc of a black hole!

2.3. Molecules at high red shift

A dramatic paradigm shift occurred in millimetre astronomy with the detection of CO emission and then continuum dust emission from the IRAS galaxy F1024+4724 at $z = 2.3$. Dust emission, CO, and CI lines have also been detected in the gravitationally lensed 'cloverleaf' quasar H1413+117 at $z = 2.5$ (Fig. 3), while millimetre-wavelength molecular absorption lines (at $z = 0.25$, 0.69 and 0.8) have been seen in the spectra of several BL Lac objects which are also gravitationally lensed (see e.g. Wiklind & Combes 1996). In these cases the molecular gas is in the lensing galaxy. It is now clear that millimetre astronomy has an important role to play in studies of the early Universe. In fact, at a recent meeting to discuss the science to be done with the LSA, the detection of millimetre continuum emission from dust was hailed as one of the best methods for finding and studying the evolution of primeval galaxies at redshifts up to 10 or more. The large amounts of interstellar dust, assumed to arise in the early bursts of star formation, will

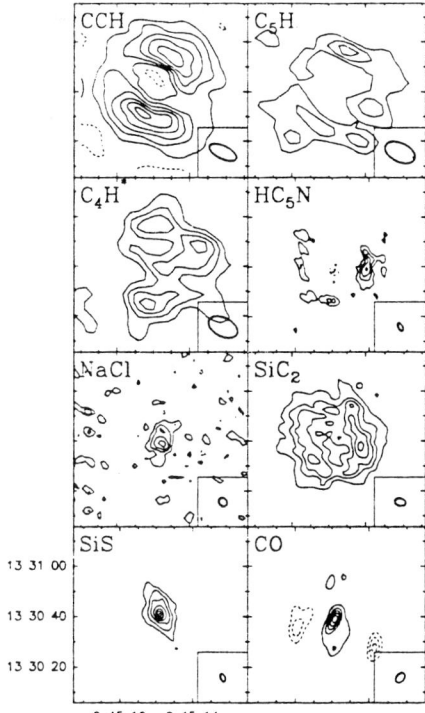

FIGURE 2. Maps of the circumstellar envelope of IRC+10216. Each box presents the average of the emission in a 5 km/s velocity interval centered on the star velocity (corresponding to a cut through the envelope by a plane perpendicular to the line of sight). The beam size is given in the lower right corner of each map. The CO observations (snapshots) miss most of the flux for this very extended envelope, but reveal the slightly elongated shape of the inner dense regions.

be detectable at these high redshifts since the far infrared spectral peak is red-shifted into the millimetre band, compensating for the distance effect. Returning to high-z line studies, both emission and absorption lines will be more easily detected with greater sensitivity. If, as seems to be the case, the emission lines are only detectable today because of amplification ($\times 10$) or focusing by lensing, the exceptional collecting area of the LSA is imperative for molecular-line studies of the parent population of high-z galaxies. Through high-resolution, high-sensitivity observations of the absorption lines, we have the prospect of studying the molecular structure and chemistry of galaxies at epochs corresponding to redshifts of 1 or 2. It is interesting to note, in the context of the talk by Braun, that it is much easier to detect gas in the early Universe at millimetre than at cm or decimetre wavelengths. Not only can one detect thermal continuum emission from dust in the millimetre band but since the emitted power in lines that have about the same brightness temperature and line width varies as $\nu^2 T_b \Delta\nu$, and hence as ν^3, a CO (3-2) line redshifted to 100 GHz emits about 30 million times more power than the corresponding H I line shifted to 400 MHz!

3. The millimetre arrays

The parameters of the three proposed arrays are described in Table 1. In Table 2, the equivalent details of the currently available arrays are shown for comparison.

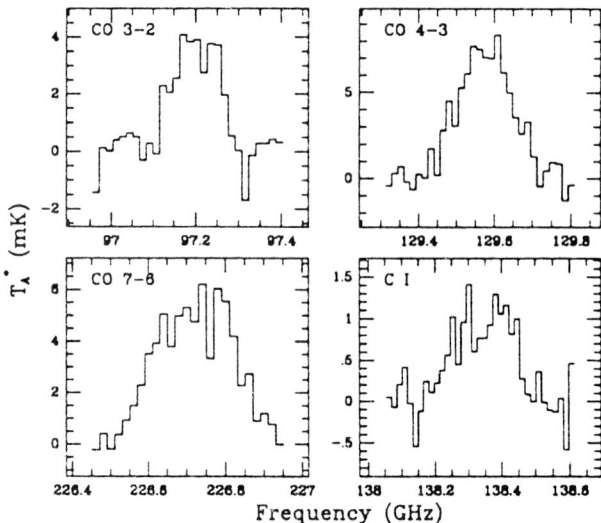

FIGURE 3. Various lines of the CO molecule and [CI] detected towards the Cloverleaf quasar at a redshift $z = 2.546$ at the IRAM 30 m telescope (from Barvainis et al. 1994).

3.1. The US Millimetre Array

The MMA, a project of the US National Radio Astronomy Observatory, NRAO, was first discussed in the mid 1980s. At that time it was proposed to build two separate instruments, a compact array of about 21 small (3–5 m) dishes, possibly mounted on a single structure of about 25 m in diameter, to provide a large field of view for mapping extended low brightness objects, and another array, similar to the VLA, consisting of about 15 to 27 larger telescopes, to provide high resolution and sensitivity. At that time it was proposed to site the array(s) close to the VLA in New Mexico. The present concept of the MMA (40×8 m dishes)was proposed to the National Science Foundation in 1990 (see also Brown 1994) and is the most developed of the three array projects under discussion. The dish size is chosen to give a relatively large field of view, bearing in mind the extended nature of most Galactic molecular clouds, and with high surface accuracy in mind. Baselines up to 3 km are proposed giving the array a resolution of about 0.2 arcsec at a wavelength of 3 mm. The only change since that time has been the decision, in response to community wishes, that the minimum operating wavelength should be in the atmospheric window centred at 0.35 mm (860 GHz). However, since the goal for the precision of the individual telescopes is ambitiously described as 25 microns rms, (equivalent to $\lambda/14$ at 860 GHz) this has, in principle, always been a possibility. Sites currently under discussion are Hawaii (4000 m) and Cerro Chajnantor (5000 m) in the Chilean Andes, with a preference for the latter (see later).

3.2. The Large Millimetre/Submillimetre Array

The LMSA was first proposed at a meeting on millimetre interferometry in Hakone, Japan in October 1992 (Ishiguro et al. 1994). The array will have 50×10-m antennae and, like the MMA, will operate down to 0.35 mm. Maximum baselines of 2 km are proposed giving a resolution of 0.3 arcsec at 3-mm wavelength. Again, sites in Hawaii and Northern Chile are under test.

	EUROPE	**USA**	**JAPAN**
Name	LSA	MMA	LMSA
Antennas	50×16 m	40×8 m	50×10 m
Total area (m^2)	10054	2010	3930
Altitude (m)	> 3000	> 5000	> 4000
Min. λ (mm)	0.8	0.35	0.35
Site testing in N. Chile	Pampa del Chino	Cerro Chajnantor	Rio Frio

TABLE 1. The planned arrays

	EUROPE	**USA**	**USA**	**JAPAN**
Institute	IRAM	BIMA	OVRO	Nobeyama
Antennas	5(6)×15 m	9×6 m	6×10.4 m	6×10 m
Total area (m^2)	884 (1060)	255	510	471
Baselines	10 (15)	36	15	15
Min λ (mm)	1.3 (0.8)	1.3	1.3	1.3
Altitude (m)	2552	1043	1216	1350
Location	Plateau de Bure	Berkeley	Owens Valley	Nobeyama

TABLE 2. Existing Millimetre Arrays

3.3. The Large Southern Array

The LSA is a European project to build a large millimetre array in the southern hemisphere and it has been under discussion since 1992 (Booth 1994; Downes 1994), although serious studies commenced only in 1995 with the formation of the study consortium. Unlike the other two arrays, the LSA has been conceived with high-sensitivity observations in mind – especially observations of molecules and dust in high-z galaxies for studies of the early Universe. Hence, it will have a collecting area of 10,000 square metres and will operate in the millimetre band (λ_{min}=0.8 mm), making use of the fact that the rotational spectrum of CO has lines at integral multiples of 115 GHz and so for all reasonable values of z, there will be a line accessible to one of the receivers in the standard millimetre bands (80–115 GHz, 130–170 GHz, 210–260 GHz & 300–360 GHz). In addition, it will have a resolution of 0.1 arcsec or better at 3 mm in order to resolve the distant galaxies, so baselines of 5–10 km are required. These parameters are also essential for the other major goals: studies of protostellar discs and stellar envelopes around evolved stars. In fact the sensitivity of the LSA will facilitate studies of stars of many types, over practically the whole range of the HR diagram. The preference is for 16-m diameter telescopes and a development of the IRAM design antenna, despite the comparatively small field of view, also because the dominant scientific drivers are small-diameter sources.

3.4. Site considerations

The next generation of millimetre arrays requires high, dry sites. Probably the best northern-hemisphere site is Mauna Kea where the JCMT and the CSA millimetre telescopes are located and where the Smithsonian submillimetre array is being built. At altitudes of 4000 m, these telescopes enjoy relatively long periods with low atmospheric opacities at millimetre wavelengths. However, even at the driest sites, the atmospheric transmission decreases dramatically with increasing zenith angle. As an example, a site with only 1 mm of precipitable water has zenith transmissions at 345, 460, 690 and 860 GHz of 0.8, 0.5, 0.25 and 0.25, respectively. It has been known for some time that some of the world's best millimetre sites lie in the Atacama desert on the coastal side of the Andes in Northern Chile where there are many large, flat tracts of land at altitudes even as high as 5000 m. Extensive water vapour measurements have been made on Cerro Paranal (altitude 2650 m), the site for the ESO VLT. These data show the precipitable water vapour content is below 1.5 mm for at least 30% of the April-November period, and below 3 mm for 60% of this period (Sarazin 1990; Martin 1990). Paranal is only 15 km from the ocean; the higher inland sites have a lower water vapour content. Serious site testing began about one year ago on two potential inland sites: at Cerro Chajnantor at an altitude of 5000 m and at Rio Frio at an altitude of 4100 m by teams form NRAO and Nobeyama, respectively. The NRAO data is available on their WWW site, presented as opacity at 225 GHz (1 mm precipitable water along the path to the zenith gives an opacity at 225 GHz of 0.057), and is very encouraging and considerably better than that for Hawaii over the same period, showing opacities of less than 0.055, or less than 1 mm precipitable water for more than 50% of the time between April and October in 1995 and less than 1.5 mm precipitable water (opacity less than 0.085) for nearly 50% of the time, November 1995 to March 1996. Although these data are considerably better than typical data for Mauna Kea, we should bear in mind that they cover only one year. In an attempt to measure the potential of the sites for interferometry, the groups have measured the relative phase of an 11.2-GHz incoming signal from a geostationary satellite, over a baseline of 300 m (see e.g. Ishiguro et al., 1990). Since the atmosphere is non-dispersive, this phase may be scaled to millimetre wavelengths. On Cerro Chajnantor, typical values for June, 1995 show an equivalent phase difference less than 26 degrees at a frequency of 100 GHz, for 75% of the time. This corresponds to 61 degrees at 230 GHz. Although phase-correction techniques will be necessary for longer-baseline work, these data are very encouraging. The data for Cerro Chajnantor suggest that it is an excellent site for a millimetre array. However, at an altitude of 5000 m, it is close to the limiting height at which people can function efficiently. Therefore, it will be very interesting to compare the site data from Rio Frio at 4100 m and the data which will be measured at the possible LSA site of Pampa del Chino at an altitude of 3500 m.

4. Summary

In this talk I have outlined the millimetre-array proposals, their scientific potential and discussed recent site data from the proposed site for NRAO's MMA. It is difficult to compare the projects fairly, since not all of their parameters are specified or known. On the basis of collecting area alone, the LSA is a much faster instrument than the other two and should reach a given sensitivity 25 times faster than the MMA (ratio of collecting areas squared) at millimetre wavelengths. However, this assumes that the array antennae have similar efficiencies, the receivers are identical in performance and that the atmospheric opacities are the same. While there is no reason to expect that the receivers will differ significantly from array to array, the current MMA antenna specification and site properties give it the edge in telescope efficiency and atmospheric opacity, bringing its 3-mm sensitivity closer to that of the LSA. The LSA performance at 3 mm is compared with that of some other instruments in Fig. 5. Although none of the projects is yet funded and all have a long way to go before completion, it is clear that they are similar and that they will be costly. Particular challenges are the telescope design for high-frequency performance, the production and efficient operation of many tens of millimetre/submillimetre receivers, many probably operating in 4-K cryostats, and the large correlators. Returning to the cost, it would seem appropriate for the advocates

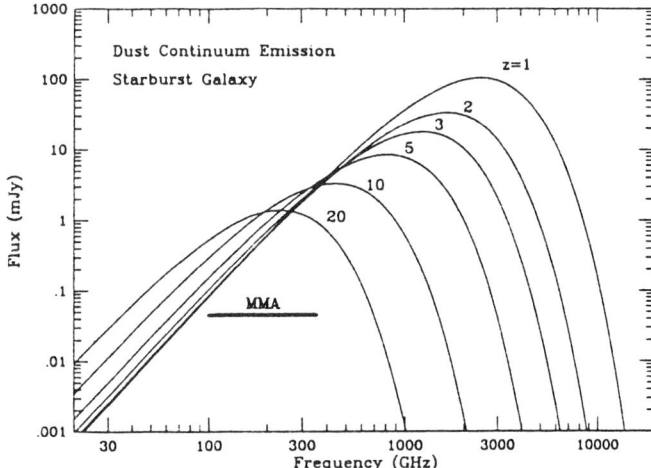

FIGURE 4. The continuum spectrum of a starburst galaxy similar to the prototypical object Arp 220 as the object is moved to higher redshift. The frequency range of the MMA and its sensitivity after ≈ 2 hours of integration is indicated. The dust emission from Arp 220-like (proto)galaxies will be detectable at any redshift.

of the three projects to come together and attempt to collaborate. Perhaps one major array in each hemisphere would be a useful outcome.

REFERENCES

BARVAINIS, R., TACCONI, L., ANTONUCCI, R, ALLOIN, D., COLEMAN, P. 1994, *Nature* **371**, 586.

BOOTH, R.S. 1994, in *IAU Colloq. 140, Astronomy with millimetre and submillimetre interferometry*, eds M. Ishiguro and J. Welch. ASP Conf. Series vol **59**, p.413.

BROWN, R.L. 1994, *ibid* p. 398.

BROWN, R.L., VANDEN BOUT, P.A. 1992, *Ap.J. Lett.* **397**, L11.

DOWNES, D. 1994, in *Frontiers of Space and Ground-Based Astronomy*, eds. W. Wamsteker, M. Longair, Y. Kondo, Kluwer, Dordrecht p. 133.

DUTREY, A., GUILLOTEAU, S., SIMON, M. 1994, *Astron. Astrophys.* **280**, 149.

ISHIGURO, M., KANZAWA, T., KASUGA, T. 1990, in *Radio Astronomical Seeing*, eds. J.E. Baldwin & Wang Shouguan, Pergamon, Oxford, p. 54.

ISHIGURO, M., KAWABE, R., NAKAI, N., MORITA, K.-I., OKUMURA, S.K., OHASHI, N. 1994, in *IAU Colloq. 140, Astronomy with millimetre and submillimetre interferometry*, eds M. Ishiguro and J. Welch. ASP Conf. Series vol **59**, p. 405.

LUCAS, R. *ibid*, p. 135.

LSA: LARGE SOUTHERN ARRAY, an *IRAM-ESO-OSO-NFRA Study Project*, October 1995.

MARTIN, R.N. 1990, *SMT Technical Memo. UA-90-2* (U. Arizona, Tucson).

MYOSHI, M., MORAN, J., HERRNSTEIN, J., GREENHILL, L., NAKAI, N., DIAMOND, P., INOUE, M. 1995, *Nature*, **373**, 127.

NATIONAL RADIO ASTRONOMY OBSERVATORY 1990, *Proposal for a Millimeter Array*. NRAO, Charlottesville, Va.

RAND, R., KULKANI, S. 1990, *Ap.J. Lett.* **349**, L43.

SARAZIN, M. 1990, *ESO VLT Site selection Working Group*, Final Rep. No.**62**.

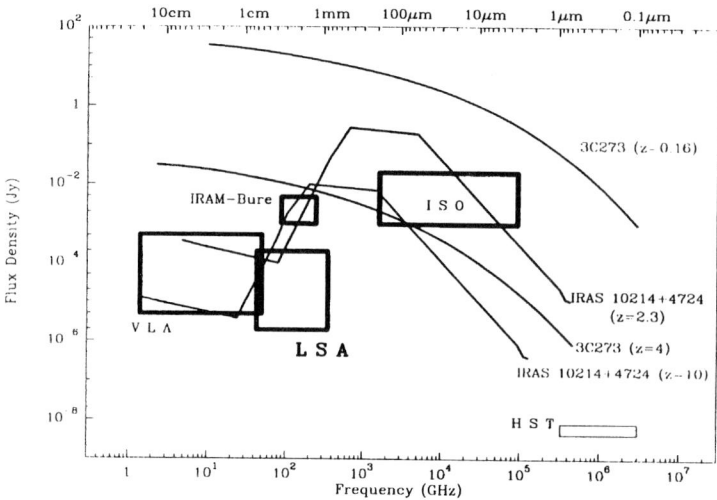

FIGURE 5. Sensitivity of the Large Southern Array compared with that of some other large instruments. The heights of the boxes correspond to r.m.s. sensitivities for spectroscopy (top) and the continuum (bottom). For the LSA, the sensitivity limits are for an 8 h integration with $T_{sys} = 100$ K for galactic spectroscopy ($\Delta\nu = 0.2$ MHz, upper border) and continuum observing ($\Delta\nu = 2$ GHz, bottom border). The curves show the observed nonthermal continuum spectrum of the quasar 3C273 at $z = 0.16$, and the flux that would be expected if 3C273 were at $z = 4$. Also shown is the observed continuum spectrum of thermal emission from dust in the galaxy IRAS 10214+4724 at $z = 2.3$, and the flux expected if the galaxy were at $z = 10$. There is good evidence that a gravitational lens magnifies the far-IR emission from 10214+4724 by a factor of 10. The curves show that even without gravitational lensing, the LSA would be capable of detecting this galaxy at high redshifts.

TACCONI, L., GENZEL, R., BLIETZ, M., CAMERON, M., HARRIS, A.I., MADDEN, S. 1994, *Ap. J. Lett.* **426**, L77.

WIKLIND, T., COMBES, F. 1996, *Molecular Lines in Absorption (and Emission) from Distant Galaxies and Quasars*, to be published in Science With Large Millimetre Arrays, ed. P.A. Shaver, Springer Verlag.

Submillimetre-wave technology for extragalactic spectral-line astronomy

By STAFFORD WITHINGTON[1]

[1] Cavendish Laboratory, University of Cambridge, Madingley Road, Cambridge, UK

The possibility of using submillimetre-wave telescopes to study the state and distribution of molecular and ionised gas in distant galaxies has received a considerable amount of attention in recent years. Indeed, the noise temperatures of submillimetre-wave receivers are now so low, 0.15 K GHz^{-1}, that it should be possible to detect and study the broad, >100 km s^{-1}, faint, <10 mK, lines associated with extragalactic sources with relative ease. In practice, however, it is found that the instability of the atmosphere and imperfections in the instrumentation make it almost impossible to integrate to mK levels at all, let alone in an efficient and straightforward manner. In this paper, I will discuss the future of submillimetre-wave instrumentation in the context of extragalactic spectral-line astronomy. The discussion is illustrated by describing a high-performance extragalactic spectral-line imaging facility (WEASEL) that is currently being developed at MRAO.

1. Scientific background

There are essentially three types of astronomy that one might want to do with a sensitive submillimetre-wave extragalactic spectral-line receiver:

• **Point-source observations.** For objects having redshifts of greater than 2, a number of far-infrared lines are shifted into the submillimetre-wave atmospheric windows. Of particular interest are transitions of C$^+$ and N$^+$, and both neutral and doubly-ionised oxygen; high-order rotational transitions of CO are also important. The 159-μm line of C$^+$ is particularly strong and is known to account for up to 1% of the total luminosity of nearby starburst galaxies. For redshifts of around 4 to 5, corresponding to the most distant quasars, the line should appear in the particularly transparent 350-GHz atmospheric window. The ability to detect and study lines from redshifted molecular and ionised gas would complement the activities of facilities like ISO and SCUBA. In future years, millimetre and submillimetre-wave lines may offer a new way of determining the redshifts and kinematics of optically unidentified, high-redshift, continuum sources.

• **Continuum observations and absorption-line studies.** Thermal continuum emission has now been detected towards a number of quasars having redshifts of greater than 4. With the advent of sensitive bolometer arrays, and colour-based optical searches, it is likely that sources having redshifts of greater than 5 will be found. The main purpose of the submillimetre-wave observations is to determine the mass and temperature of the radiating dust. Although standard observations are effective for determining whether radiation is thermal in origin or not, the very coarse resolution offered by bolometers means that there is a limit to the complexity of the functions that can be used to model redshifted continuum emission. If heterodyne instruments could be built that have instantaneous bandwidths of several tens of GHz, and spectral resolutions of several tens of MHz, then it would be possible to use more sophisticated models when studying the form of the radiating region. In fact, with sufficient sensitivity it is likely that broad absorption features would be found, and these would give important information about the nature of intervening systems.

- **Mapping extragalactic lines.** There is a considerable amount of interest in studying the spatial distribution and excitation of molecular gas in galaxies having relatively-low redshifts. Ideally, one would like to be able to measure line ratios and velocities as a function of position in a whole variety of different objects: normal galaxies, Seyfert galaxies, AGN, starburst galaxies, and elliptical galaxies. The ability to measure rotation curves and identify nonuniform gas distributions in a straightforward manner would be of considerable benefit and would allow statistical studies to be undertaken. Another area of interest concerns the possibility of detecting molecular gas in regions between interacting galaxies.

In short, any instrument that is designed specifically for extragalactic spectral-line work should be able to image faint, extended (100″–200″) objects. It is vitally important, however, not to compromise the ability to detect and study point sources. An extragalactic instrument should have a large instantaneous bandwidth, >4 GHz, but it need only have low spectral resolution, >30 MHz. The instrument must be free from systematic effects such as baseline structure. Finally, because many programmes require a significant number of objects to be observed, in order to build up statistical information, it must be possible to observe faint lines in a reliable and straightforward manner.

2. Technical considerations

In an attempt to satisfy all of the above requirements, we are developing an instrument called WEASEL, which consists of a linear array of four 345-GHz SIS detectors. These detectors will be used to take linear cuts through galaxies. This arrangement is, of course, in contrast to a galactic array which is designed to cover large, two-dimensional fields. To maintain sensitivity on point sources, the aperture efficiency of each pixel must be high. Consequently, the array undersamples the sky by a factor of 4. For the arrangement presented here, with a beam spacing of 25″, we will get a fully sampled image 100″ long after positioning the telescope 4 times. Clearly, many thousands of galaxies could be viewed with an array of this size.

At first sight, an array does not in itself improve the effectiveness with which point-source observations can be made. It must be remembered, however, that at the present time the subreflector has to be chopped and the telescope nodded to remove the sometimes rapidly varying sky emission. Although various attempts have been made to understand the spectral form of the atmospheric fluctuations, it is still not clear to what extent they can be removed on long integrations. The inability to remove the sky fluctuations completely, essentially stems from the fact that the statistical properties of sky noise are nonstationary and there is a limit to the rate at which the subreflector can be chopped. A key feature of WEASEL is to configure the spectrometer so that continuous differencing can be achieved. Another important advantage of continuous differencing is that the telescope looks at the source for essentially all of the time.

It is important that an extragalactic receiver achieves near-quantum-limited sensitivity. To achieve quantum-limited performance requires that the number of optical components between the mixers and the telescope is at a minimum. Indispensible components, such as those required for local-oscillator injection, should be cooled to low temperatures. Moreover, the unwanted sideband should be terminated in a cold, 4-K, load.

Because of the extremely low noise temperatures, it will be necessary to calibrate the receiver accurately. To date calibration has almost always been achieved by placing hot, 290-K, and cold, 80-K, loads in front of the receiver. When taking into account the various uncertainties, however, these temperatures are too high to make accurate measurements on sensitive devices. To overcome this problem, we intend to use a load that is located inside the Dewar. The load is essentially an inverse bolometer, and it will allow the radiometric temperature to be varied continuously and electrically, within a few seconds, over the temperature range 4 K to 100 K. Each mixer will have its own variable temperature load so that the differential and common-mode response can be assessed.

There is nothing in the physics of an SIS mixer that says that it cannot operate with an extremely large instantaneous bandwidth. A change in performance will not occur until the voltage scale of the DC nonlinearity is comparable with the photon energy of the intermediate frequency, and even then the change may not be detrimental. Combined with the development of very broadband IF amplifiers, there is no reason why, in the foreseeable future, receivers should not be constructed that have instantaneous bandwidths of several tens of GHz. Such receivers could be used very effectively for observing continuum emission; and even during faint spectral-line observations, the appearance of the sloping continuum baseline would give one confidence that the system is operating correctly. In the more immediate future, the instantaneous bandwidth should, as far as extragalactic lines are concerned, be at least 1 GHz. In reality, it is desirable to have much larger bandwidths so that acceptable mathematical techniques can be used to remove residual baseline structure. Moreover, the redshifts of many high-redshift objects are not known accurately, and it is impractical to search large regions of velocity space in a piecemeal fashion. An instantaneous bandwidth of 4 GHz would be very effective.

A four-element array of 4 mixers would require a 16 GHz spectrometer. Looking to the future, however, it is desirable for the technology to be extendable, at least in principle, to bandwidths of several tens of GHz. In this way completely blind searches could be made for extragalactic spectral lines. In reality, the instantaneous bandwidth is going to be set by the spectrometer and not by the mixers. Clearly, there is a fundamental problem to manufacturing wideband digital correlators with the above capacity. What is required is a spectrometer that is in some way intermediate between the very-low resolution offered by bolometers and the very-high resolution offered by digital correlators. The filter-bank spectrometer satisfies this requirement, and does not suffer from many of the inadequacies that are inherent in other approaches.

As part of the WEASEL project we are looking into the possibility of constructing a filter-bank spectrometer based on superconducting microcircuits. The advantages of using superconducting circuits are considerable, and it seems as if a complete system could be accommodated easily within a liquid helium Dewar. A key advantage of this approach is that because the spectrometer is small it could be located next to the receiver thus avoiding the numerous problems associated with bringing broad-band signals down a telescope: stability, equalisation, dynamic range, etc.

3. Technical details

In order not to lose signal on point sources, it is important that each mixer element is coupled to the telescope, and ultimately the sky, in an optimum way. For this reason, each mixer will be based on a horn-reflector antenna. That is to say, we will use a large-flare-angle corrugated horn with a with an off-axis paraboloidal mirror at its aperture. This arrangement has the advantage that, if the aperture of the horn is reasonably large, a highly-collimated beam can produced without using a plastic lens. Submillimetre-wave plastic lenses are difficult to model, they introduce dielectric loss, and they give rise to troublesome multiple reflections. Our arrangement allows a highly-collimated beam to be produced with extremely-high efficiency and with a high degree of circular symmetry. The highly-collimated beam means that an array of dielectric beam splitters can be used in the Dewar to inject local oscillator power, and each mixer can have its own calibration load.

The first stage of IF amplification is, of course, extremely important. The intention is to have the first stage of IF amplification—likely to be around 8 GHz—in the mixer block, and lower-grade amplifiers on the 4-K stage. The integrated amplifiers will be of the balanced design in order to get the large bandwidth necessary and also to avoid cumbersome isolators.

Because the mixers will be working close to the quantum limit it is important to have a high-quality calibration system. It is our intention to use a calibration load that is internal to the Dewar. Moreover, because of quantum effects and saturation, we wish to be able to sweep the temperature of the load continuously over a range of low temperatures. It turns out to be extremely difficult to design a load that has a radiometric temperature that is known to with a fraction of a degree, that can be heated with small amounts of electrical power, and that has a time constant of a few seconds. We are developing two different types of load based on the principle of an inverse bolometer. Here a resistive film is deposited on a quartz substrate which is suspended on fibreglass rods and heated in a variable manner up to 100 K. A prototype load has already demonstrated exceptionally good thermal behaviour. Its time constant is of the order of a few seconds and its temperature is known and is stable to within 10 mK over several hours.

Finally, as stated earlier, we intend to investigate the construction of a superconducting filter-bank spectrometer. We believe that we can make a 16-GHz spectrometer with a resolution of 30 MHz which can be contained easily within a liquid-helium Dewar. The incoming signal is coupled onto a long superconducting microstrip transmission line, and then a row of directional couplers, edge-coupled filters, GaAs detectors, and video amplifiers are used to detect adjacent frequency channels. Many of the components are fabricated using Nb circuit technology on thick quartz substrates. The dielectric of the microstrip lines is a thin layer of artificial oxide. There are a number of advantages with this scheme: (i) because of kinetic inductance the microstrip components will be much smaller than if they were made out of normal metals; (ii) the losses will be small and dispersion negligible; (iii) high-Q filters with almost perfect isolation can be manufactured; (iv) low-noise operation can be assured; and (v) the temperature of a helium Dewar is extremely stable and drifts due to environmental changes will be eliminated.

4. Conclusions

There is little doubt that the submillimetre-wave part of the electromagnetic spectrum can make a major contribution towards the study of distant galaxies. What is is required is a small linear array of extremely high-performance detectors. It seems that by using SIS mixers based on horn-reflector antennas, cooled optics, continuous differencing, integrated IF amplifiers, and a superconducting filter-bank spectrometer, it should be possible to satisfy all of the technical requirements. Indeed these developments represent a first step towards constructing an extremely sensitive receiver that has many tens of GHz of instantaneous bandwidth; such an instrument could be used for blind emission and absorption line searches and for relatively high resolution continuum observations where the shape of the atmospheric window is taken into account.

I would like to thank all of my colleagues at MRAO for interesting discussions regarding this project. I particular, I would like to acknowledge High Gibson for suggestions regarding the filter-bank spectrometer and Ghassan Yassin and Matthew Buffey for their work on the horn-reflector antenna.

Very deep continuum observations at submillimetre wavelengths

By E. I. ROBSON

Joint Astronomy Centre, 660 N.A'ohoku Place, University Park, Hilo, HI 96720, USA

Submillimetre continuum observations can be made through atmospheric windows up to a frequency of almost 900 GHz from high, dry ground-based sites. Single-pixel photometers employing bolometers cooled to 300 mK have mostly been used so far, and over the past decade excellent progress has been made in exploring a wide variety of astronomical investigations, prominently those involving thermal emission from cool and cold dust. However, in terms of limiting flux densities, the sensitivity has been insufficient to make significant inroads into some major areas of research. The James Clerk Maxwell Telescope is about to be equipped with SCUBA, a common-user, photon-limited, multi-waveband capability submillimetre camera, with simultaneous observations through two wavebands. This is made possible by employing 128 bolometers, cooled to below 100 mK. SCUBA will revolutionise submillimetre continuum observations and open up new studies previously impossible to undertake and will be a perfect complement to ISO.

1. Introduction

Ground-based astronomy in the submillimetre regime is restricted to observing through transmission windows in the atmosphere. These allow measurements to be made from very high, dry sites such as the 4.2 km-high summit of Mauna Kea in Hawaii, or from the dry plateau of Antarctica. The transmission windows extend from 345 GHz (0.85 mm) to 860 GHz (0.35 mm). Bolometric techniques have the advantage over heterodyne detectors in that their enormous instantaneous bandwidths give increased detection sensitivity, however a drawback of the large bandwidth is increased susceptibility to degradation of performance through sky-noise. The effect of variations in the large background from the immediate surroundings, but more critically from the sky, is reduced by fast switching the secondary telescope mirror between the source and a very nearby sky position at frequencies up to 7 Hz in the case of the JCMT. Phase-sensitive detection then allows much of the sky fluctuations to be removed, albeit at the expense of observing the source for only half the time. The previous generation of continuum detectors were single-pixel devices with a range of passband filters to select the observing wavelength and bandpass, basically single pixel photometers. The best example is undoubtedly the common-user UKT14 bolometer (Duncan et al 1990) which has been a stalwart detector since the commissioning of the JCMT. Recently, array detectors have started to appear, such as those at IRAM – operating at 1.3 mm (Neininger,N. et al., 1996), and at the CSO – operating at 0.35 mm (Hunter et al 1996). Both of these devices are now producing good science, but are aimed at a single waveband and are not optimised for photon sensitivity or optical performance. Nevertheless, they show the way for the future. The single-pixel bolometers (which are usually more sensitive than the newer arrays due to better photon coupling) have typical noise equivalent flux densities (NEFD s - 1s1s) values of around 400 mJy/Hz at 1.1 mm (detector noise limited), rising to a few Jy/Hz at 0.35 mm, where the sky is the limiting factor. Even so, these values are for excellent conditions (high atmospheric transparency and low sky-noise) and to achieve these at the telescope, even at an excellent site such as Mauna Kea, is relatively rare. Therefore, in the absence of flexible scheduling, to obtain the lowest detection fluxes requires extensive observing time and much good luck. The limiting flux values which have been achieved are around 2 mJy at 1.25 mm, 10 mJy at 0.8 mm, 80 mJy at 0.45 mm and 150 mJy at 0.35 mm. The 1.25 mm value is for IRAM, the rest for the JCMT. These levels are insufficient to make a significant impact on the study of the emission from the radio-quiet AGN population, for example, among many other areas. The continuum emission from radio-quiet AGNs has been reviewed by Hughes (1996). Although we can detect some of these AGNs (see *e.g.* Hughes, Dunlop, Robson & Gear - 1993) and observe thermal emission from dust, the sensitivity is inadequate to investigate adequately the properties of the dust and evolution with redshift. Another important study, the spatial distribution of dust in galactic and extragalactic sources is still in its infancy due to the time-consuming manner of source mapping with a single-channel device, even with on-the-fly techniques. The new arrays are just now showing how this can be improved; however, the single waveband map gives only limited scientific value as properties of the dust are still required and these are not simply deduced from a single long or short wavelength. The picture longward of 1 mm can also be confused for hot sources due to underlying contributions from free-free emission.

2. SCUBA

SCUBA will revolutionise submillimetre astronomy, just as array cameras have done in the infrared. As well as the instantaneous snapshot imaging and wide-field mapping capability, SCUBA's sensitivity is much higher than that of other systems. SCUBA has 91 pixels in the 0.45-mm camera, and 37 pixels in the 0.85-mm camera. It views the sky simultaneously through both these arrays with a field-of-view of 2.3 arcmin. The 0.45-mm camera can also operate at 0.35 mm, whilst the long-wavelength camera can also operate at 0.75 mm and 0.6 mm to obtain maximum scientific information from multi-wavelength mapping/imaging/photometry. For photometric studies an additional three pixels are provided at 1.1 mm, 1.4 mm and 2.0 mm. SCUBA comes with its own internal calibration unit (for flat-fielding) and a sky-chopping unit for atmospheric attenuation correction. The software which comes with SCUBA is very powerful and will, in effect, provide calibrated images in real-time. The bolometers are cooled to below 100 mK with a helium-three dilution refrigerator and SCUBA will be photon-noise limited in all wavebands (Gear et al. 1995). Passband filters define the wavelengths of observations and optical matching is via single-moded horns for all the photometric pixels and optimised for the 0.85-mm and 0.45-mm arrays. The apertures of the horns are, however, twice the diffraction spot size and so this leads to severe undersampling for an image which is sampled by staring at the source (while chopping). In order to obtain the diffraction-limited spatial resolution (by sampling at about half the diffraction pattern size), instead of just staring, an image is constructed by using a fast jiggle pattern of the secondary mirror to fully sample the image at the required resolution. This requires a 16-position jiggle for a fully sampled 2D image of the one array and a 64-position jiggle to fully sample both arrays simultaneously. The modes of observing are determined by the source size. For images smaller than or about the size of the FOV, the jiggle image described above is employed, while for more extended images scan-chopping is employed. For point sources, the possibilities of sky-noise removal by chopping across the array is obvious. SCUBA will operate in a flexibly-scheduled environment and comes with its own queuing software. SCUBA is now in the final laboratory commissioning phase. It will be delivered to the JCMT in April 1996 and observations for the community should

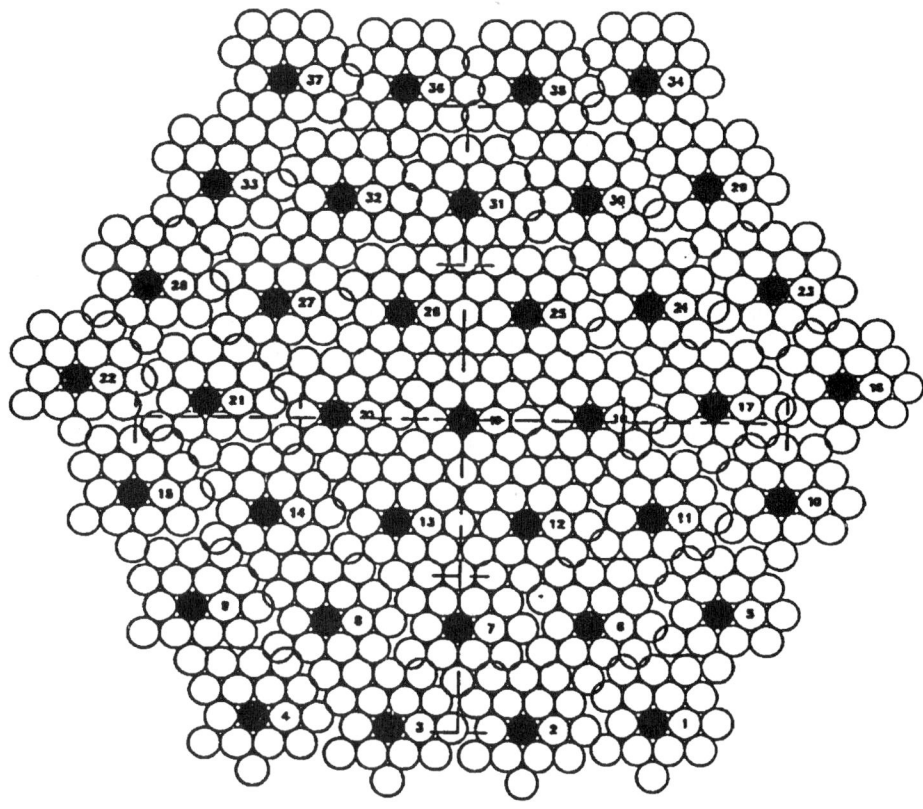

FIGURE 1. A fully sampled jiggle pattern as it appears on the sky

commence by November, following extensive Hilo laboratory commissioning (including operating-staff training) and telescope commissioning.

What about the SCUBA sensitivities? Under the very best conditions and making some (optimistic) assumptions about the sky- noise, it is expected that SCUBA will achieve NEFDs of $10\,\mathrm{mJy}\,\mathrm{Hz}^{-0.5}$ at 1.1 mm, $20\,\mathrm{mJy}\,\mathrm{Hz}^{-0.5}$ at 0.85 mm and $150\,\mathrm{mJy}\,\mathrm{Hz}^{-0.5}$ at 0.45 mm. This leads to limiting 5σ detection sensitivities of:

Wavelength	1hr	25 hours
1.1 mm	0.83 mJy	0.17 mJy
0.85 mm	1.67 mJy	0.33 mJy
0.45 mm	12.5 mJy	2.5 mJy

This will have a major impact on investigating continuum emission from the following: radio galaxies at high redshift, radio-quiet quasars, elliptical and blue galaxies, stellar photospheres, stellar envelopes, brown dwarfs, proto-stars, proto-planetary condensations, solar-system bodies. Fig. 2 demonstrates the capability of SCUBA for deep integration. In terms of mapping, the revolution is even more pronounced. SCUBA will deliver a factor of almost 10,000 increase in mapping speed at 0.45 mm, due to a factor of 10 increase in sensitivity and almost 100 detectors. This will enable statistically mean-

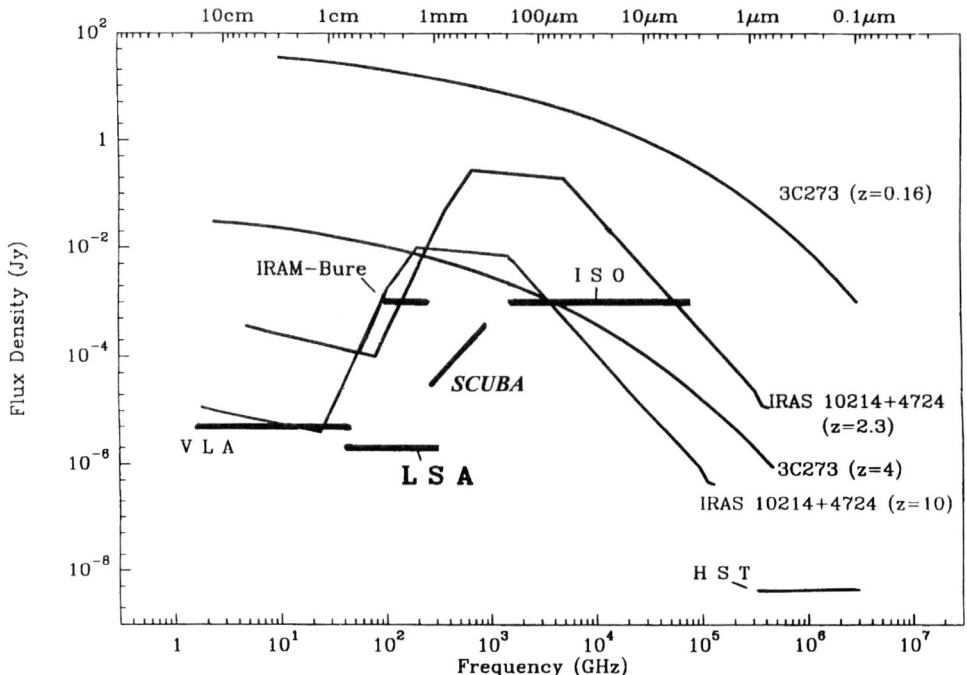

FIGURE 2. RMS sensitivities for a number of facilities. (Adapted from the Large Southern Array Study document).

ingful samples for the condensations, the so-called class zero sources, within molecular clouds to be obtained, rather than the handful of sources we know today. The impact of dust on outflow sources is another area that will benefit from SCUBA observations. In terms of extragalactic astronomy, the distribution and temperature of dust in relatively nearby galaxies, ordinary galaxies, close-by active galaxies (such as NGC5128) and starburst galaxies can all be probed effectively. The SCUBA and ISO fields of view are extremely complementary.

There is yet another area in which SCUBA will bring about a revolution: blank-field/cosmological studies. Studies of the S-Z effect, the determination of H_o, galaxy source-counts and proto- galaxies will all be tackled by major programmes on the JCMT. The future is undoubtedly rosy. Further information about the JCMT and SCUBA can be found on the WWW at http://www.jach.hawaii.edu.

REFERENCES

BEARMAN, P. W. & GRAHAM, J. M. R. 1980. *J. Fluid Mech.* **99**, 225

DUNCAN, W.D., ET AL. 1990. *MNRAS* **243**, 126

GEAR, W.K., ET AL. 1995. *ASP Conf.Ser.* **75**, eds., Emerson, D.T. & Payne, J., p215.

HUGHES, D.H., DUNLOP, J.S., ROBSON, E.I. & GEAR,W.K. 1993. *MNRAS* **263**, 607

HUGHES, D.H. 1996, *MNRAS*, in press

HUNTER, T.R., ET AL. 1996. *BAAS* **27**, 108, p1441.

NEININGER, N., ET AL. 1996. *IRAM Newsletter* **24**, 21

Cold dust in galaxies at 1.2-mm wavelength

By RICHARD WIELEBINSKI

Max-Planck-Institut für Radioastronomie, Auf dem Hügel 69, 53121 Bonn, Germany

Observations of cold dust in galaxies can best be done at short-mm wavelengths with continuum bolometers. A series of bolometers were developed by Ernst Kreysa of the MPIfR and are used in the 30-m radio telescope of IRAM on Pico Veleta. By now we have an operational system with 19 channels at $\lambda=1.2$-mm wavelength. This system has now been successfully used for two seasons. A number of nearby galaxies have now been mapped showing that cold dust is more extended than the CO gas, while the H I extends even further. A new bolometer with 37 channels is under construction. Multibeam bolometers will be available in the 10-m Heinrich-Hertz Telescope on Mt. Graham in Arizona at submm wavelengths.

1. Introduction

Radio astronomy was first successful at metre wavelengths and slowly worked itself towards cm and later mm wavelengths. At short cm wavelengths we observe a mixture of thermal (free-free) emission from H II regions and nonthermal emission from supernova remnants, diffuse regions (old supernova remnants?) and halos of galaxies. Above 200 GHz the nonthermal component is very weak and a new component is observed, namely the cold dust emission. This component has a very steep spectrum with a positive spectral index so that at submm wavelengths and in the FIR dust emission dominates.

2. The development of multibeam bolometers

In general, radio continuum observations at the frequency of 200 GHz and above use broadband bolometers. The superheterodyne systems have too narrow bandwidth to detect the weak continuum emission. A sensitive bolometer developed by MPIfR was described by Kreysa (1990). This bolometer was optimised for the 30-m radio telescope of IRAM at Pico Veleta. It was quite clear from the developments at cm wavelengths that multibeaming offers great advantages. In particular, the reduction of sky noise can be achieved by appropriate software. To follow these possibilities the next step was a seven-beam bolometer described by Kreysa et al. (1993). This step confirmed the spectacular possibilities of multibeaming and led to the development of a 19-beam system. The next step envisaged by Ernst Kreysa is 37-beam bolometers. This method of step-by-step development has offered observational capabilities which were gradually expandable.

3. Some results

I wish to describe some of the results obtained in the field of galaxies. In fact some of the objects mentioned by Dr. Robson as prime targets for SCUBA have already been done at $\lambda=1.2$ mm. We have mapped M82 and the nucleus of NGC 253 with a single-channel bolometer (Krügel et al. 1990). The galaxies NGC 891 (Guélin et al. 1993), NGC 3627 (Sievers et al. 1994, NGC 4631 (Braine et al. 1995) and M51 (Guélin et al. 1995) were mapped with the seven-channel bolometer. With the 19-beam bolometer the galaxies NGC 4565 (Neininger et al. 1996), NGC 5907 and NGC 3079 have been studied. The 1995/96 winter season should lead to further exciting results.

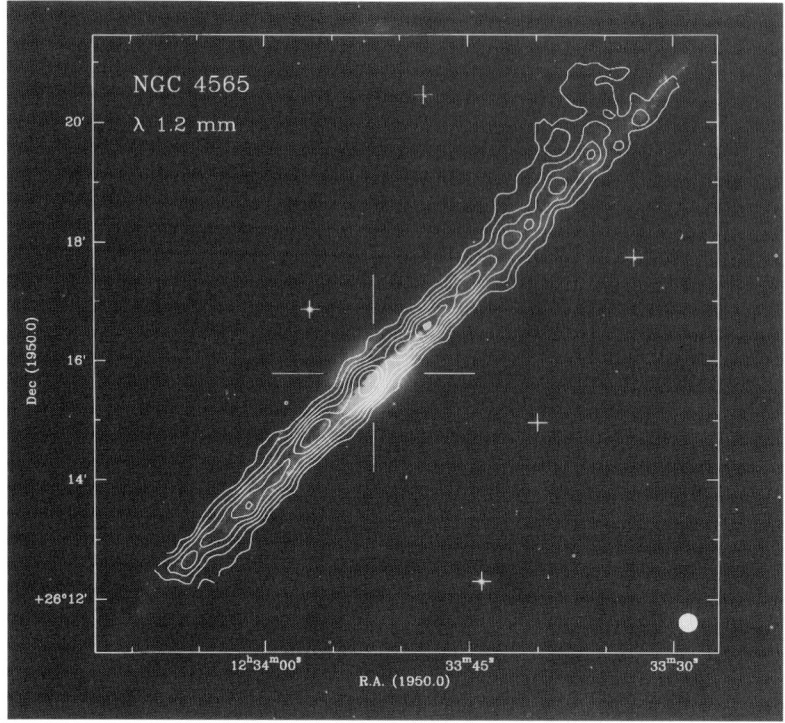

FIGURE 1. NGC 4565 observed at $\lambda = 1.2$ mm (from Neininger et al. 1996)

From these observations of galaxies it was possible to estimate the dust mass and temperatures with an unprecedented accuracy. Furthermore, in the edge-on galaxies it was possible to show that the cold dust mimics the CO distribution in the central regions but then follows the H I distribution out to the optical limit. Maps of 'our' galaxies at 450 μm with identical angular resolution (JCMT/SCUBA combination) should give unique data on the constitution of cold dust in galaxies.

REFERENCES

BRAINE, J., KRÜGEL, E., SIEVERS, A., & WIELEBINSKI, R. 1995. *A&A* **295**, L55

GUÉLIN, M., ZYLKA, R., MEZGER, ET AL. 1993. *A&A* **279**, L37

GUÉLIN, M., ZYLKA, R., MEZGER, ET AL. 1995. *A&A* **298**, L29

KREYSA, E. 1990. In: *Proc. 29th Liège Int. Astrophys. Coll. From Ground-based to Space-borne Sub-mm Astronomy*, ESA SP-314, 265

KREYSA, E., HALLER, E.E., GEMÜND, H.-P., HASLAM, C.G.T., LEMKE, R., & SIEVERS, A.W. 1993. In *Proc. 4th Int. Symp. on Space Terahertz Technology*, 692

KRÜGEL, E., CHINI, R., KLEIN, U., ET AL. 1990. *A&A* **240**, 232

NEININGER, N., GUÉLIN, M., GARCIA-BURILLO, S., ZYLKA, R., & WIELEBINSKI, R. 1996. *A&A* in press

SIEVERS, A.W., REUTER, H.-P., HASLAM, C.G.T. ET AL. 1994. *A&A* **281**, 681

The Lovell Telescope upgrade

By R. J. DAVIS[1]

[1]NRAL, University of Manchester, Jodrell Bank, nr Macclesfield, UK

It is intended to rebuild the Lovell telescope using the experience gained in the design of the radio properties of the 32-m telescope in MERLIN. The aim is to increase the frequency range of the present instrument to at least 22 GHz and increase the overall sensitivity by typically a factor of thirty at 5 GHz. Such improvements together with the use of the telescope for pulsars, H I, MERLIN and VLBI will make this telescope a major contributor to astronomical research into the next millenium. Analysis shows that such performance would make the telescope first equal in the world.

1. Introduction

Engineering studies indicate that the fatigue life of the Lovell telescope with some modifications is good. These studies have been principally of the elevation rotation parts, and the foundations. I will concentrate on the radio design in this talk. The principal parts to the upgrade are a new surface of aluminium, tetrapod legs and a deformable subreflector and/or primary reflector. The surfaces may be dual shaped similar to the DSN dishes and our own 32-m telescope. Calculations indicate that an overall aperture efficiency of 85 percent should be achievable including blockage. For a 250-ft (76.2-m) telescope this corresponds to 1.41 K/Jy. The other half of the calculation is the system temperature. This can be split into two halves: the receiver part and the telescope part. Measurements from the 32-m telescope indicate that a design aim of 10 K for the telescope part is achievable; receiver temperatures do currently achieve 4 K and state-of-the-art receivers can now achieve sub-1 K (GBT memos) at L-Band and thus 10 K overall system temperature should be possible. This should give an overall system performance of 7 Jy. The present frequency range is limited by the panels and their setting, and the distortions in the structure. This distortion may be predominantly astigmatic and as in the Greenbank 140-ft telescope a deformable subreflector may work well. However for more complex deformations, controlling the shape of the main reflector may be preferable. It is intended to drive all the motors independently with tachometer and delta tachometer feedback to achieve the required pointing control of the telescope axes.

2. Fatigue Life

With such a rebuild project, the most important consideration is the integrity of the existing structure. To this end a number of engineering studies have been made. I will mention some of them here. The foundations have been studied and measured by AEA Engineering. Some repairs and increased drainage were recommended and some of this work has been carried out. The overall integrity is good and the deviations on the track, loaded and unloaded, show arcsecond attitude stability is achievable. Parts of the track, however, will need to be replaced or relaid. The Elevation Rotating Structure (ERS) has been analyzed in detail by Acer Special Structures. A detailed stick model of the telescope has been made together with many detailed measurements of the structure. This has resulted in a stress analysis which, where measurements exist, agrees well with the actual structure. The overall results look good and some strengthening will be required.

3. Extend Frequency of Operation to at least 22 GHz

The aim here is to extend the useable frequency range up to at least 22 GHz *i.e.* the highest frequency currently used by MERLIN. The plan is to remove the existing steel surface which is made from large steel panels of $\sim 7 \times 4$ m and replace them with a steel frame of the same area covered with 4 or 6 aluminium panels. These new composite panels should be accurate to 0.1 mm. These composite panels will then be attached to the backing structure with adjusters which could be manual or motorized. The central tower will be replaced by a new tetrapod which will support a secondary mirror which could either be a Gregorian or Cassegrain system.

3.1. Pointing

By far the biggest problem in using giant telescopes at high frequency is that of pointing. Specifications approaching an arcsecond are required. To this end Comsat RSI have been contracted to examine the pointing control of the Lovell Telescope. They have established that the structure is such that the two orthogonal axes are such that the two drives appear to be independent. In other words driving in azimuth does not, to first order, affect elevation and vice versa. This is fundamental for accurate control and shows an excellent design in the way that the loads and torques are transferred through the complex structure of the Lovell Telescope. They have simulated the servo system as it is now. The biggest problem at present is the elevation drive. Recent improvements in the control computer have given significant improvements but the system still shows a degree of instability. The motors are driven in groups. Each group has its motors in series. The simulations show the same instability that the actual telescope experiences. The solution appears to be to control all the motors through the appropriate whiffle trees and feedback. Oscillations in the system are a problem and it is hoped that these can be ironed out with delta tachometer feedback on top of the more usual PID system and current feedback. The simulations indicate that arcsecond servo control may be possible. The other half of the pointing problem is knowing where the telescope beam actually is. One can mount an encoder but different parts of the structure can deform such that this encoder is no longer telling one the required information. Thus a servo system may keep the formal errors at a subarcsecond level but this is of little value if the beam is pointing somewhere else. Work is underway to study better ways to mount such systems.

3.2. Dish Deformation

Up to now the dish deformation has been measured using radio holography. This was achieved using the two largest telescopes at Jodrell Bank, the Lovell and the Mk II. A broad-band interferometer was employed with a bandwidth of 400 MHz and continuum radio sources were used to make the holograms. In this way it was possible both to measure the present panels and their setting and also the way the structure deforms as a function of elevation. The results indicate that the main deformation is astigmatic with a magnitude of ~ 20 mm. More detailed measurements may show deformations on a finer scale and certainly more measurements will be required. If the simple astigmatism proves to be the only problem then one way to correct this would be to deform the secondary mirror. S Von Horner has shown that this works well both theoretically and practically in the case of the Greenbank 140-foot telescope. However the use of a deformable secondary is limited both for correcting non-astigmatic deformations, but also temperature and wind deformations. It is the primary surface which is deforming and thus in principle it is the primary which should be put right, otherwise the illumination is interfered with to some degree. The further measurements could be made with holography but optical techniques would give the absolute shapes whereas holography only gives the best-fit shapes to a best-fit paraboloid.

4. Improve Sensitivity by 30 at 5 GHz

The essential aim is to improve the sensitivity of the telescope by a factor of 30 at 5 GHz. This will make the performance of the telescope first or at least first equal in the world and close to the expected performance of the Green Bank Telescope (GBT).

4.1. *Gain*

The plan is to make the surfaces dual-shaped, similar to the DSN and our own 32-m telescope. A conventional Cassegrain or Gregorian system has a paraboloid primary mirror and a hyperboloid or ellipsoid for the secondary mirror. The dual-shaped system alters the distribution of power on the surfaces to optimize efficiency. Figures of 55 percent are typical for an unshaped system whereas simulations show that 85 percent efficiency should be achieved with the Lovell. This results in a telescope performance of 1.41 K/Jy. The main disadvantage of dual-shaped systems is the loss of field of view. A conventional unshaped system has smooth surfaces and one can offset feeds by up to 25 diameters without undue loss of performance. Thus in principle 2500 horns could be installed. The shaped system however does have deviations from the analytic curves in its surfaces and at most only 3 touching horns along one axis may be used. Thus, as a 2-dimensional array, a hexagonal ring with one in the centre making 7 is possible. Here the outer 6 have their gain reduced by 1 db compared to the centre feed. This performance is still better than the unshaped systems. One clearly uses shaped systems where forward gain is the most important. Examples of this are the VLA, VLBA, AT, DSN, MERLIN 32-m, MERLIN E systems telescopes. Examples of paraboloids are Effelsberg, Parkes, GBT *i.e.* the largest dishes used in radio astronomy. The Lovell telescope we see as very much a dish for MERLIN, VLBI, and space VLBI. There is also much interest in single-telescope operation but it is agreed that the seven horn systems for multifeed operation is a good compromise. At L-Band anything more than a 7-feed system is difficult to entertain from the secondary due to its size. Currently it is planned that the GBT will operate a 3-feed system at L-Band.

For the future, focal-plane arrays offer interesting possibilities. Here with a large array of small horns one can synthesize a large horn and by suitable processing, the shaping can be removed and converted in software back to a parabola/hyperbola system. Thus the large field of view of an unshaped system is restored. At the moment this is difficult to achieve due to the very large number of horns and receivers required and the extremely large number of correlators required to do the appropriate synthesis.

4.2. *System Temperature*

By far the biggest increase in sensitivity can be achieved by careful attention to the system temperature. The system temperature can be broken down into two parts: a receiver contribution and a telescope part. With modern lattice-matched pseudomorphic HEMT technology, receivers with virtually zero receiver temperature can be made at L-Band and at 5 GHz. The telescope part consists of 2.7 K from the cosmic microwave background and, at 5 GHz, 2 K from the neutral atmosphere in the beam. The rest is from radiation which is not from in the beam. It comes from scattering from feed legs, from panel gaps from the edge of the subreflector and any other pieces of metal which can interact with the receiver by scattering in ground radiation. With most telescopes this is of the order of 20 K in total. Careful attention to this problem indicates that figures as low as 8 K total for the telescope contribution can be achieved and a design aim for better than 10 K overall telescope contribution is our aim. An example of this can be seen with our 32-m MERLIN telescope where, for example, we have made sure that the central blocked radiation from the feed finishes up in the cold sky and the feed legs are carefully designed so as not to scatter ground radiation into the system. Here we achieve a receiver contribution of 12 K at 35 degrees elevation. This should reduce to 9 K at 55 degrees elevation but as yet it does not: it remains up at 12 K. The cause for this has not been found to date, but presumably there is a little more spill-over past the edge of the main bowl than expected and when the bowl tips up and the in-beam atmosphere radiation goes down, the effect of this spill-over goes up by the same amount.

4.3. *Gain / System Temperature*

Thus we plan to achieve overall system temperatures of 10 K. By comparison J. Lockman (private communication) intends to achieve single-figure system temperatures at L-Band for the GBT. With all these design considerations taken together we should achieve an overall sensitivity or gain / system temperature, G/T, of 7 Jy. This figure means that a 7-Jy source doubles the power out of the receiver. This comprises 10 K divided by 1.41 K/Jy. This exceeds the present performance at L-Band by about 8 and at 5 GHz by 30. It exceeds the present Effelsberg performance by 3 at L-Band and 6 at 5 GHz. 7 Jy is the sort of performance hoped for on the GBT. Thus, if these design aims are achieved, the telescope sensitivity will be level with the best in the world. To reiterate; this is done partially by high gain but more by very careful attention to overall system temperatures.

5. Multifrequency Rapid Operation

Flexible use of this telescope is vital and it is intended that all useable frequencies can be electrically selected in a minute and also that these giant multifeed systems at long waves and smaller ones at short wavelengths can be brought in in 10 minutes or so. Exactly how this can be done is not clear. The different frequencies could be selected using an offset carousel as in the case of our 32-m telescope. The offset is arranged such that simple rotation of this carousel will bring any receiver onto the axis of the instrument. The problem then is how to locate the giant multifeed systems. The short-wavelength multifeed systems would fit on the carousel. This may have to be moved out of the way to let the waves access the giant multifeeds.

6. Conclusions

Although the Lovell telescope is 40 years old next year and was originally designed for a shortest wavelength of 1 m, modern analysis, in particular the understanding of fatigue together with advances in radio design, suggests that the telescope can be brought up to the sort of performance expected from state of the art instruments of the future. This particular instrument is important because of its strategic position in the MERLIN array and in the EVN and for space VLBI.

The Giant Metrewave Radiotelescope

By G. SWARUP, S. ANANTHAKRISHNAN,
C. R. SUBRAHMANYA, A. P. RAO,
V. K. KULKARNI AND V. K. KAPAHI

National Centre for Radio Astrophysics, TIFR
Pune University Campus, Pune 411 007, India.

The Giant Metrewave Radio Telescope (GMRT) under construction in India is a major new facility in the field of radio astronomy. GMRT consists of 30 fully-steerable parabolic dishes of 45-m diameter each. Twelve of the dishes are located in a central array of about 1 km × 1 km in size and the other eighteen along three arms of an approximately Y-shaped array, providing a maximum baseline of ∼ 25 km. Signals received by the 30 antennas are brought together to a central laboratory using coherent local oscillators and optical fibre links. GMRT will operate in the following frequency bands protected for radio astronomy in India : 37.75-38.25 MHz, 152-155 MHz, 230-235 MHz, 322-328.6 MHz, 608-614 MHz and 1400-1427 MHz. However, the front-end amplifiers have bandwidths varying from about 20 MHz to 40 MHz for the above frequency bands but cover 1000-1430 MHz for the highest frequency range. Narrower bandwidths can be selected in the IF and baseband systems. It has been possible to build the light weight 45-m dishes quite economically using a novel concept, nicknamed SMART (Stretched Mesh Attached to Rope Trusses).

One of the important scientific objectives of GMRT is to search for neutral hydrogen clouds prior to the formation of galaxies and clusters in the Universe. The signals are expected to be extremely weak and may occur somewhere in the frequency range of 150-600 MHz. GMRT will, however, be a versatile instrument for investigating a variety of other astrophysical problems concerning the sun, radio stars, pulsars, HII regions, supernova remnants, the Galactic centre, nearby galaxies, radio galaxies, quasars and cosmology.

Twenty-four of the thirty dishes have been taken over by TIFR after the installation of the mechanical drives; the remaining antennas are expected to be completed by May 1996. Astronomical observations, calibrations, on-line programming and debugging have been done using four antennas since September 1993. Electronics systems have been installed on eleven more antennas recently and it is hoped that the first astronomical maps will be available within a couple of months. The full telescope is expected to be ready by late 1996.

1. Introduction

The Giant Metrewave Radio Telescope is being set up in India by the National Centre for Radio Astrophysics (NCRA), Pune, which is a part of the Tata Institute of Fundamental Research. It will provide a large collecting area over a wide frequency range of about 38 to 1430 MHz. The two most important scientific objectives that have been kept in mind while designing GMRT are (a) to detect the highly-redshifted 21-cm line of neutral hydrogen from protoclusters or protogalaxies, and (b) to search for and study a large number of rapidly-rotating pulsars in our Galaxy. However, the specifications of GMRT will make it a highly versatile instrument for investigating a variety of other astrophysical problems concerning objects in our solar system, the Milky Way, nearby galaxies, clusters of galaxies and the very distant radio galaxies and quasars.

GMRT consists of thirty fully-steerable parabolic dishes of 45 m diameter each, located over a region of about 25 km (Swarup et al. 1991). It is now in an advanced stage of completion and is expected to become operational by end-1996. The array configuration of GMRT is briefly described in Section 2 and the major design features of the GMRT

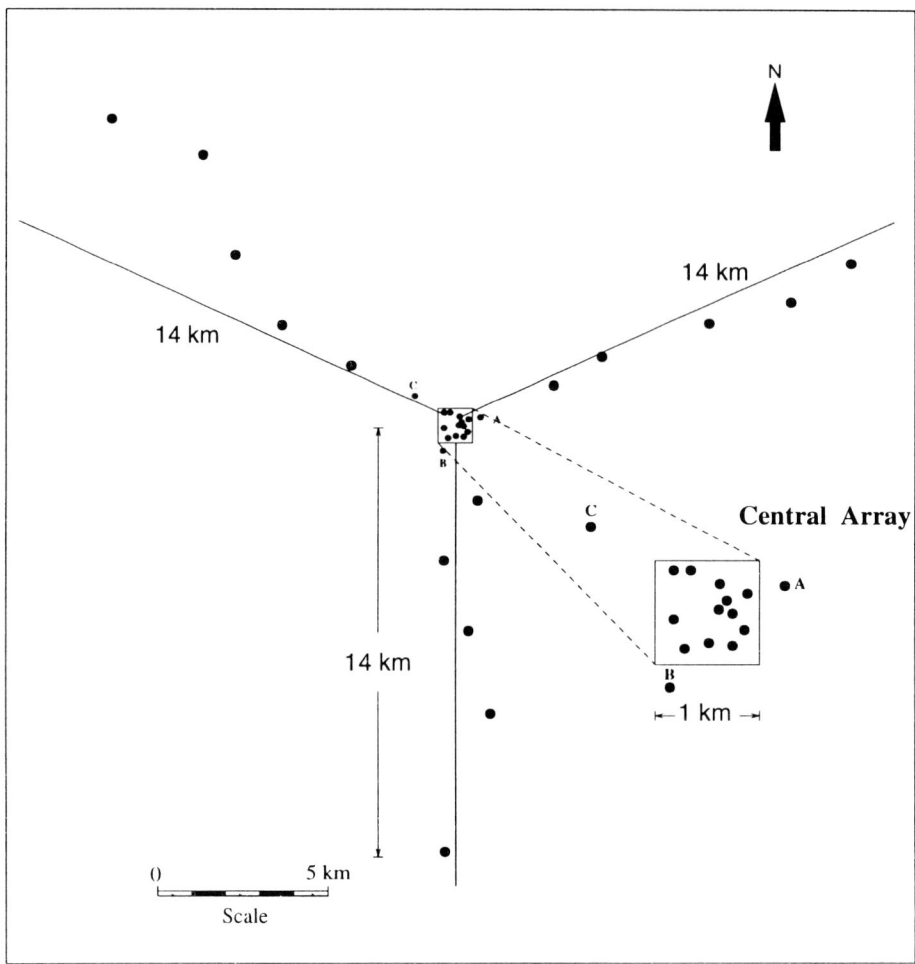

FIGURE 1. Array Configuration of the thirty 45 m diameter dishes of GMRT

dishes and of the electronics system are summarised in Sections 3 and 4 respectively. In Section 5, we give the expected system parameters and results of some recent performance tests. A few examples of the astronomical capabilities of GMRT are mentioned in Section 6.

2. Array Configuration

GMRT operates as an "Earth Rotation Aperture Synthesis Radio Telescope" (Thompson et al. 1986). Its array configuration has been chosen so as to be sensitive to both compact and broad features of celestial radio sources. Twelve antennas are placed somewhat randomly in a central array with a maximum baseline of about 1.1 km. The other 18 antennas are distributed along the three 14-km-long arms of a Y-shaped configuration (Fig. 1). With its 30 antennas, GMRT would measure 435 Fourier components of the brightness distribution across a radio source at any given instant and roughly a million components in about a 12-hr period of observation.

FIGURE 2. A photograph showing a 45-m diameter GMRT dish

3. Design Features of the GMRT Dishes

The antenna system forms a major part of the cost of a radio telescope. Since the galactic background temperature at high latitudes varies as $\sim 50\ \lambda^{2.7}$ K, where λ is the wavelength in metres, the system temperature of a radio telescope is dominated at the longer wavelengths by the Galactic background temperature rather than the receiver temperature. Hence it becomes necessary to construct a large collecting area for a sensitive metrewave radio telescope. After investigating many alternative possibilities, a novel concept was developed, nicknamed SMART (Stretched Mesh Attached to Rope Trusses), resulting in considerable economy in the cost of the 45-m-diameter dishes (Swarup, 1990; Swarup et al. 1991). In this concept, the reflector surface of the antenna is made of low solidity wiremesh, which is supported by rope trusses attached to the backup structure of the dishes. This design cuts down forces due to wind load on antennas. A 45-m-diameter parabolic dish was preferred for the elements of the GMRT, rather than a more economical parabolic cylindrical antenna or a cluster of several smaller dishes, say 4 dishes of 22.5-m diameter, in order to avoid complexity of the electronics system and also to have a relatively small primary-beam size to minimize the effects of the non-isoplanaticity of the ionosphere.

The backup structure of the 45-m parabolic dish consists of 16 radial parabolic frames made of tubular steel and connected to a 12-m-diameter central hub (Fig. 2). The outer edges of the radial frames are connected to a triangular rim truss and are also interconnected using guy ropes for rigidity. Circumferential stainless steel wire rope trusses of 4 and 2.5-mm diameter are connected to anchor blocks of adjustable height which are welded to the parabolic frames at spacings of 1.2 metre (Fig. 3). These rope trusses are suitably adjusted and tensioned to give a roughly parabolic curvature to the top wire ropes. Wiremesh panels made of 0.55-mm-diameter thin stainless-steel wires are then stretched on the rope trusses to form a series of plane facets approximating the paraboloidal reflecting surface. The mesh has a size of 10 mm × 10 mm in the central

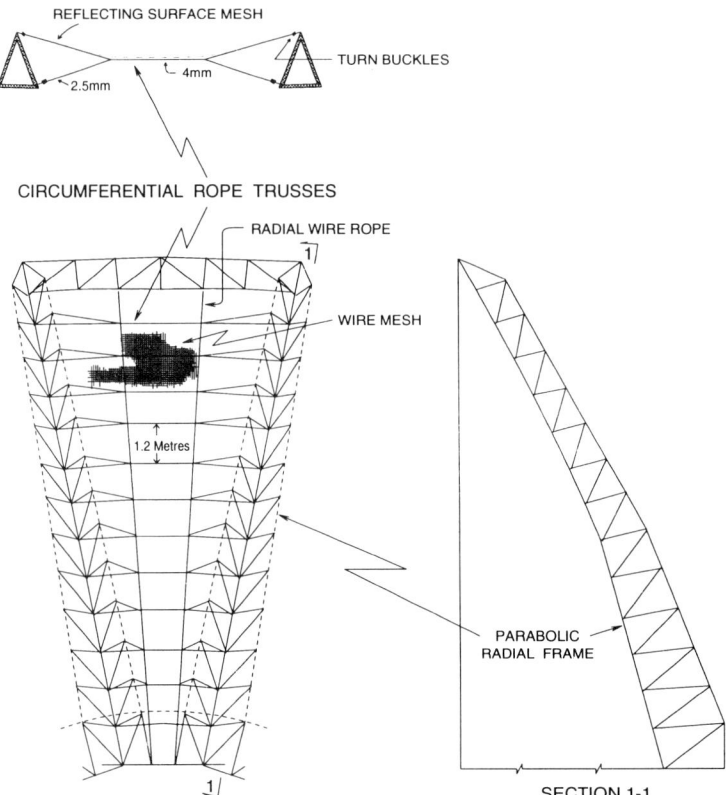

FIGURE 3. A section of the GMRT dish illustrating the SMART (Stretched Mesh Attached to Rope Trusses) concept

one-third, 15 mm × 15 mm in the middle one-third and 20 mm × 20 mm in the outer one-third aperture of the dish. The expected reflectivity of the wire mesh at 1420 MHz is about 95 per cent and the contribution of the ground radiation is thus only about 5 per cent. The measured value of the root-mean-square deviation of the surface from a true paraboloid is about 10 mm for the completed 45-m dishes.

The central hub of the parabolic dish is connected to a cradle at 4 points. The cradle is supported by two elevation bearings on the top of a U-shaped yoke. A bullgear is connected to the cradle which allows rotation of the 45-m dish in elevation from about 17 deg to 110 deg from the horizon. The yoke is placed on a 3.6-m-diameter slewing ring bearing, with a built-in gear-wheel which forms part of the azimuth drive system. The slewing ring bearing is supported on a concrete tower of 12-m height. A counter-torque system, consisting of a pair of 6-hp servo motors each, is connected to the elevation bullgear and similarly for the azimuth slew-ring gear through a pair of planetary gear boxes, providing a total gear reduction of about 25000:1 and 18000:1 respectively. Seventeen-bit encoders are connected to both elevation and azimuth axis. Measured rms tracking accuracy of the antennas is about 20 arcsecond which is negligible compared to the antenna beam-width at 1420 MHz.

The total weight of the parabolic dish including yoke is only about 85 tonnes. This is much lower than the typical weight of about 250 tonnes for the structural parts of a 25-m dish, operating at cm and decimetre wavelengths.

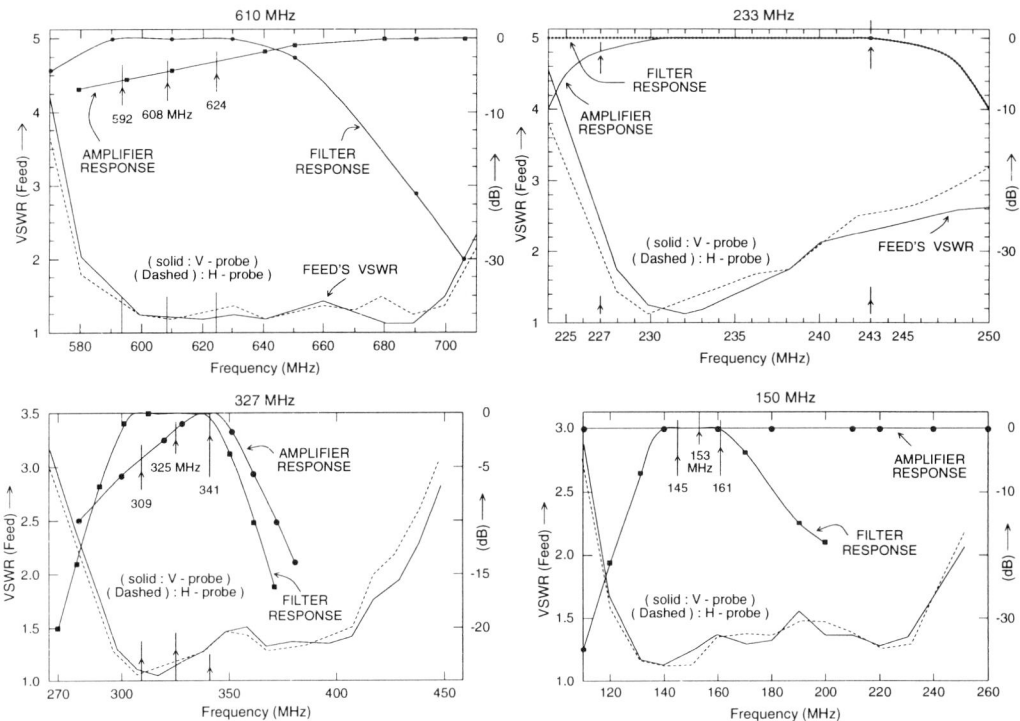

FIGURE 4. Voltage Standing Wave Ratio (VSWR) for two linear dipoles or probes and amplifier and filter responses for feeds operating at 150, 233, 327 and 610 MHz.

4. Electronics System

GMRT is to operate in six different frequency bands as follows : 50 ± 13, 153 ± 16, 235 ± 8, 325 ± 16, 610 ± 16, and 1000-1430 MHz. The 50 and 150-MHz primary feeds consist of two orthogonal pairs of dipoles at spacings of 0.5 λ, placed in a quad-formation. However, while the 150-MHz feed is placed above a ground plane, the 50-MHz feed is backed-up by linear rod-reflectors. All the primary antenna feeds, except the 50-MHz feed, are mounted near the focus of the 45-m dish on a rotating cage supported by a quadrupod structure. The 50-MHz feed is permanently fixed to the quadrupod structure. The 325-MHz feed consists of 2 orthogonal half-wave dipoles in a cross formation with a beam forming ring placed above a ground plane having a diameter of about one wavelength (Kildal 1982). Dual concentric coaxial feeds are used at 235 and 610 MHz. A broad-band horn has been developed by the Raman Research Institute to cover the frequency range of 1000-1430 MHz.

The performance of the 150, 235/610 and 325-MHz feeds are shown in Figs. 4. It is seen that the 150-MHz feed provides frequency coverage over a bandwidth of about 1.8:1 but the bandwidth of the other feeds is limited. Over the next few years, it is proposed to replace some of these feeds with broadband feeds in order to provide a near-continuous frequency coverage from about 130 MHz to 1430 MHz.

The outputs of all the feeds (except the 1.4-GHz feed) are connected to polarisers giving Right-handed (RH) and Left-handed (LH) signals. The 1000-1430 MHz feeds provide two orthogonal linear polarizations. The feeds are followed by low-noise amplifiers with RF filters. The RF outputs are brought to a LO-IF system placed at the bottom of the concrete tower of each antenna.

The local oscillators (LOs) are phase-synchronized to a master oscillator placed in the central electronics building (CEB) located near the centre of the GMRT array. The RF signals received by each antenna are down-converted to 32-MHz wide IF signals at 70 MHz using the above LO. The IF signals for the two polarizations are then frequency translated to 130 and 175 MHz using a set of second LOs. The two IF signals are then combined and transmitted to the CEB on optical-fibre links. The optical-fibre links are also used for transmitting reference signals at 106 MHz, 201 MHz and 97.5 ± 1 MHz from the CEB to the different antennas for the purpose of phase synchronization of the local oscillators, and also for sending the telemetry signals required to control and monitor all the GMRT antenna and receiver systems. At the CEB, a third Local Oscillator translates the 130/175-MHz IF signals (of 32 MHz bandwidth each) back to 70 MHz and these are then down-converted using a 4th LO giving two 0-16 MHz baseband signals for each of the two polarizations. The receiver bandwidth can be restricted using saw filters placed in the IF chain at the antennas to either 5.5, 16 or 32 MHz and finally in 9 binary steps from 64 kHz to 16 MHz in the baseband system. Attenuators are placed in both the RF and IF systems to allow observations of the Sun and other strong sources.

The baseband outputs from each of the four outputs of the 30 antennas are then digitised by a 4-bit sampler. A 4-bit system has been chosen, rather than the usual 2-bit sampler, in order to ensure that any narrow-band interference signals from man-made transmitters, after being sampled, do not produce intermodulation products in the passband higher than -45 dB of their input levels. The samplers are followed by delay lines and Fast Fourier Transform machines giving 256 complex spectral outputs over a bandwidth of 16 MHz for each of the two sidebands and two polarizations for each antenna. The dynamic range of FFT for intermodulation is expected to be about 35 dB.

The signals from the FFT machines of the two side-bands are fed to two separate multiplier and accumulator (MAC) systems, since correlation between the sidebands is not required. Each of the MAC systems forms all the cross products between the 256 channels in each of the polarizations of the 30 antennas giving a maximum of $(30 \times 31/2) \times 256 \times 2 = 238{,}080$ RR and LL products including self-products. The MAC system supports a polarization mode which gives the RR, LL, RL and LR products but with half the frequency resolution (128 frequency channels). Each of these products can be averaged in an accumulator for integration times selectable from 64 ms to 10 s. The correlator system uses about 1500 VLSI chips developed by the National Radio Astronomy Observatory for the Very Long Baseline Array (VLBA).

5. System Parameters

The expected system parameters of GMRT are summarised in Table 1. The antenna efficiency and system temperatures have been measured for several completed GMRT antennas and these have been found to be close to those given in Table 1. The measured antenna efficiency including cable losses is about 55 per cent at 327 MHz and 38 per cent at 1420 MHz. Further measurements are being made to verify various system parameters.

6. Astronomical Objectives

The design of GMRT has been motivated by a need to set up a powerful facility in radio astronomy to investigate a variety of astrophysical problems that are either exclusive to the metrewave region or which will complement the existing facilities at

TABLE 1. Some estimated system parameters for GMRT

	Frequency (MHz)					
	38	150	233	327	610	1420
Primary beam (deg)	13	3.1	2.0	1.4	0.8	0.32
Synthesized beam :						
Total array (arcsec)	80	20	13	9	5	2
Central compact array (arcmin)	28	7	4.5	3.2	1.7	0.7
System temperature (K), Total T_{sys}	10,280	580	250	110	100	70
RMS noise in image (μJy)†	1,420	55	24	11	10	12

†For assumed bandwidth of 16 MHz, integration time of 10 h, and natural weighting and 2 polarizations

shorter wavelengths. As already mentioned, one of the important scientific aim of GMRT is to search for the 21-cm line from primordial clouds of neutral hydrogen in the redshift range of about 3 to 8, in an attempt to determine the epoch of galaxy formation. Such a discovery would be of fundamental importance to our understanding of the formation of structure in the Universe. Attempts will be made also to determine the variation of HI content of clusters of galaxies with redshift by observing in the frequency range of 1000-1430 MHz.

Pulsars would form another very important topic of study with GMRT. Apart from detailed studies of individual pulses for elucidating pulsar emission mechanisms, GMRT would be a powerful instrument for undertaking a search for hundreds of new pulsars. The optimum frequency for carrying out such searches is 327 MHz but searches towards the galactic plane may be carried out also at 600 and 1400 MHz. Particularly important would be searches for pulsars with milli-second periods, those in binary systems and in globular clusters. Timing of milli-second pulsars can be an effective probe of the theories of gravitation and also of internal structure of the neutron stars. Observations of pulsar scintillations can provide valuable information about the interstellar medium.

Apart from the above-mentioned observational programmes, the special features and capabilities of GMRT will make it a powerful tool for undertaking a variety of other astrophysical studies, a few of which are mentioned below.

- HI studies of our Galaxy and of external galaxies out to cosmological distances through emission as well as absorption.
- Relic radio sources and ageing effects from observations of the steep-spectrum diffuse emission associated with the population of old relativistic electrons.
- Non-thermal halos around spiral galaxies and in some rich clusters of galaxies.
- Large-scale deep surveys of the sky at metre wavelengths, to catalogue millions of radio sources for a variety of statistical and cosmological studies.
- Search for the Deuterium line at 327 MHz which is of considerable cosmological interest.
- Studies of a variety of objects in our Galaxy and nearby galaxies such as HII regions, supernova remnants, transient radio sources, etc.
- Study of recombination lines from our Galaxy as well as from external galaxies.
- High-time-resolution studies of solar radio bursts and studies of solar wind through interplanetary scintillations.

7. Conclusion

GMRT will provide high sensitivity, good spectral and polarization capability, excellent sky coverage and reasonably good angular resolution at six different frequency bands from 38 MHz to 1430 Hz. Its sensitivity will be comparable to that of the VLA at 21 cm but is expected to be about 4 to 8 times better at 327 MHz. Further, it will be operating at several decimetre and metrewave bands. GMRT's sensitivity will be comparable to that of the Arecibo Radio Telescope, but with much higher angular resolution and declination coverage. GMRT is likely to be completed by late 1996. It will be used for investigating a variety of astrophysical problems which are best studied at the longer radio wavelengths.

8. Acknowledgements

Successful near completion of the challenging project of GMRT is due to the sustained and dedicated efforts of a large team of engineers and scientists of the Tata Institute of Fundamental Research. The engineering teams have been led by S.C. Tapde, M.K. Bhaskaran, N.V. Nagarathnam, M.R. Sankararaman, T.L. Venkatasubramani, A. Praveen Kumar, R. Balasubramanian, G. Sankarasubramanian and A. Dutta. V. Balasubramanian and A.J. Selvanayagam of RAC, Ooty, have also made valuable contributions. The Raman Research Institute is building the 21-cm front-end receiver system and a major part of the pulsar back-end. The engineering design of the 45-m dishes has been done by M/s. Tata Consulting Engineers. The project has also benefitted from discussions and contributions by many other scientists in India and abroad.

REFERENCES

KILDAL, PER-SIMON AND SKYTTEMYR, S.A. 1982, *IEEE Trans. on A & P*, **AP 30**, 529.

SWARUP, G. 1990, *Indian J. Radio & Space Physics*, **19**, 493.

SWARUP, G., ANANTHAKRISHNAN, S., KAPAHI, V.K., RAO, A.P., SUBRAHMANYA, C.R. & KULKARNI, V.K. 1991, *Current Science*, **60**, 95.

THOMPSON, A.R., MORAN, J.M. & SWENSON, G.W. JR. 1995, *Interferometry and Synthesis in Radio Astronomy*, John Wiley & Sons, New York.

Status and Plans for the Effelsberg 100-m telescope

By W. REICH

Max-Planck-Institut für Radioastronomie, Radioteleskop Effelsberg,
D-53902 Bad Münstereifel-Effelsberg, Germany

The Effelsberg 100-m radio telescope has been in operation since 1971. Its performance has been improved all the time. Today the telescope is in routine operation in the frequency range from 1.3 GHz up to 43 GHz. Successful VLBI observations have been made even at 86 GHz, where its collecting area is comparable with that of the largest mm-radio telescopes. Highly stable cooled HEMT amplifiers, a number of multi-feed systems, adequate backends and adequate computing facilities are the basis for single-dish observations with unprecedented sensitivity and angular resolution.

However, during the last years severe problems with the track and azimuth-drives have showed up and require a major repair. This takes place in spring 1996. For the next years an improvement of the aperture efficiency at high frequencies is expected from a replacement of the outer low-quality panels of the telescope.

---

1. Introduction

The Effelsberg 100-m telescope of the Max-Planck-Institut für Radioastronomie, Bonn, is located in a valley of the Eifel mountains at about 40 km to the west of Bonn. It was commissioned in 1971 and was designed for observations in the frequency range from 0.4 GHz up to 10 GHz, but soon it became clear that useful observations could be made at much higher frequencies. The telescope is a multi-purpose instrument for continuum, polarization, line, VLBI and pulsar observations. A large number of receivers and backends is available to meet the special requirements of the different observational needs. The telescope is open to the astronomical community and a large fraction of observing time is allocated for outside observers. Even after 25 years of operation the demand for observing time with the 100-m telescope exceeds substantially the available telescope time.

2. Operation

The 100-m telescope is located at an altitude of 350 m in the Eifel mountains, which is not an ideal site for a telescope operating in the mm-wavelength range. To make optimal use of clear nights we run outside of VLBI sessions two schedules in parallel: "Pian A" for high-frequency and "Plan B" for low-frequency observations, respectively. A decision for the night is made in the late afternoon, based on the weather prediction, by the observer scheduled for high-frequency observations. Subsequently, the selected receiving system is installed, checked and the corresponding on-line computer set up; the backend needed for the scheduled observations is activated by the telescope operator.

Observations with the 100-m telescope are possible either from the prime focus (PFC) or the secondary focus (SFC). For PFC observations a single receiver box needs to be manually positioned, while the SFC receivers can be selected by a computer command. For this flexibility and to make use of the large focal plane a number of new receiving systems have been installed in the SFC.

Focus	Frequency [GHz]	HPBW [']	Aperture Efficiency [%]
SFC	4.85	2.40	55
SFC	10.5	1.15	42
SFC	14.7	0.85	35
PFC	23.0	0.55	24
SFC	32.0	0.45	23
PFC	43.0	0.40	16

TABLE 1. HPBW and aperture efficiencies (diameter 100 m) for selected receiving systems at the Effelsberg 100-m telescope

3. Aperture efficiency

The telescope surface consists of aluminium panels for the inner diameter of 79.4 m. The outer area has one ring of perforated aluminium panels and three rings of wire-mesh panels. In particular the wire-mesh panels have a substantially lower surface accuracy compared to the panels of the inner area, where after a holographic adjustment in 1984 a surface accuracy of about 0.4 mm (rms) for the elevation range between 10° and 50° has been achieved. The surface accuracy decreases to about 0.6 mm (rms) at an elevation of 80°. The measured aperture efficiencies (full diameter) and the angular resolution are listed for a number of presently available receivers in Table 1. There is a steady decrease of the effective area with increasing frequency as is expected and the elevation dependence of the telescope gain increases. However, the available collecting area is larger than for any other fully steerable telescope in the world.

Observations in the high-frequency range are regularly scheduled. In particular studies of the various ammonia transitions, SO line observations at 30 GHz or 43 GHz observations of SiO lines should be mentioned. Polarized emission from the complex Sgr A region has been detected for the first time at 32 GHz (Reich, 1994). Wielebinski et al. (1993) have reported on pulsar observations up to 35 GHz, which are the highest frequencies where pulsars have been observed so far.

The telescope is meanwhile regularly used at 86 GHz for VLBI observations. Although the gain of the telescope depends strongly on elevation, in its optimum range around 30° elevation and illuminated to 80 m (taper −14 dB) a beamwidth of about 10'' has been measured and its aperture efficiency (for 100-m diameter) is about 8% (D. Graham, 1996, priv. comm.). This collecting area is comparable with that of the largest mm-radio telescopes.

4. Multi-feed systems

A list of available receivers at the Effelsberg telescope can be found on the World Wide Web: http://www.mpifr-bonn.mpg.de. Almost all of them are our own developments.

Of particular importance are the multi-feed systems for their ability to speed up mapping observations and to reject the influence of weather effects. These systems are all installed in the SFC cabin. Table 2 lists some of their main parameters. These systems are used for sensitive continuum-mapping observations and all except the 14.7-GHz system have in addition IF-polarimeters for each feed for sensitive observations of linearly-polarized emission. The 14.7 GHz system is rotatable to compensate for the variation of the parallactic angle during an observation. All multi-feed systems rely on the application of "software beam switching" (Morsi and Reich, 1986), which requires extremely stable receivers. The 10.55-GHz system was the first multi-feed system where two total-power

Frequency [GHz]	Bandwidth [MHz]	Feeds	Channels	HPBW [']
4.85	500	2	8	2.40
10.45	300	4	16	1.15
14.7	1000	4	4	0.85
32.0	2000	3 (+6)	12 (+24)	0.45

TABLE 2. Multi-feed systems installed at the SFC of the Effelsberg 100-m telescope. The 14.7-GHz 2×2 feed system at the central position is exchangeable with other systems, as for instance six additional 32-GHz feeds.

receivers with cooled HEMT-amplifiers are connected to each feed's left- and right-hand circularly-polarized component. The system and the data reduction have been described in detail by Schmidt et al. (1993). Reich (1995) has shown that the system is confusion limited (0.08 mJy) rather than stability limited by the total-power receivers. The data reduction is based on the restoration method described by Emerson et al. (1979), where the antenna pattern of the different feeds is required to be rather similar. Observations have shown that this is in fact the case (Schmidt et al., 1993).

5. Backends

A number of modern backends are available to cope with the present observational demands. Continuum measurements are normally made with a simultaneous registration of linear polarization by a broad or narrow-band IF-polarimeter. Up to 80 channels can be recorded by a digital continuum backend. A 1024-channel autocorrelator is available for line observations and a new 8096-channel autocorrelator will be added soon. VLBI MK-IIIA and VLBA type observations are scheduled for about 30% of the available telescope time. Pulsar observers use a four-channel backend connectable to every receiver. They have available in addition a dedicated polarimeter, a pulsar de-disperser or a 480-MHz filterband for a pulsar search at high frequencies.

6. On-line computer

Meanwhile the fourth generation of on-line computer hardware is installed at the 100-m telescope. The control software is basically the same at Effelsberg, the 30-m IRAM telescope and the 10-m Heinrich-Hertz telescope on Mt. Graham. At Effelsberg there are now two VAX 4000-505A in operation, each with 128 Mb RAM. In particular we have set up a configuration that ensures that in the case of a hardware failure there is redundancy, and that observations are continued with a minimum loss of time.

For post processing the raw data are automatically converted and stored at UNIX-based SUN workstations. The Effelsberg site is connected with its home institute at Bonn by a fast data link.

7. Problems

The first fracture of the telescope track happened in winter 1987 and two more failures occurred since then. It turned out that the grouting material was damaged and the 3200-t weight of the telescope caused local depressions of 1 mm or more. In addition a number of broken wheel bearings of the azimuth drives have been noticed, which produce velocity-, temperature- and time-dependent hysteresis-like effects and reduce the pointing accuracy substantially. In February 1996 the telescope was taken out of operation to replace the azimuth rail by a new one with a slightly modified cross-section. All wheel bearings will be exchanged. Calculations have shown that a tilt of the wheels by 54' should improve the driving performance of the telescope.

8. Next steps

A finite-element study of the effect of wind pressure on the telescope structure by the VERTEX company has shown that this load was overestimated at the time when the telescope was constructed. This result is important in view of the replacement of the outer damaged wire-mesh panels. Because the wind pressure is less severe we plan for an exchange for perforated aluminium panels similar to those already installed for one ring. These panels will have an rms surface accuracy of the order of 0.5 mm. After an adjustment of the panels based on the results of holographic measurements the aperture efficiency is expected to be improved, in particular for the frequency range between 10 GHz and 40 GHz.

Other activities at Effelsberg are related to improving the pointing and to correcting for focal changes due to temperature effects. Four inclinometers and a number of temperature sensors have been installed. Recently we have successfully tested an infra-red ranging system (accuracy 0.1 mm) aimed at improving the focussing.

I am grateful to Richard Porcas for his support.

REFERENCES

EMERSON, D. T., KLEIN, U. & HASLAM, C. G. T. 1979. *A&A* **76**, 92

MORSI, H. M. & REICH W. 1986. *A&A* **163**, 313

REICH, W. 1994. In *The Nuclei of Normal Galaxies* (ed. R. Genzel & A. I. Harris). Kluwer Academic Publisher, pp. 55–62

REICH, W. 1995. *Multi-feed systems for radio telescopes* (ed. D. T. Emerson & J. M. Payne). ASP Conference Series, vol. 75, pp. 171–178

SCHMIDT, A., WONGSOWIJOTO, S., LOCHNER, O., REICH, W., REICH, P., FÜRST, E. & WIELEBINSKI, R. 1993. *Kleinheubacher Berichte (URSI)* **36**, pp. 99–108.

WIELEBINSKI, R., JESSNER, A., KRAMER, M. & GIL, J.A. 1993. *A&A* **272**, L13

Renovating the Nançay radio telescope: the FORT project

By W. VAN DRIEL[1], J. PEZZANI[1] AND E. GERARD[2]

[1] Unité Scientifique de Nançay, Observatoire de Paris, 92195 Meudon, France

[2] ARPEGES, Observatoire de Paris, 92195 Meudon, France

The Nançay decimetric radio telescope is a transit instrument with the equivalent collecting area of a 94-m diameter round dish. Its 'hoghorn' feed system is being replaced by an optimized system consisting of corrugated horns in a dual-reflector offset Gregorian mirror configuration. By mid-1998 it will thus have an average system temperature of 25 K in the wavelength range of 8.5–30 cm (including the CH, OH and H I spectral lines), providing a 6 times better observing time efficiency at 18/21-cm wavelength and even a ten-fold gain at 9-cm wavelength.

We intend to define a number of scientific long-term Key Projects for the renovated telescope. At present we are considering a number of themes based on ongoing projects at the Paris Observatory (comets, evolved stars, pulsars and large-scale structures in the Universe), but a final choice will undoubtedly be made only after an international call for proposals.

1. The Nançay decimetric radio telescope – its present status

The 30-year old Nançay decimetric radio telescope is actually still the third largest instrument operating at decimetric wavelengths (9–21 cm), with the equivalent collecting area of a 94-m round dish. It is a transit telescope of the Kraus/Ohio State design, with a very high focal (f/D) ratio, consisting of a fixed spherical mirror, 300 m long and 35 m high, a tiltable flat mirror (200×40 m), and a focal carriage moving along a 90-m long curved rail track, which allows the tracking of a source on the celestial equator for about 1 hour. Its southern declination limit is $-39°$.

The telescope became operational in 1965, and is operated by the Paris Observatory, with financial support from the French National Scientific Research Council (CNRS) and the Région Centre, where the Nançay Radio Observatory is located on a terrain owned by the Ecole Normale Supérieure – in the rural heartland of France, 200 km due south of Paris.

Nançay was recently measured to be still one of the most 'radio-quiet' radio astronomy sites, in terms of unwanted interference, partially due to the extended, radio-wave-absorbing forest surrounding it. In our fight against radio interference, we have enclosed a number of buildings in Faraday cages, installed an antenna above tree-top level monitoring the radio spectrum, and started developing on- and off-line hardware and software for the elimination of parasitic line emission (Weber, 1996).

The present focal system of the telescope consists of three 'Hoghorn' feeds operating at wavelengths of 9, 18 and 21 cm, allowing observations in the CH, OH and H I spectral lines. These feeds, of vintage 1970s design, are no longer adequate in terms of efficiency and noise temperature, since the efficiency in the vertical polarisation is about 25% less than in the horizontal one, and there is a significant contribution to the system temperature from the ground and the vegetation surrounding the highly elongated telescope reflector.

2. Renovating the telescope – the FORT project

Though the telescope has a large collecting surface, its focal system is no longer state-of-the-art, as mentioned above, but preliminary studies made during the late 1980s showed it could be improved considerably, by about a factor 2 in sensitivity. Thus, in a review of French astronomy policy a few years ago, the decision was made to equip the Nançay radio telescope with a new, optimized focal system. This decision was motivated by the scientific return in the form of long-term research projects that can be carried out with the renovated telescope (see hereafter), for the moderate cost of the renovation (about 3 million US Dollars, co-financed by the Paris Observatory, the CNRS and the Région Centre.).

The renovation was dubbed the FORT project, a French acronym for 'Optimized Focal System for the Radio Telescope' – pronounced as 'for' in English, it is a name with a twist, as it also means 'strong' in French.

The aims of the renovation are:

1. to reduce the system temperature to an average of 25 K in the required wavelength range of 8.5 to 30 cm – implying a six-fold gain in observing time efficiency at 18/21-cm wavelength (OH and H I lines) and even a ten-fold gain at 9 cm (CH line),

2. to equalize the efficiency in the horizontal and vertical polarisations, and to increase the overall efficiency by about 20%, while allowing observations of all four Stokes parameters, and

3. to allow observations with about equal efficiency in the 8.5–30-cm wavelength range.

The actual 'Hoghorn' feeds will be replaced by an optimized, dual-reflector offset Gregorian configuration (see Figure 1), using two corrugated horns, one for shorter wavelengths (8.5–19 cm) and the other for longer wavelengths (15–30 cm). The performance specifications of the new focal system design were validated, both in amplitude and in phase, through laboratory tests of a 1:20 scale model.

The Division of Radiophysics of CSIRO in Australia has designed and is constructing the Gregorian subreflector system and corrugated horns for the optimized focal system, while GIAT Industries of France will construct the focal carriage and the new 90-m-long rail track; the carriage will measure about 10 m on each side, and is expected to be installed on the Nançay site for testing in March 1997.

The duration of the FORT project is expected to be 3 years, ending mid-1998. The actual focal system will remain in operation as long as possible for scientific observations; the telescope will be shut down intermittently over a period not exceeding 6 months during 1997 and 1998, with a maximum contiguous stop of 4 months.

3. Future scientific key programs

We intend to define a number of scientific Key Projects for the renovated Nançay decimetric radio telescope, which will be awarded a considerable fraction of the telescope time. The radio observations are expected to be complemented by observations at other wavelengths and by high-resolution radio synthesis observations. The Key Projects are expected to start by the end of 1998 and to last for a few years, at least.

A final choice has not been made yet, and will undoubtedly be made only after an international call for proposals. At present, we are considering the following themes, based on ongoing long-term international projects with the Nançay radio telescope: Large-scale structures in the Universe, Evolved stars, Millisecond pulsars and Comets.

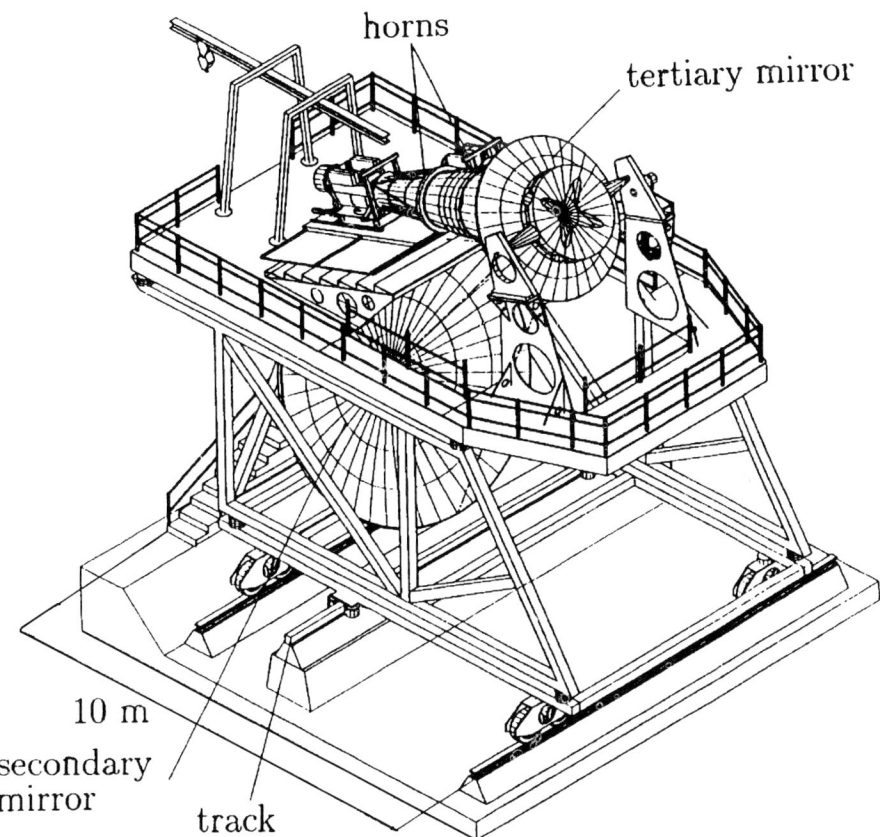

FIGURE 1. The new, optimized focal system for the Nançay decimetric radio telescope: a CAD representation of the dual-reflector offset Gregorian reflectors and the two corrugated horns mounted in the focal carriage on its rail track – scale: the largest mirror will be about 5 metres in diameter and the overall dimensions of the carriage will be about 10 m on each side (Courtesy M. André, SERT, Paris Observatory).

1. Large-scale structures in the Universe

Goal: observe 10 to 20,000 spiral galaxies in the 21-cm H I line in about 5 years, sampling the Universe out to a distance of about 200 million light years.

For studies of the large-scale mass distribution in the Universe and the Hubble constant, we plan to observe large galaxy samples in the 21-cm H I line, to provide a more accurate and more bias-free determination of the Tully-Fisher relation, and to use this relation to determine galaxy distances. The galaxies could be selected, for example, from among the 500,000 objects expected to be detected in the DENIS near-infrared southern hemisphere survey presently being performed at ESO.

2. Millisecond Pulsars

Goals: Detection of new pulsars, and frequent observations (once every few days) of a network of known pulsars – both timing and flux density measurements can be made; participation in the World Pulsar Array.

Pulsars, with their extreme long-term rotational stability ($\geq 10^{-14}$ s/s), can be used to study the stability of the atomic time scale and general relativity, as well as to search for background gravitational waves and extra-solar planets.

3. Evolved stars

Goal: observe the circumstellar envelopes of a thousand evolved stars in the 18-cm OH lines, combined with high-resolution interferometric studies (with EVN, MERLIN) to analyse the spatial maser structure.

For a study of their gaseous envelopes (important mass loss, dust and gas clouds, intense OH line masers), the physics of interstellar masers (pumping, variability, polarisation), as well as of the kinematics (expansion) and dynamics (acceleration) of the envelopes, the structure of the magnetic field, stellar evolution scenarios, and the spatial distribution of evolved stars.

4. Comets

Goals: to monitor faint, periodic comets in the 18-cm OH lines, and to detect more 'new' comets from the Kuiper belt.

For a study of the water production rate and the onset of sublimation in comets, water expansion velocity and anisotropic outgassing, the magnetic field (Zeeman effect) and the OH hyperfine line ratio. So far, 53 comets have been observed at Nançay, 5 of which in 1995 alone – i.e., two thirds of all comets ever detected in the 18-cm OH lines.

Other possible interesting long-term projects well-suited for the transit-type Nançay radio telescope could be, for example:

- monitoring radio counterparts of X-ray and γ bursts in X-ray binaries,
- magnetic field measurements in the H I and OH lines of a variety of objects
- blind H I line surveys – search for, e.g., gas-rich low surface brightness galaxies,
- surveys in the 9-cm CH line of Galactic objects and nearby galaxies,
- VLBI observations with ground based or space radio observatories,
- search for extraterrestrial civilisations for the HRMS/Phoenix (ex-SETI) program,

and ... whatever else our dynamic, and often unpredictable, science will come up with.

The Unité Scientifique de Nançay (the Nançay Radio Observatory) of the Observatoire de Paris is associated to the French Centre National de Recherche Scientifique (CNRS) as USR B7040, and receives financial support of the Conseil Régional of the Région Centre in France.

REFERENCES

GRANET, C. 1995. Ph.D. thesis, Université d'Orléans, France

GRANET, C., JAMES, G. L., PEZZANI, J. 1995. In *Proc. JINA (Journées Internat. de Nice sur les Antennes, 8-10 Nov. 1994)*

VAN DRIEL, W. 1996. In *Proc. XVth Moriond Astrophysics Conf. - Clustering in the Universe* (eds. C. Balkowski et al.), in press. Editions Frontières.

WEBER, R. 1996. Ph.D. thesis, Université de Paris XI Orsay, France

High sensitivity and high dynamic range observations with the WSRT

By A. G. de BRUYN[1,2]

[1]NFRA, Postbus 2, 7990 AA, Dwingeloo, The Netherlands

[2]Kapteyn Astronomical Institute, Postbus 800, 9700 AV, Groningen

In order to fully exploit the steadily improving thermal noise levels in synthesis array data we have to improve the dynamic range. Various atmospheric and instrumental errors start affecting images when the data have a potential dynamic range exceeding 10,000:1. In the case of the WSRT at 21cm the total intensity images are limited, on-axis, by closure errors. Off-axis a whole range of effects become important. Polarization images are affected by closure errors and variable polarization leakage terms. We describe the procedures developed to diagnose and remove such errors with data on the 22-Jy sources 3C84 and 3C147. We may expect to produce thermal noise-limited images with a dynamic range of 1,000,000:1.

1. Introduction

The thermal noise levels achievable in aperture synthesis imaging have steadily gone down over the years as a result of improvements in receiver technology, increasing bandwidth and increasing integration time. When observing very bright sources these low noise levels require a correspondingly high dynamic range.

Only 10-15 years ago a dynamic range (DR) of 10,000:1 was a record achievement (Noordam and de Bruyn, 1982). Since then much higher DR has been achieved at the WSRT, VLA and, recently, the VLBA. With the WSRT the technique of selfcalibration routinely is capable of reaching a DR of 100,000:1. Here we define DR as the ratio of peak brightness to rms noise. It should be noted, however, that the rms noise generally varies with location in the image and is generally higher near the brightest source.

There is a whole range of atmospheric and instrumental errors that limit the achievable dynamic range. These effects depend strongly on frequency and probably vary from one synthesis array to another. The discussion below reflects the situation in WSRT imaging. At low frequencies (e.g. 92 cm) time-variable ionospheric effects are the main obstacle to high dynamic range imaging because the wide fields of view invalidate the basic assumption of position-invariant phase distortions in selfcalibration (non-isoplanaticity). Strong background sources at the edge of the fields also cause instrumental off-axis effects (due to polarization effects and bandpass mismatches). At 21 cm the atmosphere (troposphere and ionosphere) is probably the least damaging and pointing and instrumental closure errors become the limiting factor. At higher frequencies, pointing and rapid atmospheric phase fluctuations become an additional limiting factor. Finally it should be noted that it has long been known that a serious DR limitation in imaging *extended* sources is set by the CLEAN deconvolution algorithm. Even if the data is good enough to image point sources at 10^5:1, the on-source DR in extended sources will rarely get better than 10^4:1.

The current thermal noise of the WSRT at 21 cm, in a 12-hour synthesis, is about 50 μJy which means that in a 6×12-hour synthesis a potential DR of 10^6:1 is available on a 22-Jy source. The science that can be done at these DR levels has hardly been explored. One application that motivates us to try to reach these levels is the study of Thomson-scattered haloes around bright radio sources in clusters and at high redshift (e.g. Syunyaev 1982). Because we are dealing with Thomson scattering these haloes should be highly polarized unless destroyed by Faraday depolarization. Such haloes carry a wealth of information on the evolution of radio sources, anisotropic emission, the intergalactic medium and intergalactic magnetic fields.

In this talk I will show you that we have every reason to expect to actually reach a level of 10^6:1 or more! The corrections to be applied to the data to reach these levels are, in fact, fairly straightforward. This gives us hope that in the future, when the WSRT continuum sensitivity will be improved by about a factor 5, we may still be 'thermal noise-limited', even around strong (point) sources.

2. Observations of 3C147 and 3C84

Observations of 3C147 and 3C84 were obtained in the winter of 1994/1995 at a wavelength of 21 cm. The continuum backend (O'Sullivan, 1984) was used with 8 bands of 5 MHz. A full 6×12 hour synthesis was made of 3C84, a radio source with very complex structure imaged with the WSRT many years ago (Sybring 1993; de Bruyn and Sybring 1996). The array was reconfigured once a week to provide continuous $u-v$ coverage with 12-metre baseline increment from 36 to 2760 metres. This pushed the grating response to a radius of 60′, enabling us to synthesize the whole primary beam. The quasar 3C147, a compact source with a flux density of 22 Jy, i.e. equally bright as 3C84, was observed for only 1×12 hour, mainly to calibrate instrumental behaviour on a 'point source', and to check the image fidelity as a function of distance from the source. Because the haloes that we are interested in are believed to extend over more than 10′ the observations were done on winter nights, to avoid contamination by, highly polarized, signals from the sun on the shortest baselines.

2.1. The WSRT

The WSRT continuum backend provides 87 interferometers for the parallel-hand correlations and 40 interferometers for the cross-hand correlations. From these signals we form the four Stokes parameters. The earlier 21-cm observations were made with the dipoles oriented in position angles 0° and 90°. The sum and difference of the parallel hand correlations then yield Stokes I and Q, while the combination of the cross-hand signals yields Stokes U and V (see e.g. Hamaker et al, 1996; Sault et al., 1996). In the new observations we rotated all dipoles by 45° to swap the signals from Q and U. By doing this we hoped to make an (instrumentally) independent confirmation of a large-scale polarization signal in the Perseus cluster that we had seen in our earlier observations.

2.2. Errors in parallel-hand signals

A range of small, but very localized, instrumental errors is obvious in the data (e.g. 'instrumental ghosts'). Although they are small (at a level below 0.01%) they were clearly limiting the dynamic range locally. In the Perseus cluster data the radio sources 3C83.1 (NGC1265) and 3C83.1A cause serious problems because of their strength and location at the 5dB point of the primary beam. These problems, however, appear to be mainly due to pointing errors and therefore only affect (to first order) the Stokes I image. Such errors do not exist in the 3C147 field, which is free from strong background sources. The dominant errors in the 3C147 Stokes I and U images are, in fact, simple closure errors. These errors appear to be constant in time and could be eliminated by solving for constant, multiplicative, gain and phase errors following a telescope-based selfcalibration. About 60% of the interferometers have closure errors <0.03% (in gain) and <0.01° (in phase). Figure 1 gives an impression of the magnitude of these closure errors for the standard set of 40 baselines (i.e. excluding the redundant baselines). There is only a

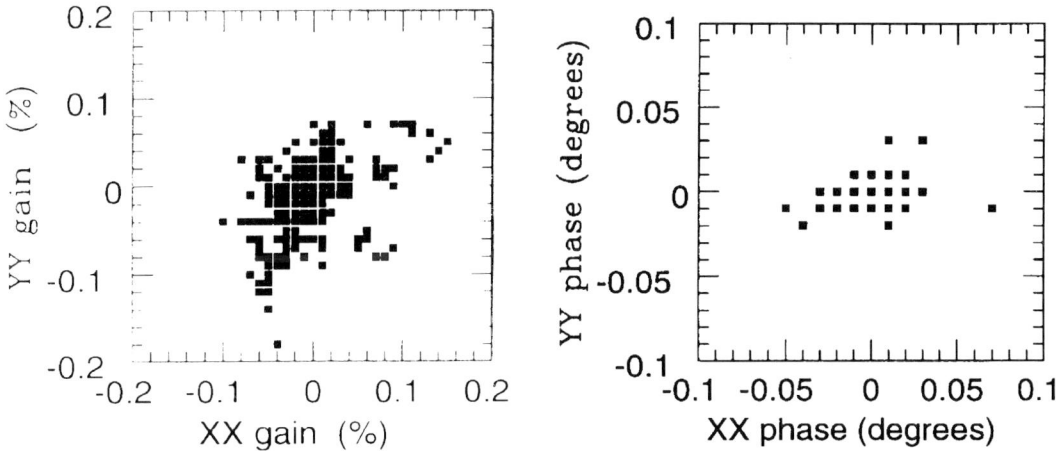

FIGURE 1. Amplitude (left) and phase (right) closure errors for the standard 40 interferometers of the WSRT in the 3C147 dataset: the XX errors are plotted against the corresponding YY values. All 8 bands are plotted but due to truncation at the 0.01% level many points fall on top of each other, most of them close to the zero point.

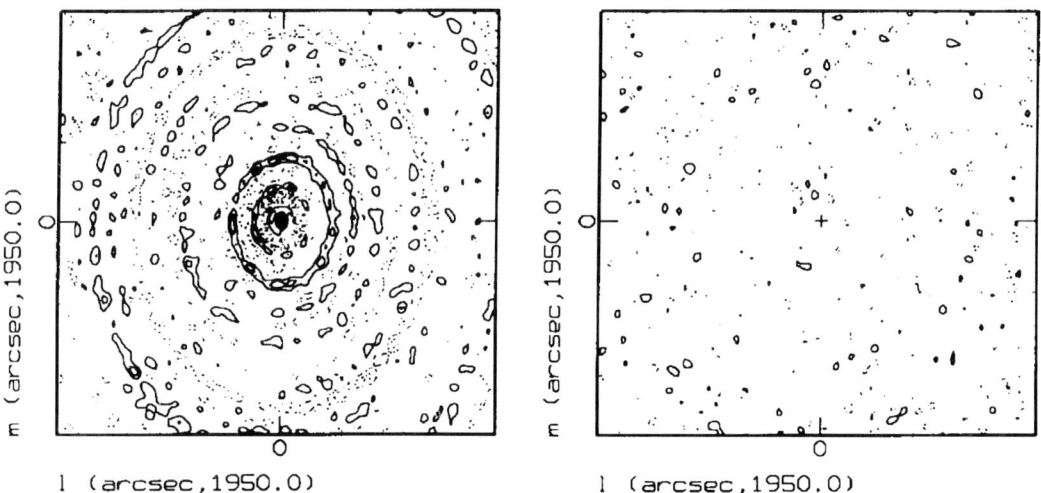

FIGURE 2. Left: Stokes U image of 3C147 with ringlike errors resulting from stable complex closure errors. Contours are in steps of ± 100 μJy (2σ). Right: The U image after removal of these errors.

small degree of correlation between the two orthogonal polarizations and the errors in the Stokes I and U images are therefore only weakly correlated. On the other hand, the errors in the various frequency bands are highly correlated. The closure errors cause elliptical structures around the brightest source in the image. Figure 2 shows the Stokes U image before and after correcting for these closure errors. Figure 3 shows the I image after correcting for closure errors. Both the U and I images, after closure corrections, have formal dynamic ranges of 22 Jy/50μJy or 440,000:1.

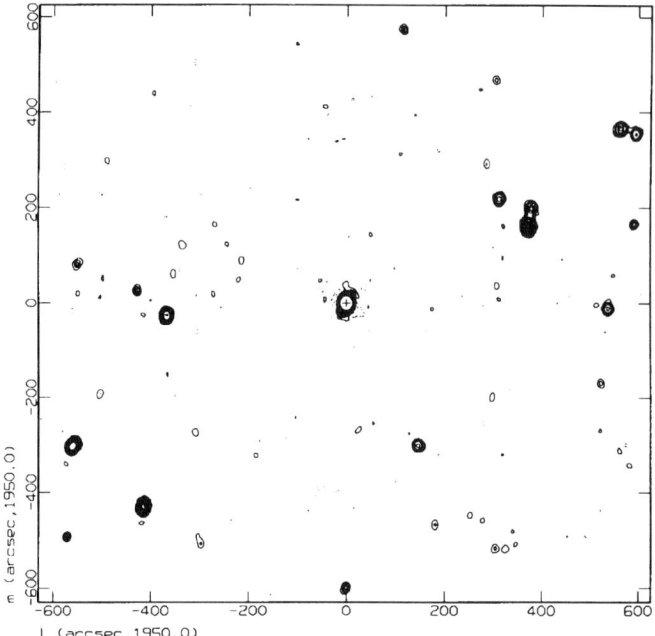

FIGURE 3. Stokes I image of 3C147 after correcting for complex closure errors. The central source was 22 Jy. The lowest contours are at ± 200 µJy, or 0.001% of the central source.

2.3. Errors in cross-hand signals

Following the standard polarization calibration (e.g. Sault et al 1996) the Stokes Q and V images were dominated by errors at a level of 0.05% of the peak brightness in Stokes I. After studying these images, and the leakage terms causing them, it was deduced that the dominant error was caused by small (0.05°–0.15°) but systematic rotation of the telescope feeds during a 12-hour observation. This rotation was different for each telescope, but repeated very well in the six 12-hour observations. The frontends equipped with the heavier cooled receivers showed the largest and most systematic rotation. We therefore solved for the dipole orientation and feed ellipticity as a function of time (in 1-hour intervals) using the source itself. This is probably best described as a polarization selfcal. Here we made the assumption that the central source is unpolarized (an assumption probably good to a level of 0.1%). This led to a spectacular improvement of the Stokes Q and V images which now reached the thermal noise level of 50 µJy per beam everywhere in the synthesized field.

3. Results

The data of 3C147 have been fully reduced. The total intensity image shows that 3C147 may have some very weak extensions in position angle –20°. The Stokes U and Q images, when smoothed to a resolution of about 1', revealed low level but highly significant polarization in a few regions well separated from 3C147. No unambiguous localization of these structures could be made due to be fact that we have only a single 12h synthesis which is self-confused.

We have not yet finished the reduction of the 3C84 data. The effects of the strong off-axis emission of NGC1265 (which is both intrinsically and instrumentally polarized) need to be properly taken care of in the total intensity and polarization selfcalibration. However, we are confident that we will also reach the thermal noise (at 22 μJy) in the polarization of this dataset. This implies a DR of 10^6:1. We can already confirm the detection of large scale polarization structure in the Perseus cluster. Its brightness is typically only 20–30 μJy per 15″ beam. Possible explanations for this emission are Thomson scattering, from the bright radio sources embedded in the cluster, or highly polarized diffuse synchrotron emission from our galaxy.

4. Conclusions

We have shown that, with some simple tools, a DR of up to 1,000,000:1 can be reached in data obtained with the WSRT continuum backend at 21 cm. However, we should hasten to add that currently this can only be reached when the dominant source is a point source and is located near the centre of the field. The standard polarization dynamic range can be increased enormously, to reach thermal noise levels, by using the central source and solving for time-variable leakage terms (polarization selfcal). Closure errors are the current limit to reliable imaging. Their stability in time gives hope that for many applications they will not present a fundamental limit.

REFERENCES

DE BRUYN, A.G. & SYBRING, D. 1996. submitted to A & A.
HAMAKER, J.P., BREGMAN, J.D. & SAULT, R.J. 1996. *A&AS* **117**, 137
NOORDAM, J.E. & DE BRUYN, A.G. 1982. *Nature.* **299**, 597
O'SULLIVAN, J.D. 1984. In *Indirect Imaging*, Cambridge University Press. **299**, 597
SAULT, R.J., HAMAKER, J.P. & BREGMAN, J.D. 1996. *A&AS* **117**, 149
SYBRING, D. 1993. Ph.D thesis, State University of Groningen.
SYUNYAEV, R.A. 1982. *Sov. Astron.Lett.* **8**, 175

MERLIN Phase 3

By R. J. COHEN

University of Manchester, Nuffield Radio Astronomy Laboratories, Jodrell Bank, Macclesfield, Cheshire SK11 9DL, U.K.

The current status of the MERLIN array is summarised. MERLIN uniquely supplies the baseline coverage between the VLA and VLBI Networks. A development plan is underway to enhance the frequency coverage, frequency flexibility and mapping fidelity of the instrument, and to provide better integration with other arrays. Funded developments include improvements to the frequency flexibility of three telescopes, and a new frequency band 12.0–15.4 GHz. Highest on the list of unfunded developments is an upgrade to, or replacement of, the Defford telescope. In the longer term a large capital investment will be needed to increase the bandwidth and the sensitivity of the array. The sensitivity envisaged is about 5 microJy per beam. This will give an astrometric precision better than 0.1 milliarcsec. MERLIN Phase 3 will have the angular resolution and sensitivity to resolve thermal sources such as PN and HII regions in external galaxies.

1. Introduction

MERLIN is a long-baseline-interferometer with baselines which uniquely fill the gap between the VLA and VLBI networks. The instrument came into operation in 1980 (Davies, Anderson & Morison 1980), and underwent a major upgrade ten years later. The performance of the original instrument has been reviewed by Thomasson (1986), and the Phase 2 upgrade was outlined by Wilkinson (1993). The present paper summarises current status of the Phase 2 instrument after four years of operation, and outlines plans for a Phase 3 development.

2. Current Status

The MERLIN network has baselines from 6 to 218 km, which overlap the longest VLA baselines and the shortest VLBI baselines. It currently has five frequency bands: 150–152 MHz, 406–410 MHz, 1.4–1.7 GHz, 4.6–5.1 GHz and 21–24 GHz. The corresponding angular resolution ranges from 1.4 arcsec at 151 MHz down to 8 mas at 24 GHz. At the most sensitive frequencies of 1.4 and 5 GHz the angular resolution (0.13 and 0.04 arcsec respectively) is well-matched to that of the HST. The instantaneous bandwidth is 32 MHz, and is usually configured as 16 MHz in each of left and right hand circular polarizations. Full polarization capabilities are provided for line and continuum observations. The instrument has the sensitivity to image thermal sources such as planetary nebulae, novae, hot stellar winds and compact HII regions (e.g papers by Bryce, Eyres, and Hughes, in these proceedings).

Observations are usually made by phase-referencing, except at 22–24 GHz where suitable calibrator sources are too few, given the array sensitivity. The astrometric precision achieved is currently 2 mas, and a target of 1 mas seems realistic for the near future (Garrington, these proceedings). Several flexible observing modes have been implemented. MERLIN is a sparse array, and a full 12-hour track is normally required to achieve good baseline coverage. If the full sensitivity of a 12-hour track is not essential then multi-source tracks can be used to image several sources at once. This involves cycling around the different targets on a timescale of about half an hour. The technique of multi-frequency synthesis (MFS) is routinely used to generate new baselines for continuum measurements by observing at several frequencies across an observing band. Frequency-switching is also used to observe several different spectral lines during a single track, and bandpass-switching can be used to provide narrow-band measurements on the target source but broad-band measurements on the phase-calibrator source.

MERLIN has already achieved most of the scientific goals set out in the original Phase 2 proposal. Particular highlights in stellar astronomy include the first direct measurements of the expansion of Nova Cygni 92 (Pavelin et al. 1993), measurement of proper motions of 44 pulsars and the subsequent deduction of their high birth velocities (Harrison, Lyne & Anderson 1993; Lyne & Lorimer 1994), and imaging of thermal and non-thermal emission from the nebula around HMSge and comparison with HST data (Eyres et al. 1995). Extragalactic highlights include the imaging of more than 20 young supernova remnants in the M82 starburst (Muxlow et al. 1994), high resolution mapping of neutral hydrogen absorption against the complex radio nucleus of the Seyfert NGC4151 (Mundell et al. 1995), imaging of the jet of 3C273 and comparison with HST images (Bahcall et al. 1995), and the ongoing search for gravitational lenses, which has so far yielded 10 new discoveries (e.g. Patnaik et al. 1993).

3. Frequency flexibility

The funding of MERLIN as a national facility includes modest support for developments of the array. The top priority is to improve the frequency flexibility. At the present time only one frequency is available at a time, and it may be years before a particular band is scheduled again. This restricts the science that can be done. We have therefore begun a programme of upgrades to the MERLIN telescopes. The newest telescope, the 32-m dish at Cambridge, is already frequency-flexible. It has a suite of receivers mounted on a carousel at the secondary focus. By rotating the carousel any receiver can be brought on axis and into operation on a timescale of one minute. We have begun a programme to install similar carousels on the three E-systems telescopes. The telescope at Pickmere was upgraded during summer of 1995, and similar upgrades are scheduled for the Darnhall and Knockin telescopes during summer 1996. The work involves a new vertex cabin which protrudes into the bowl, the rotating carousel, and strengthening of the vertex cabin floor. The secondary focus will be used for frequencies of 1.4 GHz and above. Frequencies below 1 GHz will require prime-focus operation and it will be some time before they can be included in the frequency-flexibility plan.

With four frequency-flexible telescopes we will be able to offer the first dual-frequency session in Semester 1997B (October 1997). This will be at 5 and 22 GHz. The Defford telescope does not operate above 5 GHz, so it will be possible to switch between a six telescope array at 5 GHz and a five telescope array at 22 GHz on a timescale of one day (the time to change over the home station Mark II telescope). Design work is proceeding on ways to upgrade the Mark II itself. The Defford telescope is the oldest in MERLIN, and will require substantial modification or more likely complete replacement to give it similar frequency coverage and flexibility to the other five telescopes. This could not be achieved within the current funding profile, and will require a separate grant application.

In order to pay for these developments it will be necessary to decommission the Mark III telescope at Wardle, which will cease operation in March 1996. Wardle was built as a dedicated telescope for long baseline interferometry (Palmer & Rowson 1968), and was in a sense the first MERLIN telescope. However it has the slowest drive speeds and it does not operate above 1.7 GHz. Furthermore it is on a site which has become an industrial estate entirely unsuitable for a radio telescope. The closure of the Wardle telescope will not have a major impact on the science programme.

4. New frequency band 12.0–15.4 GHz

MERLIN provides baselines intermediate between those of the VLA and VLBI, but in order to capitalize on this there must be matching frequency coverage to enable combined array observations. At present these can be done only in the bands 1.4–1.7, 4.6–5.1 and 21–24 GHz. Our long-term strategy is to provide matching frequency coverage in all bands where there is a strong scientific interest.

The first new frequency band we will develop for MERLIN is 12.0–15.4 GHz, matching the VLBA receivers. This will provide a sensitivity close to that at 5 GHz, and three times better than that currently achieved at 22 GHz. The angular resolution will be 13 mas. The band will be particularly valuable for stellar research and other high sensitivity projects. For example it will be possible to image novae within about one month of outburst at 15 GHz, compared with three months at 5 GHz. The receivers will also cover the important methanol maser line at 12.2 GHz, as well as lines from rotationally excited OH and formaldehyde at 13.6 and 14.5 GHz respectively. In the first instance only a five-telescope array will be available. The 15 GHz receiver will be mounted in the new carousels, so joint high frequency sessions at 5, 15 and 22 GHz may available by 1998/9.

5. Future development plans

A frequency-flexible replacement for the Defford telescope has already been mentioned as highly desirable. In fact it is the most pressing development not yet funded. Considerable ingenuity has been expended in attempts to obtain a replacement dish at low cost, for example by acquiring a satellite communications dish which is no longer needed for that work. These plans foundered for lack of capital. However funds for the project are likely to become available before the end of the century. With a six-telescope array of frequency-flexible telescopes operating to 31 GHz and 43 GHz, MERLIN could make a major impact.

The drive to higher sensitivity continues. At this meeting we have heard of the European Mark IV tape-recorder system, which will be the first 1-GHz VLBI system in the world, and will be ready next year. We have also heard plans for similar bandwidths on other VLBI systems, and on the VLA. Clearly MERLIN must join this race if its baselines are to provide data of suitable sensitivity to combine with data from the other arrays. We are looking at ways of achieving this technically.

At the present time the radio astronomy bands from the MERLIN outstation telescopes are relayed to Jodrell Bank over microwave links for real-time correlation. The microwave link system could be upgraded to provide instantaneous bandwidths of order 100 MHz, but it is doubtful if 1 GHz of bandwidth could ever be achieved this way. If we wish to retain real-time correlation then the only suitable link at the present time is optical fibre. There is no way we could pay the commercial rate for a 1-GHz fibre link, but there are possibilities for collaborative research and development which we are exploring. A single fibre could carry not only the radio astronomy band but also the timing signals to keep the array coherent, and the site monitoring including TV pictures.

An available bandwidth of 1 GHz will have profound implications for the way MERLIN operates. All observations will be multi-channel as at present, but each baseline will provide a wider range of spatial frequencies, and MFS will be achieved by default. This will greatly improve the continuum mapping fidelity, and the possibilities for rejecting interference. The sensitivity of the system will be of order 5 microJy/beam. It will thus be possible to use much fainter phase-reference sources. Phase-referencing will be possible at all frequencies, and at most frequencies there will be several suitable sources within one degree of the target. The astrometric precision will increase to better than 0.1 mas. Many spectral lines of methanol, silicon monoxide and other molecules will be accessible simultaneously, and a flexible correlator will be needed to deal with them, for example simultaneously mapping different SiO maser transitions around red supergiants while imaging the stellar continuum.

The rms noise level for continuum measurements with a 1 GHz band corresponds to a brightness temperature of 50K. Thus targets could include solar system objects for the first time with MERLIN. Particularly spectacular advances are anticipated in imaging hot cores, jets and winds in star-forming regions. The angular resolution at 43 GHz will be 4 mas, which will enable MERLIN to image Hii regions and planetary nebulae in nearby galaxies in the same way it currently images non-thermal sources (SNRs).

The name MERLIN first appeared two years after the instrument was commissioned (Norris, Diamond & Booth 1982). The acronym Multi Element Radio Linked Interferometer Network may need modification to reflect any move from radio links (RL) to optical fibres (OF or FO?). Likewise the MERLIN logo which took some years to produce depicts a telescope which is soon to be decommissioned. Should we retain it for historical interest? Should we enshrine the University of Manchester's own logo within the MERLIN logo? These are difficult questions. In view of the long gestation period I urge all MERLIN users to start thinking now about a new name and a new logo for MERLIN Phase 3.

MERLIN has always been a team effort. It is a pleasure to thank the many Jodrell Bank colleagues and MERLIN users who have helped to form the views expressed here.

REFERENCES

BAHCALL, J. N., KIRHAKOS, S., SCHNEIDER, D. P., DAVIS, R. J., MUXLOW, T. W. B., GARRINGTON, S. T., CONWAY, R. G. & UNWIN, S. C. 1995. *ApJ* **452**, L91

DAVIES, J. G., ANDERSON, B. & MORISON, I. 1980. *Nature* **288** 64

EYRES, S. P. S., KENNY, H. T., COHEN, R. J., LLOYD, H. M., DOUGHERTY, S. M., DAVIS, R. J. & BODE, M. F. 1995. *MNRAS* **274** 317

HARRISON, P. A., LYNE, A. G. & ANDERSON, B. 1993. *MNRAS* **261** 113

LYNE, A. G. & LORIMER, D. R. 1994. *Nature* **369** 127

MUNDELL, C. G., PEDLAR, A., BAUM, S. A., O'DEA, C. P., GALLIMORE, J. F. & BRINKS, E. 1995. *MNRAS* **272** 355

MUXLOW, T. W. B., PEDLAR, A., WILKINSON, P. N., AXON, D. J., SANDERS, E. M. & DE BRUYN, A. G. 1994. *MNRAS* **266** 455

NORRIS, R. P., DIAMOND, P. J. & BOOTH, R. S. 1982. *Nature* **299** 131

PALMER, H. P. & ROWSON, B. 1968. *Nature* **217** 21

PATNAIK, A. R., BROWNE, I. W. A., KING, L. J., MUXLOW, T. W. B., WALSH, D. & WILKINSON, P. N. 1993. *MNRAS* **261** 435

PAVELIN, P., DAVIS, R. J., MORRISON, L. V., BODE, M. F. & IVISON, R. J. 1993. *Nature* **363** 424

THOMASSON, P. 1986. *Q. Jl. Roy. Astr. Soc.* **27** 413-431.

WILKINSON, P. 1993. MERLIN – Phase 2 In *Sub-arcsecond Radio Astronomy* (ed. R. J. Davis & R. S. Booth), pp. 422-424. Cambridge University Press, Cambridge, U.K.

The VLA upgrade project

By R. A. PERLEY

National Radio Astronomy Observatory, Socorro, NM, USA

The National Radio Astronomy Observatory has formally organized the Very Large Array Upgrade Project. The initial goal of the project is to prepare a plan by mid-1997 for a comprehensive upgrade of the Very Large Array. The plan will propose extensive improvements in sensitivity, frequency coverage, correlator capability, spectral resolution and dynamic range, instantaneous bandwidth, spatial resolution, and imaging fidelity of objects on all angular scales. The upgrade represents an opportunity for outstanding scientific return for a very modest investment.

1. Introduction

The Very Large Array was formally dedicated in October, 1980. At that time, the VLA was a state-of-the-art instrument, providing an enormous improvement in resolution, sensitivity, flexibility, speed, spectral capacity, and image fidelity over existing instruments. These advantages remain with the VLA to this day – there is no instrument like it, and there is none to surpass it under construction or in planning – it will remain the pre-eminent cm-wave radio telescope in the world well into the next century.

However, the VLA was designed with the technology of the 1970s and 1980s. Although the instrument at that time incorporated the best that could then be offered, great advances in technical design and capacity have occurred since then which, if implemented on the VLA, could vastly improve its performance. Improvements of over a factor of ten in sensitivity, spatial resolution, and spectral resolution can now easily be obtained at very modest cost, using these new technologies.

Ideally, improvements of the type which will be studied for the upgrade would have been implemented more-or-less continuously as part of instrumental support budgets. However, the VLA, like many (and probably most) scientific instruments, has suffered considerably from insufficient infrastructure support and thus has been hard pressed to extend its capability since its dedication. This is not to say that no improvements have been made. Improvements in sensitivity have been made in three bands (21, 2, and 1.3cm), two new bands (3.6cm and 90cm) have been installed, while two other bands (7mm and 400cm) are partially installed. With regret I note that the great majority of the cost of these new bands has been borne by other agencies than the National Science Foundation.

Because of the very slow pace of improvements, we have now reached a point where it seems best to regard the upgrade as a project, rather than as a list of incremental improvements. There is an advantage in this – the many diverse elements of the plan can be viewed as parts of a cohesive unit. In addition, parts of the upgrade too large to regard as incremental units can be proposed as part of a single package.

The VLA has enormous untapped potential, which will be unlocked by an upgrade. It is necessary to emphasize that virtually all current observing on the VLA is limited by the instrument in:
- Sensitivity
- Spectral resolution, flexibility, or agility
- Frequency availability
- Spatial resolution (both on large or small scales)

All of these limitations can be greatly reduced – in most cases by factors greater than ten – by the improvements to be proposed in the upgrade plan. And because the upgrade will combine elements of both the VLA and the VLBA, it is best to think of the resulting facility as a new instrument, whose power will considerably exceed that of either the VLA or VLBA.

2. Key Elements of the VLA Upgrade Project

The VLA Upgrade Project falls naturally into two phases. Phase A deals with improvements located at the VLA site itself, while Phase B concerns resolution enhancement using both VLBA antennas and new antennas sited at intermediate distances between the VLA and VLBA. These two phases are not necessarily to be considered as sequential. I discuss each phase in turn.

2.1. Phase A

There are a large number of improvements and enhancements contained under this phase of the project. In summary:
- Sensitivity Improvements. The sensitivity of the array will be increased by a factor of 2.5 to 35 by a combination of three improvements:
  ○ Retrofits of existing receivers to utilize 'VLBA-style' systems in which the cooled amplifier and polarizer for each band are located at the base of the horn. Also, replacement of older noisier amplifiers with new devices. And, completion of the Q-band system (from the current 13 to a full complement of 28 receivers).
  ○ Increasing the instantaneous bandwidth from the current 200 MHz to 2 GHz.
  ○ Improving antenna efficiency through panel adjustments using holographic measurements. This programme is already well advanced, with the 43-GHz efficiency now doubled on the adjusted antennas.
- Increasing the Tuning Bandwidth. The new receivers will each have a full bandwidth ratio (defined as the highest tunable frequency divided by the lowest) of at least 1.5:1, and perhaps as wide as 2.0:1.
- Adding New Bands. The current VLA has a limited frequency coverage. The upgrade will propose adding two or three new bands which when combined with the new wide-bandwidth receivers described above would vastly increase the frequency coverage. When combined with the larger subreflector described in the next section, continuous frequency coverage from below 1 GHz through 50 GHz is feasible. The new bands would be 'S' band (2.2 to 3.4 GHz), and 'K_a' band (26.5 to 40 GHz), and possibly an upper 'L' band (1.5 to 2.2 GHz).
- Extend L-band down to 1.0 GHz. There is great scientific interest in redshifted HI. The VLA upgrade will propose to obtain optimum performance in the range 1.0 to 1.3 GHz by installation of a new, high efficiency feed, plus possible replacement of the current subreflector with one which is approximately 25% larger. The larger subreflector has the useful spinoff advantage of permitting more feed horns to be installed around the secondary focus feed ring, thus permitting continuous frequency coverage. A 50% larger subreflector option will also be explored.
- Expansion of 'P' Band. The upgrade will consider options for obtaining continuous coverage from ∼ 250 MHz to the lower edge of the improved L-band system through use of a number of deployable scalar feeds, or perhaps a phased array located on the secondary focus feed ring.

Wavelength (cm)	VLA now	VLA upgrade
50	X	20
30	X	5.3
20	6	2.7
11	X	1.9
6	6.7	1.3
3.6	4.4	1.3
2.0	20	1.7
1.3	37	2.6
0.9	X	2.0
0.7	75	3.0

TABLE 1. VLA Upgrade Project: Point Source Sensitivity

- Fibre-Optic Transmission System. To conduct the 2 GHz-wide signal from the antennas to the correlator, the current waveguide-based IF transmission system will be replaced with a modern fibre-optic system.
- An Expanded Correlator. A new correlator, providing up to 8192 channels, will provide vastly increased flexibility for spectral line observing. This machine will be matched with a new IF system which would provide up to 16 independent channels from one or two bands.
- A Compact ('E') Configuration. To permit fast, accurate observations of extended low-surface-brightness emission, a new, compact 'E'-configuration will be considered.

Table 1 shows the current and expected sensitivity of the VLA. The quoted values are 1 sigma in microJy, assuming full bandwidth, and 12 hours integration. Full bandwidth is assumed to mean 2.0 GHz for the six highest frequency bands, 1.0 GHz at 11cm, 500 MHz at 21cm, and 100 MHz at the 50-cm band.

2.2. Phase B

This phase of the project would increase the resolution of the array by a factor of about eight. This will be done by adding the two closest VLBA antennas (at Pie Town and Los Alamos) to the array, plus adding four to six new, fixed antennas at distances intermediate between the VLA's maximum scale (35 km) and the VLBA's minimum scale of approximately 400 km. This new proposed arrangement has come to be known as the 'A+' configuration. Signals from these six to eight antennas would be conducted to the new VLA correlator using fibre-optic lines. For the closer antennas, we envision a dedicated fibre which we would either own, or lease from the local telephone company. For further antennas, we must use commercial fibre. Discussion with the telephone companies on the technical challenges are now beginning. The new antennas would be part of both the VLA and the VLBA. Their use would be determined by the configuration the VLA is in, and by the project. A fairly obvious division of their use would be to have the new antennas, and the innermost VLBA antennas be part of the VLA during its A-configuration when the project specifically needs resolution on 300-km baselines. At all other times (B, C, and D-configurations, and when the A-configuration project does not need baselines over 35 km), these new antennas would be part of the VLBA, thus greatly increasing its sensitivity and imaging capability.

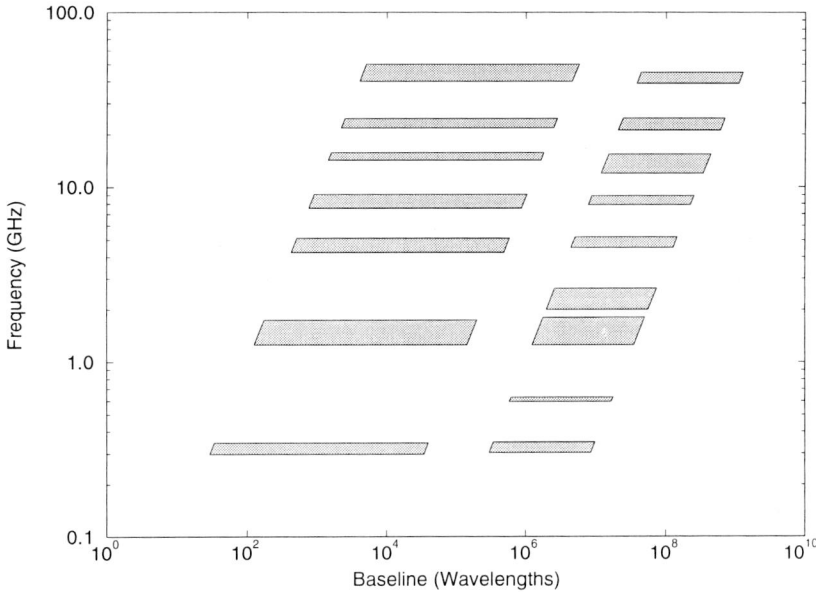

FIGURE 1. The frequency-resolution coverage of the VLA and the VLBA. The important features are the large gaps, both in frequency and in resolution. Most important for imaging is the large slanted gap separating the VLA from the VLBA. This gap lies in the critical range of approximately 2 to 0.02 arcseconds resolution.

3. Comments on Aspects of the Development Plan

3.1. Coverage of the Spatial-Frequency Plane

The flexibility of an instrument can be defined in many ways. Obvious ones are the speed of imaging, or the speed at which frequencies can be changed. In these aspects, the VLA is clearly very flexible. However, another, and equally important indicator of flexibility is the region in the resolution-frequency plane that a telescope can access. Figure 1 shows the current coverage of the VLA and VLBA in this plane. Notable are both the large gaps in frequency, and the well-known 'VLA-VLBA' gap in resolution. Both classes of gap present serious limitations to science accessible with these instruments:

• Frequency gaps. It is clearly preferable to provide continuous frequency coverage. Studies of molecular and atomic line-forming regions greatly benefit by the availability of multiple transitions. Figure 2 shows the distribution of astrophysically important molecular spectral transitions, and demonstrates the usefulness of wide frequency coverage. Although nearly all molecular transitions occur at frequencies above 10 GHz, a powerful argument for extending continuous frequency coverage below that comes from studies of lines from moderate to highly redshifted objects. Continuous, or nearly-continuous frequency coverage will also benefit spectral studies, and Faraday rotation studies.

• Resolution Gaps. The slanting gap between the VLA and VLBA is due to the complete lack of spacings between 35 and ∼200 km in these instruments. This 'VLA-VLBA' gap is a serious impediment to scientific studies of objects in the crucial resolution

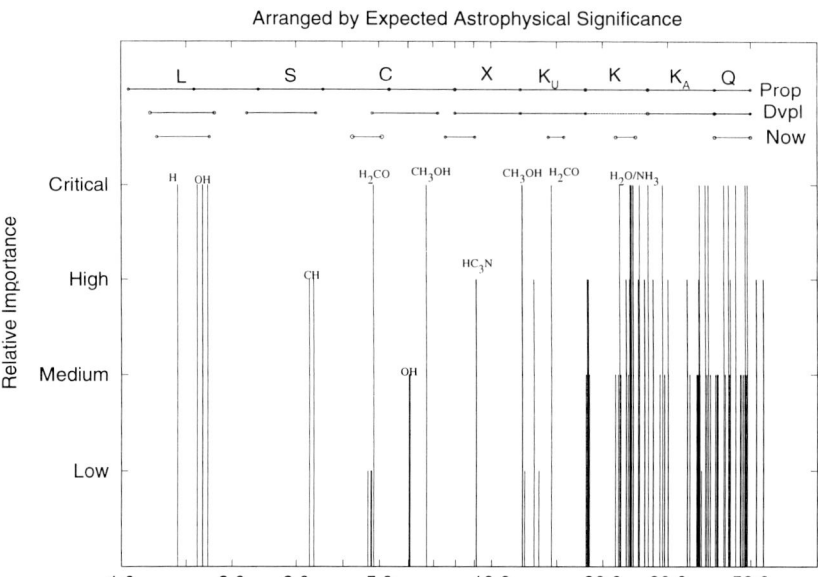

FIGURE 2. A representation of important molecular transitions between 1 and 50 GHz. The height of the lines conveys the expected astrophysical significance of that transition. Across the top of the figure are displayed the frequency coverage of the current VLA, a modified version of the proposal in the 'VLA Development Plan', and that described in this report which gives continuous frequency coverage

range of of 20 to 2000 mas. Although MERLIN does cover this gap, and is a fine instrument, it does not have the sensitivity, speed, or imaging fidelity of an expanded VLA which incorporates the current 27 antennas with 6 to 8 additional antennas distributed around the VLA. MERLIN has, however, clearly demonstrated the richness of the science available in this resolution range. The expanded VLA would provide the same resolution as MERLIN, but with many times the sensitivity and imaging fidelity, as well as the already existing advantages in frequency flexibility and speed.

Figure 3 demonstrates the coverage in the frequency-resolution plane given by an upgraded VLA which incorporates the existing 27 antennas, the two closest VLBA antennas, and 4 new antennas situated in New Mexico and Arizona. Looking ahead to the day when the entire VLBA can be made to be 'on-line', the VLA and VLBA could potentially become a single instrument, resulting in the frequency-resolution coverage displayed in Figure 4. Note that for this 'ultimate' combined operation, there will be possible constant resolution over an extremely wide frequency range – a most valuable technical capacity absent in all current instruments.

3.2. Low-Frequency issues

There are compelling scientific reasons to improve the VLA's performance at frequencies below 1.3 GHz, which is about the current lower limit for L-band. However, it is difficult to provide optimum performance below that frequency. A prime-focus system would

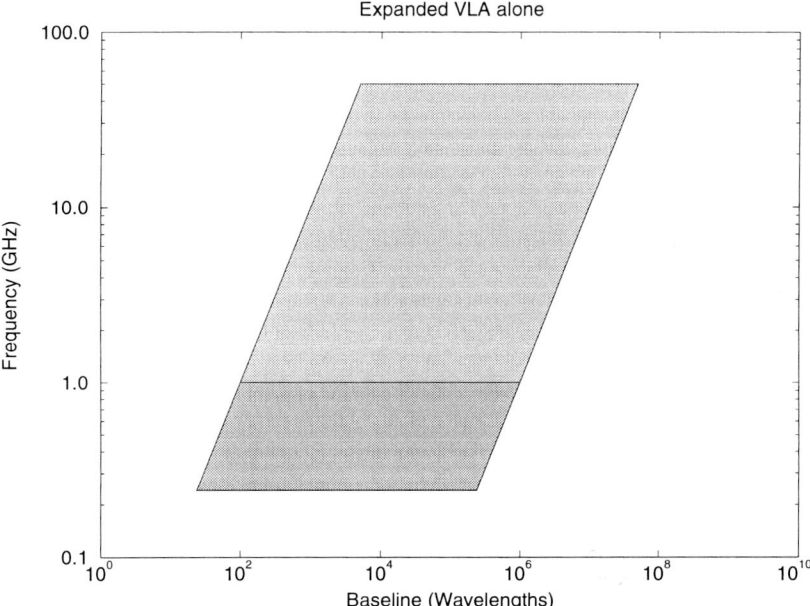

FIGURE 3. The frequency-resolution coverage of the VLA with the complete upgrade. The new antennas in the A+ configuration would extend the resolution coverage by one order of magnitude, while the new wideband receivers and new frequency bands would fill in all the frequency gaps. The horizontal line separates the regime of Cassegrain operation (above) from prime-focus operation.

need to be cooled, thus adding much weight at the focus area, degrading the pointing stability of the antennas. And, because the primary-secondary system is shaped, the primary reflector is not a paraboloid – there is thus no single defined focus with the result that a prime focus feed will be less efficient than a secondary focus system. A feed at the secondary focus is preferable, but is not physically feasible due to the size of the subreflector. Efficient illumination of the current subreflector requires a feed of about eight wavelengths aperture. Providing such a feed for frequencies below 1.2 GHz would require elimination of one or more higher-frequency bands – most probably the S-band (2.7 GHz) system. This is a trade-off we wish to avoid. A better, although more expensive, solution is to replace the subreflector with one perhaps 25% larger. This will permit both the low-frequency limit for Cassegrain operation to be lowered by a similar factor (to approximately 1.0 GHz), and also will permit ten frequency bands and continuous frequency coverage from 1.0 to 50 GHz. A more radical idea is to increase the subreflector to the maximum size defined by the unpanelled area of the primary reflector – an increase by approximately a factor of 1.5. This would permit a further lowering of the lowest frequency practical for Cassegrain operation. However, it would also require modification of the quadruped feed legs. Issues such as these are currently being examined. However, increasing the subreflector size considerably complicates solutions to the problem of obtaining frequency coverage below the limit where operation from the secondary focus becomes impractical. Below this limit (about 1.2 GHz with the current

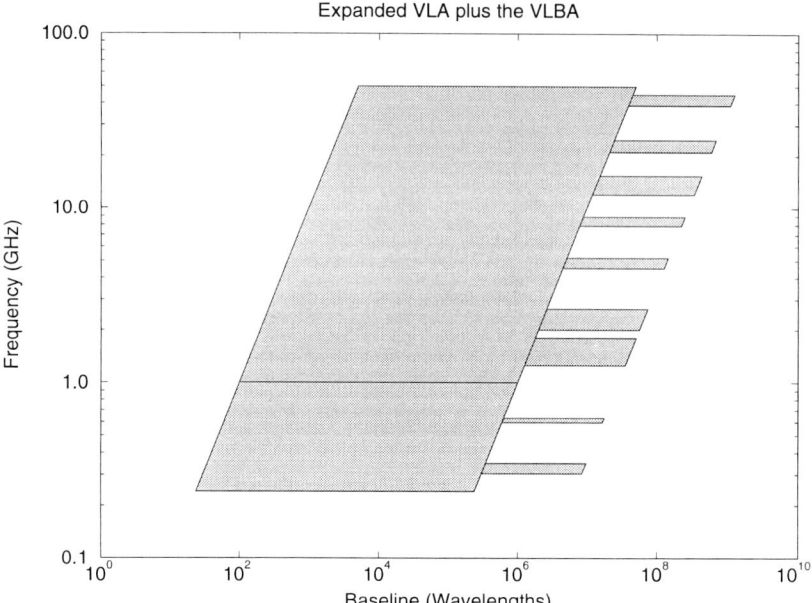

FIGURE 4. The frequency-resolution coverage of the VLA when the full VLBA is added. Over seven orders of magnitude in resolution is accessible with such a combination. Although this complete combination is not part of the VLA Upgrade proposal, we must not rule out its possibility with a decade or so.

subreflector, and about 1.0 GHz for a modestly enlarged subreflector), the only reasonable way is to employ a primary focus system. Since the focus lies behind the subreflector, and the subreflector cannot be withdrawn to expose this focus, effective operation in the range from 400 MHz to the Cassegrain lower limit will require major structure changes to either, or both the quadruped legs or the subreflector. And the larger the subreflector, the more difficult it will be to remove it for prime focus operation. These are difficult issues, requiring trade-offs and compromises to be made. It is entirely possible that the disadvantages of the larger subreflector for higher-frequency operation, combined with the difficulty of installing a low-frequency system, will require a separate low-frequency array to be designed. Issues such as these will be debated over the coming months.

3.3. *Radio-Frequency Interference*

There can be no doubt about the worsening observing situation due to the increased use of frequency space for communications. Particularly difficult for radio astronomers are the satellite-ground links, and the rapidly increasing use of cellular telephones. At the VLA, we have begun an 'environmental survey', to monitor the 'damage' from the radio astronomers' point of view. The impact of these increased communications is easy to see, although they do not yet impact current VLA bands heavily. Figure 5 demonstrates an early result of this monitoring system.

But with our plans to have a wider-band system with nearly complete frequency coverage, the impact of this telecommunications traffic cannot fail to be large, especially in

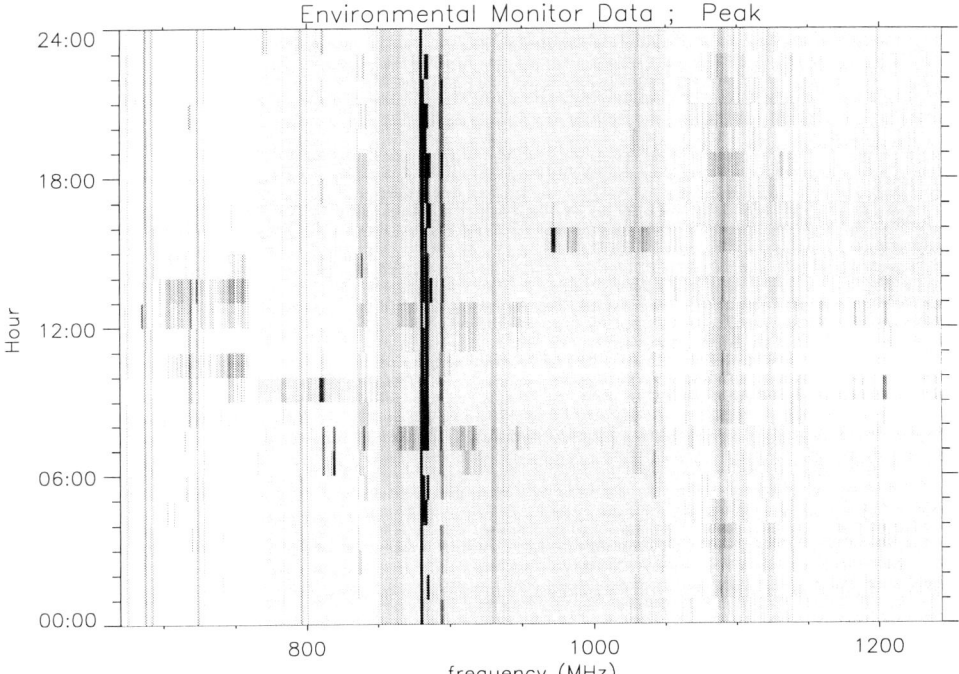
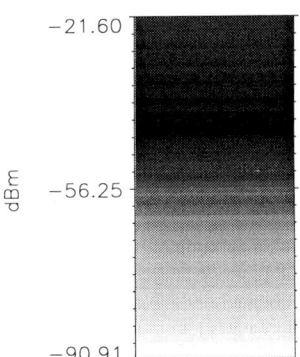

FIGURE 5. A grey-scale representation of the current RFI environment at the VLA site, from a low-sensitivity monitoring programme underway at the VLA. This plot shows the observed power spectrum from 675 through 1250 MHz, for one 24-hour period. The noise floor is near -90 dBm, and the greyscale saturates near -50 dBm. The prominent vertical bar near the centre is due to cellular telephones. The weaker grey bars between 680 and 805 MHz are from UHF television. The FAA air traffic control radars are at 1030 and 1090 MHz. The weaker vertical grey bar near the centre is due to pagers.

the lower frequency areas used by cellular phones, and in the frequency bands allocated for satellite links.

A pragmatic view is that it is hopeless to resist the increased use of these bands, and that we must learn to live with, within, and around RFI. Thus, the engineering challenge is to design robust systems which resist saturation, have multiple IF channelization to permit avoidance of particularly bad frequencies, permit flexible filtering to stop strong narrow-band signals, and employ multiple spectral channel correlators with apodization to permit removal of RFI which lies within the correlated band. And all of these must come with fast time-response.

4. VLA Upgrade Operational Issues

A major factor which favours an upgrade of the VLA, when compared to a new facility, is that the upgrade builds upon an existing facility, and thus utilizes the existing infrastructure. Operating an upgraded VLA will cost very little more than the cost of running the existing facility.

Indeed, the operational cost of running the 'Phase A' aspect of the plan may well be **less** than the current budget, since aging equipment, and its increased maintenance cost, will be replaced with modern equipment. There will clearly be an increased cost associated with maintaining the new antennas which are a major part of 'Phase B' – however we expect these costs to be modest as all these stations are within a few hours' drive of Socorro, which already houses the staff for the maintenance of the entire VLBA.

5. Costs of the VLA Upgrade

A major goal of the upgrade study is to produce a realistic estimate of the costs of such an upgrade. These studies are now in progress. However, I can at this time give an *approximate* guess of the costs of major parts of the upgrade. Bear in mind that not all the following items are likely to be implemented, and that 0 changes in some of them will certainly take place. The estimates below include labour costs.

Sensitivity Improvements to existing bands, including the instantaneous bandwidth increase	$4.5M
Adding Ka and S bands, plus completion of Q-band	$5.1M
Extending L-band to below 1.2 GHz	$3.3M
P-band improvements	$5.5M
New LO/IF transmission system	$10M
New Correlator	$7.5M
'E'-Configuraton	$6M
A+ Configuration	$6M per antenna (fully equipped)
Optical fibre connect	unknown

It will be seen that the complete costs of all proposed aspects of the upgrade will be in the neighborhood of $75M. In constant dollars, this is about half the cost of the original instrument.

The timescale for implementation of the upgrade plan is quite unknown. Current studies (described below) will go on for more than the next year. Funding for all, or even part, of the upgrade will be greatly influenced by the progress, and fate, of the pending MMA proposal. It seems unlikely that the NRAO could simultaneously handle both the MMA construction and the full-fledged version of a VLA Upgrade. An advantage of the upgrade is that its implementation could be drawn out in a way which minimizes overall impact both to observing and to the observatory's engineering and technical capabilities. However, in my view, this is not the preferred path to implementation of the upgrade. It remains, however, as a viable alternate path.

6. VLA Upgrade Activity

Current work towards the VLA Upgrade include the following:
- The project is now formally organized with a Project Scientist (Rick Perley), and Project Engineering Director (Dick Sramek).
- Approximately eight working groups are in place. Each has a defined engineering design role.
- An advisory panel, consisting of seven senior NRAO personnel, is in place, and meets bi-monthly.
- A VLA Upgrade memo series has begun.
- The initial goal of the VLA Upgrade is the preparation of a design study by mid-1997. This document will be constructed so as to permit a quick conversion to a formal proposal, should the situation merit it at that time.

At this time, the design study is severely hampered by two important problems – lack of money, and lack of manpower. The NRAO's current budget status is such that essentially no development money is available to fund special studies, or to test various proposals. And the same budget status prevents hiring individuals for the upgrade study, or even to replace key indivduals who leave. There is no foreseeable break in this gloomy situation. On the other hand, the NRAO is not devoid of talent or interest in development of the upgrade. Most of the skilled and talented individuals who designed, built, and tested the VLA are still with the observatory, and all of them have a great interest in seeing the instrument improved. We are assured that a reasonable fraction of their time will be devoted to assisting the preparation of the upgrade plan. Our situation is such that we must, for the foreseeable future, 'borrow' such resources, until better budget times come. The key is work to prepare a plan which can be implemented quickly when the time is right.

7. Summary

The VLA Upgrade must be viewed as far more than an incremental improvement to the current VLA's capability. Implementation of the upgrade plan will result in a new instrument, which combines both the VLA and VLBA, but whose power will far exceed either, or the sum, of its parents. Because the upgraded VLA would build upon a considerable existing infrastructure, the cost of running the upgraded instrument will be extremely small, when compared to the scientific return. It is even possible that the operational costs will be less than current. The combination of vastly increased observing capability with essentially no increased operational costs results in a scientific 'good deal' which, in my opinion, is better than any proposed upgrade, or new start, currently being discussed.

A high-sensitivity second-generation space VLBI mission

By J. S. ULVESTAD[1], L. I. GURVITS[2,3], R. P. LINFIELD[1]

[1] Jet Propulsion Laboratory, California Institute of Technology, Pasadena, CA 91109, USA

[2] Joint Institute for VLBI in Europe, P.O.Box 2, 7990 AA, Dwingeloo, The Netherlands

[3] Astro Space Center of P.N.Lebedev Physical Institute, Moscow, Russia

We are studying a concept for a second-generation Space VLBI mission, based on a 30-m inflatable antenna. This mission, named ARISE (Advanced Radio Interferometry between Space and Earth), will be a factor of 50–200 more sensitive in VLBI mode than VSOP at 5 and 22 GHz, and will have the ability to make VLBI observations at 43 and 86 GHz. It will also make 60-GHz single-dish observations to study the distribution of molecular oxygen in the Galaxy.

1. Toward Better Angular Resolution and Higher Sensitivity

Very Long Baseline Interferometry (VLBI) from the Earth has been employed for 30 years to achieve angular resolution of 1 milliarcsecond and finer. However, there are many astrophysical problems for which the milliarcsecond-scale resolution is not sufficient. The resolution of VLBI is limited to approximately λ/B, where λ is the wavelength of the observation and B is the size of the VLBI array. There are two possible paths to higher resolution. One is to reduce the observing wavelength, and the other is to increase the baseline length by placing a radio telescope in Earth orbit. These two paths are complementary: they may provide equivalent resolution at very different wavelengths, where different parts of a radio source (and different physical mechanisms) may be dominant.

The first successful Space VLBI observations were made using TDRSS (Tracking and Data Relay Satellite System) in the mid-1980s (Linfield et al. 1990 and references therein). These observations, at 2.3 and 15 GHz, detected VLBI fringes on projected baselines longer than two Earth diameters. The scientific results were interesting in their own right, but perhaps a more important result was the demonstration that the technique of VLBI could be extended to an orbiting element in a straightforward way, even if that orbiter did not have a highly accurate on-board clock.

From the late 1970s through the early 1990s, at least four different dedicated space VLBI missions have been considered. QUASAT (Schilizzi 1988) and the International VLBI Satellite (IVS, Pilbratt 1991) were proposed missions that did not receive final approval. The VLBI Space Observatory Programme (VSOP), led by the Institute of Space and Astronautical Science in Japan, is now proceeding toward a launch in September 1996 (Hirosawa & Hirabayashi 1995). The RadioAstron mission, led by the Astro Space Center in Russia, also was approved and is currently scheduled for launch in the late 1990s (Kardashev & Slysh 1988). These two missions are similar in character and in system sensitivity, with frequency coverage up to 22 GHz, 10-m-class radio telescopes with system temperatures near 100 K, and data-acquisition rates of 128 Megabit/s.

ARISE (Advanced Radio Interferometry between Space and Earth), a next-generation space VLBI mission currently being studied by NASA, would make significant advances in scientific capability over VSOP and RadioAstron (see Ulvestad et al. 1995a). The first advance is in frequency coverage, where ARISE would cover frequencies as high as 86 GHz. The second is in sensitivity, where a detection threshold of a few milliJansky

	VSOP(1996)	VLBA(2005)	ARISE(2005)	Gain: ARISE/VSOP
D	8 m	25 m	30 m	3.75
ϵ	41%	52%	60%	1.2
$T_{\rm sys}$	200 K	60 K	10 K	4.5
$\Delta\nu$	128 Mbit/s	8 Gbit/s	8 Gbit/s	8.0
Δt	150 s	100 s	150 s	1.0
7σ limit	450 mJy	7 mJy	3 mJy	162.3

TABLE 1. 22-GHz sensitivity on baseline to a VLBA telescope

Band	4.8 GHz	22 GHz	43 GHz	86 GHz
ARISE Diameter	30 m	30 m	30 m	30 m
ARISE Efficiency	0.6	0.55	0.4	0.1
ARISE $T_{\rm sys}$	8 K	10 K	20 K	40 K
VLBA Diameter	25 m	25 m	25 m	25 m
VLBA Efficiency	0.72	0.52	0.36	0.15
VLBA $T_{\rm sys}$	30 K	60 K	80 K	100 K
Data Rate	2 Gbit/s	8 Gbit/s	8 Gbit/s	8 Gbit/s
Coherence Time	650 s	150 s	60 s	15 s
7σ limit	1.4 mJy	2.1 mJy	8.6 mJy	85 mJy

TABLE 2. ARISE sensitivity on baseline to a VLBA antenna in 2005

or better is desired on a baseline between ARISE and a 25-m ground telescope. This sensitivity on space-ground baselines would be roughly two orders of magnitude better than for VSOP and RadioAstron. These sensitivity and frequency improvements would open up important new scientific areas that are accessible to neither VSOP and RadioAstron, nor to the ground-only VLBI performed at frequencies up to 86 GHz. If ARISE is approved, it is anticipated that it would be launched between 2005 and 2010.

The system parameters needed to provide two orders of magnitude improvement over VSOP at 22 GHz are shown in Table 1. Improvements are necessary in all areas, with most parameters pushed near their limits. It is possible that the effective integration time can be extended considerably by techniques currently being explored in millimetre VLBI observations from the ground. If this improvement is realized, it may compensate for any inability to reach the goals for the other observing parameters.

The overall sensitivity of a baseline between ARISE and a VLBA antenna at the anticipated observing frequencies of ARISE is summarized in Table 2. Note that in this table, as in Table 1, improvements in the VLBA systems, particularly in the sustainable data rate, are assumed. Comparison of the ARISE parameters to the VLBA parameters shows that, to first order, ARISE requires placing an antenna with VLBA-like properties in orbit. The lower system temperature possible for the orbiter would be balanced by the fact that the correlated flux density will be lower on space-ground baselines, with the net effect being that the space-ground baselines would have signal-to-noise ratios (SNRs) comparable to the ground-ground baselines.

2. ARISE Science

A wide variety of science could be done with ARISE. Much of this science is similar to that for the previously proposed IVS mission (Pilbratt 1991), although ARISE would make use of a much less massive antenna and spacecraft, and would have significantly better sensitivity. Here, we describe briefly the most important scientific goals that could be addressed by ARISE, but not by either the first-generation Space VLBI missions or by ground VLBI observations.

2.1. Continuum Sources

The combination of high sensitivity, high resolution, and high observing frequency would enable ARISE to image the bases of the VLBI jets in AGN radio cores where they are optically thin, rather than having observations constrained to the $\tau = 1$ surface. Within the inner parts of jets, ARISE at 22 GHz may be able to image sources with multiple resolution elements across the jets. This would show the magnetic field structure across the jets and also would help reduce beam depolarization, which seems to be an important limiting factor in ground VLBI observations.

The high sensitivity of ARISE would enable investigations of AGN with weak radio cores, including distant quasars, lobe-dominated radio galaxies, IRAS galaxies, and Seyfert galaxies. It would be possible to make statistical studies at the highest resolution of the characteristics of source samples which are free from bias by the effects of relativistic beaming.

There are several other classes of weak continuum VLBI sources that would be accessible to ARISE. Of particular interest are radio stars. Very few radio stars are strong enough to be seen with VSOP and RadioAstron even during outbursts. With ARISE, investigations of a number of radio stars in their quiescent states can be made.

2.2. Water Masers

In spectral-line mode, the improvements in antenna size and system temperature, and perhaps integration time, would enable a sensitivity improvement of a factor of about 20 over VSOP and RadioAstron. This sensitivity would permit access to many more extragalactic masers. In addition, the very accurate orbit knowledge achieved with a GPS receiver aboard ARISE (predicted to be 10 cm or better) would permit measurements of the distance to galaxies using maser spot motions. Distance measurements of galaxies as far away as 100 Mpc, well beyond the local deviations from the Hubble flow, could be made; these measurements would be important in tying down the base of the extragalactic distance scale.

2.3. 60-GHz Line Observations

The Earth's atmosphere is completely opaque to radiation in the vicinity of 60 GHz, due to absorption by a band of molecular oxygen lines. Therefore, observations in this frequency regime can only be made by a radio telescope in space. Single-dish observations using ARISE at 60 GHz would provide an important scientific addition to the mission. Dozens of star-forming regions in the Galaxy could be studied, with the distribution of temperature and molecular-oxygen density mapped throughout those regions. Models of the physics of star formation would receive an important boost from the opening of this new observational regime.

3. ARISE Mission Characteristics

The general characteristics of the ARISE mission are not finally determined, but certain parameters are well constrained. The spacecraft must be lightweight; ARISE would have a total mass of only 1.0–1.5 tons. The receivers must be cooled down to ~ 12 K at all observing frequencies. At this ambient temperature, the receiver contribution to the total system temperature can be as low as 0.2 K/GHz, only about four times the quantum limit. The leading candidate refrigeration system is a hydrogen-sorption cooler, under active development for a number of space applications.

The desired orbit for ARISE should be selected only after scientific results are available for VSOP and RadioAstron. Currently, we are assuming a nominal orbit with a 5,000-km perigee altitude and a 40,000-km apogee altitude as the baseline. The perigee height is fixed by the desire to have some overlap between ground-space and ground-only baselines. The apogee height will be a tradeoff behind the desires for image quality (implying a fairly low orbit) and high resolution (implying a fairly high orbit).

The key element of the project is a 30-m-class antenna. It must be reliably deployable, light weight, low cost, and have a surface accuracy of 0.35 mm or better for $\lambda/16$ performance at 43 GHz and $\lambda/10$ performance at 86 GHz. A number of candidate antennas for an advanced space VLBI have been discussed (Freeland 1991). Mechanical deployable antennas in the 30-m class with the required accuracy would be unacceptably expensive and far too massive. An affordable and feasible alternative for the near term is an inflatable antenna technology, as has been considered for the QUASAT project and currently is being developed by L'Garde, Inc. (Cassapakis and Thomas 1995, Ulvestad et al. 1995b). With the parameters mentioned above, the antenna could be fitted into < 3-m diameter launch fairing and would have a mass of < 300 kg.

The L'Garde antenna consists of three major elements: inflatable struts and a torus that provide structural support, and a 6-μm thick canopy which forms the reflecting surface. This reflecting surface will have its shape maintained by gas pressure of approximately 10^{-5} atmospheres. The inflatable antenna will be a paraboloid (probably off-axis), with the spacecraft (and the feeds) located at the prime focus. The inflatable antenna concept to be used for ARISE is scheduled for space testing as part of NASA's IN-STEP (In-Space Technology Experiment) program on-board a Space Shuttle flight scheduled for May 1996; this experiment is described by Veal and Freeland (1995).

REFERENCES

CASSAPAKIS, C. & THOMAS, M. 1995. in *AIAA 1995 Space Progr. & Techn. Conf.* **95–3738**

FREELAND, R. E. 1991. in *Technologies for Advanced Very Long Baseline Interferometry Missions in Space* (ed. G. S. Levy) *JPL D–8541* **3**, 33

HIROSAWA, H. & HIRABAYASHI, H. 1995. *IEEE AES Systems Magazine*, June 1995, 17

KARDASHEV, N. S. & SLYSH, V. I. 1988 in *Impact of VLBI on Astrophysics and Geophysics*, eds. M. J. Reid & J. M. Moran, Kluwer Acad. Press, 433

LINFIELD, R. P. ET AL. 1990. *ApJ* **358**, 350

PILBRATT, G. 1991. in *Radio Interferometry: Theory, Techniques, and Applications*, eds. T. J. Cornwell & R. A. Perley, ASP Conf. Series, **19**, 102

SCHILIZZI, R. T. 1988. in *The Impact of VLBI on Astrophysics and Geophysics*, eds. M. J. Reid & J. M. Moran, Kluwer, 441

ULVESTAD, J. S. ET AL. 1995a. in *33rd AIAA Aerospace Sci. Meeting* **95–0824**.

ULVESTAD, J. S. ET AL. 1995b. in *AIAA 1995 Space Progr. & Techn. Conf.* **95–3794**.

VEAL, G. & FREELAND, R. 1995. in *AIAA 1995 Space Progr. & Techn. Conf.* **95–3739**.

Giga-bit VLBI storage system for the next generation

By JUNICHI NAKAJIMA[1], HTOSHI KIUCHI[1],
YOSHIHIRO CHIKADA[2], MAKOTO MIYOSHI[2],
NORIYUKI KAWAGUCHI[2],
HIDEYUKI KOBAYASHI[3],
AND YASUHIRO MURATA[3]

[1] Communications Research Laboratory, 893 Hirai Kashima-city Ibaraki Pref. 314 JAPAN

[2] National Astronomical Observatory, 2-21-1 Osawa Mitaka-city Tokyo 181 JAPAN,

[3] Institute Space and Astronautical Science, Yunodai Sagamihara Kanagawa Pref. 229 JAPAN

1. Introduction

A joint Japanese group (NRO, ISAS and CRL) has started to develop new VLBI terminals. This system will standardize high-speed recording of a total of 1024 M-bits per second with minimum configuration. In this paper, the terminal components are introduced briefly.

Although VLBI observations have achieved remarkable progress in resolution, the minimum sensitivity has stayed around 1 Jy. To perform high-sensitivity VLBI, which enables detection of weak distant QSO continuum emission, we must make better use of the receiver bandwidth with advanced digital storage techniques. The only attempt so far has been made by the Mark-IV group (Whitney 1991). From the minimum VLBI sensitivity $S \propto \sqrt{T_{sys1}T_{sys2}}/\sqrt{A_1 A_2 B \tau}$, the receiver temperatures T have drastically improved in this decade and high-sensitivity observations have been dependent on receiver performance. On the other hand, it is difficult to increase the radio telescope size A. The integration time τ is limited especially in mm-wave observations. To maximize the bandwidth B, we have been developing both High Speed Sampler and Giga-bit VLBI Data Storage. These devices are realized by professional commercial instrument modifications and additional VLBI interfaces. Figure 1 presents the configuration of the next-generation VLBI terminal.

2. High speed A/D sampler and mm-receiver

The high-speed sampler is designed to sample a data rate of 1024/512/256MHz over 1/2/4ch with 1/2- bit selection. The total output data rate is limited by the VLBI sampler interface to 2048Mbps. The wideband acquisition will be achieved with the compact 7mm-wave-SIS VLBI receiver at Kashima. The receiver 5-7GHz IF is also used for the K4-burst sampling mode proposed by Kawaguchi (1991). The optically connected IF is converted with the Nobeyama VLBI wide-band video converter, and is fed to the high-speed sampler.

For the sampler part, we modified a stable DSO (Digital Storage Oscilloscope) A/D head. The DSO, Tektronix TDS784 has a feature of 1G/4ch continuous data sampling. The 4-channel 1/2bit VLBI data are picked up at the DSO buffer memory MSB (Most Significant Bit) side and transferred to the VLBI interface with the 128-MHz clock bus. The sampling clock is generated from an external hydrogen maser reference. At the interface, 1G/4ch data are selected and slowed down to the 32-MHz cycle by de-multiplexing. The VLBI interface has 8 ID-1 (8×256Mbps) connector outputs. After modification, the

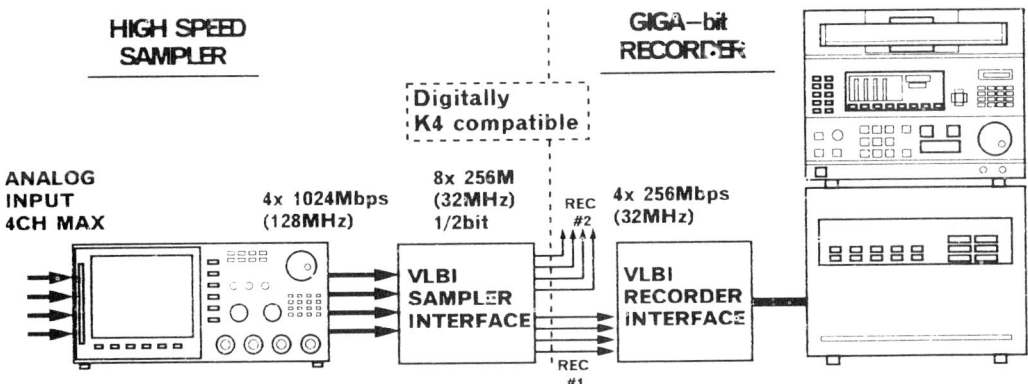

FIGURE 1. Configuration of next-generation VLBI terminal; both modified instruments have a multiple 32MHz 8parallel interface. This is similar to ID-1 and partially compatible with other K-4 instruments

DSO's oscilloscope functions remain. This sampler is not only intended for VLBI, but also for digital spectrometer front-end applications. The price of the DSO is US$30K.

3. Giga-bit(1024Mbps) Recorder

The Giga-bit storage part will establish 1024Mbps recording. Full utilization of the sampler-digitized output, 2048Mbps(1024M sample per sec., 2bit) recording, will be possible with a sampler and two recorders. An apparent concept difference from the Mk-IV system is in the small number of wideband channels. This reduces redundancy in the current VLBI Video Converter and related items. It also reduces problems associated with the channels. Table 1 shows a comparison of the Mk-IV, the K-4 and the new GBR-1000 recorder.

To achieve this unprecedented performance in VLBI, we are modifying a HDTV-VCR (High-Definition TV Video Cassette Recorder) designed for broadcasting purposes. The GBR-1000 (Figure 2), developed by TOSHIBA Co.,Ltd. is designed for a D6 standard multi HDTV format VCR without data compression; Noda (1994), Endo (1994). The recorder's original data rate is 1.2Gbps. Initial performance of the user data rate is less than 1Gbps. VLBI raw data are encoded as part of the HDTV frame data and recorded through the 16 helical-scan rotary heads. The error rate is less than 10^{-10} after speed-up modifications. The data from the VLBI sampler interface are connected to the VLBI recorder interface via the 4 ID-1-like (32MHz, 8parallel) cables. The system can almost be regarded as 4 parallel K-4 recorders digitally and its total data rate will be 1024Mbps. The VLBI data from the sampler are recorded without a time code. Figure 3 illustrates simplified recorder functions. During observations, the recorders in each VLBI station record the same numbered HDTV-frame data on the tape. This feature is realized with the recorder UTC keeping an interface with 1PPS input. In consequence, there is no need to refer to the timecode on the tape to achieve the playback synchronization during subsequent correlation processing. The buffer in the front end of the correlator removes lags within a frame. Tape for the GBR-1000 is a D6 SMPTE standard cassette.

format	data rate (Mbps)	duration (minute)	number of head	magnetic tape thickness(μ)	recording time identification
Mark-IV	1024	67.5	64fixed	16	TC†
K-4	256	50/61	4rotary	16/13	TC/TSS‡
GBR1000	1024	52/64	16rotary	13/11	TSS

TABLE 1. Comparison of Mark-IV, K-4 and GBR-1000 recording. †TC indicate the Time Code replaces VLBI data. ‡TSS indicate the time is replaced to Track Set Sync code ID in tape track. Conventional K-4 compatible to Mark-IIIA have time code in data. New K-4s employed in VSOP terminals use TSS ID. Mark-IV specification is from Whitney(1991).

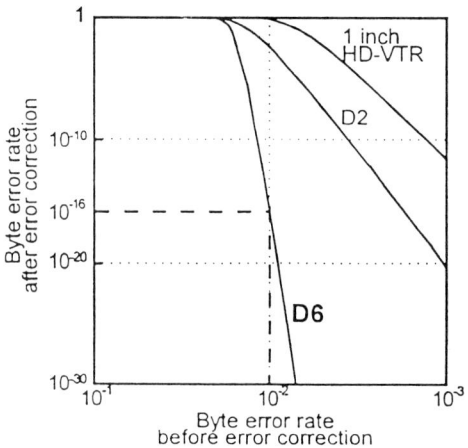

FIGURE 2. *Right*: Photo of the GBR-1000. Upper Transport, tape deck unit and Lower Control are separate units of 55kg each. Portable VLBI experiments are possible as well as K-4. An additional small VLBI interface is being developed to receive the data from the high-speed sampler *Left*: Original error rate of the recorder. An error rate of 10^{-10} is expected after the speed-up modification.

Using D2 (K4 cassette) housing, 19-mm metal particle tape of 11-micron thickness could record for a duration of 1 hour. It is easy to integrate an automatic cassette change system. Following the VSOP terminal TCU (Timing Control Unit) functions described by Kawaguchi et al. (1994), the REC/PLAY/STOP commands are programmable with action time through the GPIB. Although this recorder does not contain a longitudal recording format, playback speeds of 1/1, 1/4, 1/8, 1/16 and 1/32 are possible and it can be adapted to the existing new K-4 and other types of correlator. The price of a GBR-1000 is US$ 300K.

4. Conclusions

A high-speed sampler and VLBI interface have developed and are undergoing final tests. A giga-bit recorder and recorder interface have been designed. The next-generation Japanese VLBI terminal is composed of a modified high speed DSO and a modified HDTV-VCR. These reliable instruments will be employed as the Japanese VLBI standard terminal in the future. In addition to modifying our VLBI interfaces, there are

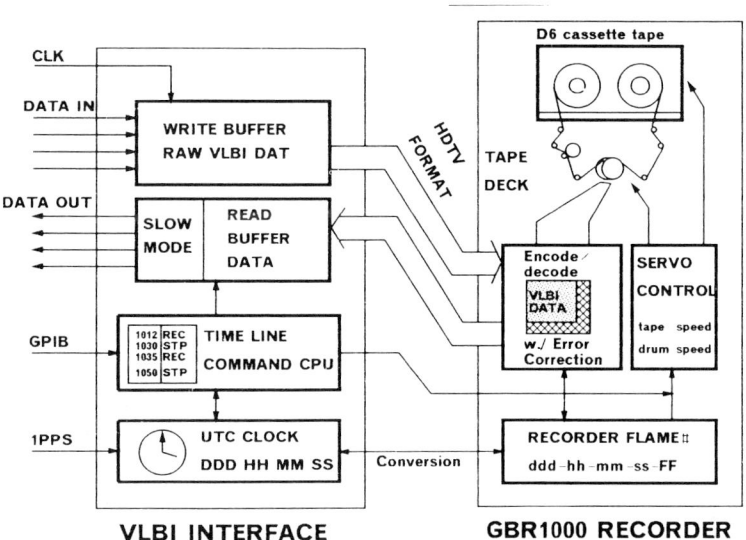

FIGURE 3. The GBR-1000 and interface VLBI recorder function. The interface keeps the UTC clock. Every station participating in an observation records the same-numbered data frame simultaneously at a certain time. VLBI data are processed and they are recorded with error correction. Recorder actions are programmable with time definition. A control program of the observation site and multi-baseline correlation becomes easy using this function.

possibilities for application to other astronomical purposes, remote sensing and other fields.

5. Acknowledgment

Part of this project was conducted between the National Astronomical Observatory and Communications Research Laboratory. Thanks are due to the technical advice and support of Mr.Kizu, Mr.Yamamoto and staff in TOSHIBA Imaging and Information division. We also grateful to Mr.Iwasa of SONY-Tektronix solution system division for his trial production and technical support of the compiler.

REFERENCES

ENDO, N., NODA, C., & KIZU, S., 1994. In *IEEE denshi Tokyo* , No.33, p. 236

KAWAGUCHI, N., KOBAYASHI, H., MIYAJI, T., MIKOSHIBA H., TOJO, A., YAMAMOTO, Z., & HIROSAWA, H., 1994. In *Proc. Int. Symp. VLBI technology, Progress and Future Observational Possibilities*, ed. Sasao, T., et al. p. 26. Terra Scientific Publishing Co.

KAWAGUCHI, N., 1991. In *Frontiers of VLBI, Proceedings of the International VSOP symposium and mm-wave VLBI Workshop*. ed. Hirabayashi, H., et al. p. 75, 269. Universal Academy Press.

NODA, C., ENDO, N., & KIZU, S., 1994. In *Proceedings of 1994 European SMPTE Conference*.

WHITNEY, A.R., ROGERS, A.E.E, CAPPALLO, R.J., HARGREAVES, J.E., HINTEREGGER, H.F., AND SMYTHE, D.L., 1991. In *Proc. AGU Chapman Conference on Geodesic VLBI*, ed. W.E.Carter, Monitoring Global change, Number NOS137 in NOAA Technical report, Rockville,MD, U.S. Dept. of Commerce, p.7–14.

The Concept of the Square Kilometre Array Interferometer

By ROBERT BRAUN

Netherlands Foundation for Research in Astronomy, Postbus 2, 7990AA Dwingeloo, NL

1. Introduction

The breadth of astrophysics which is addressed by radio astronomical observations is amply demonstrated by the range of contributions in this volume. Important physical insights continue to be made into the nature of both condensed and diffuse objects which lie at distances ranging from within the solar system to the recombination surface of the bubble which defines our visible universe.

An underlying limitation to the continued success of our quest for greater physical insight is embodied in the title of this meeting: "High Sensitivity Radio Astronomy". In one way or another, all of the various research directions in radio astrophysics are limited by our current instrumental sensitivities. Only by ensuring the continued access to order-of-magnitude improvements in our capabilities can we ensure a continued high rate of discovery. This statement is put into perspective by considering how our capabilities have evolved to their present level. In Figure 1 we have plotted the continuum sensitivity after one minute of integration for many radio telescopes as they came on-line. Substantial improvements to the performance from post-construction upgrades to the receiver systems are also indicated in the figure for some of the facilities. An exponential improvement in system performance, over at least 6 orders of magnitude can be seen between about 1940 and 1980. Instruments like the Lovell telescope, the WSRT and the VLA have become available on a schedule which maintained a high rate of discovery. The Arecibo telescope stands out as a major leap in sensitivity performance at a relatively early date. This particular example serves to illustrate that the single parameter we've chosen to examine in Figure 1, the continuum sensitivity in a short integration, doesn't necessarily tell the entire story. Additional parameters, like sky coverage, spatial resolution and survey speed also play a significant role in defining the total system performance. Even so, a disturbing trend seems to have developed in the period since about 1980. There appears to be a significant saturation of performance obtained with traditional radio telescope technology.

The time may now be ripe to explore a radically different technology in the decimetric radio band; one in which the economies of mass production are applied to high-performance, yet extremely low-cost amplifiers and digital electronics, while the dependence on large mechanical components is minimized. Just as dinosaurs were superseded by mammals, we should consider moving to distributed networks of smaller, yet more intelligent components. Hopefully, this transition can be realized without a major catastrophe first befalling the entire field. Rather than beginning with a mass extinction, we can hope for a fruitful period of coexistence of current and next-generation radio astronomy instrumentation.

The provisional name that has been given to such a next-generation instrument, is the "Square Kilometre Array Interferometer", or SKAI. This name embodies the fact that on the order of a square kilometre of collecting area will be required to provide the leap in sensitivity depicted in Figure 1.

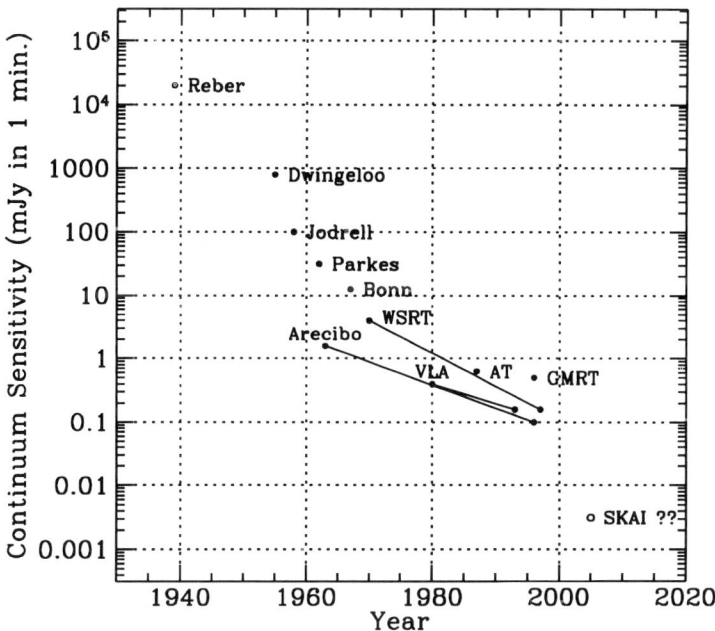

FIGURE 1. The time evolution of radio telescope sensitivity. The continuum sensitivity after one minute of integration is indicated for a number of radio telescopes as they became available. Solid lines indicate up-grade paths of particular instruments.

2. Scientific Drivers for the Array

Other chapters in this volume demonstrate the critical need for higher sensitivity in all research areas of radio astronomy. Scientific breakthroughs can be foreseen in understanding the:
• kinematics and evolution of both nearby and very distant galactic systems utilizing emission and absorption in the HI 21-cm line
• kinematics and evolution of circum-nuclear starburst systems utilizing molecular megamaser emission
• origin and evolution of galactic disks and halos utilizing radio continuum emission
• origin of pulsar emission and its utilization as a probe of extreme states of matter
• evolution and physical properties of normal stars
• AGN and jet phenomenon utilizing the highest possible angular resolution

Rather than reiterating those discussions here, we will demonstrate the impact of a next-generation radio observatory with only a few specific examples and comments related to understanding the evolution of galaxies. A much more comprehensive document outlining the general scientific motivation for SKAI is being prepared for distribution later in 1996.

The 21-cm line of neutral hydrogen remains the most valuable tracer of neutral gaseous mass in astrophysics. Even though neutral gas becomes predominantly molecular at high densities, the 21-cm line emission of the atomic component allows the total gaseous mass to be estimated to better than about a factor of two, even in the most extreme cases. This is in marked contrast to, for example, the luminosity of carbon monoxide emission lines originating in the molecular component. For the same total gaseous mass, the CO emission lines can vary in luminosity over a factor of about 10^4 depending on the

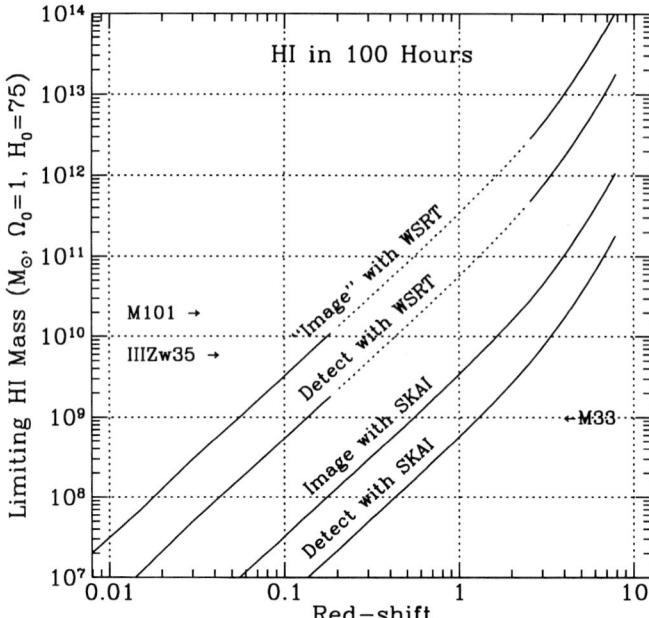

FIGURE 2. We show the detection (5σ in 50 km s^{-1}) and imaging (5σ in six channels of 50 km s^{-1}) limits of atomic gas mass as a function of redshift with current and next-generation instrumentation.

abundance of heavy elements and the intensity of the radiation field to which they are subjected.

The current and next-generation capabilities for detecting HI in emission are contrasted in Figure 2. While we must now be content to image galaxian gas masses out to redshifts of only 0.1 to 0.2, this would become possible to redshifts greater than 2, opening the way to truly understanding the evolution of gaseous into stellar mass.

Such a capability is perhaps better appreciated pictorially. In Figure 3 we have illustrated how the well known nearby system M 101, would appear to SKAI if placed at redshifts of 0.2, 0.45 and 0.9. Fairly detailed kinematic studies would be possible to at least $z = 0.5$ and cruder studies to $z > 1$. Each field of view would of course simultaneously sample literally hundreds of field galaxies rather than only the one target galaxy depicted.

An illustration of the enormous impact this capability would have on cosmological studies is given in Figure 4. In this figure the detection threshold of SKAI is overlaid on the predicted mass distribution from a recent simulation of structure formation in the early universe (Katz et al. 1996). Even the hundreds of detections indicated in this figure understate the true potential of the instrument, since the simultaneous frequency sampling would be at least 20 times deeper than that illustrated. Such large unbiased samples should allow a clear distinction to be made between competing theories for the nature of our universe.

Important advances should also follow from the utilization of both megamaser emission and non-thermal continuum emission for the study of galactic disks. These possibilities are illustrated in Figures 5 and 6. Megamaser emission could be employed to obtain sub-arcsecond (or milli-arcsec with VLBI, or even micro-arcsec with space VLBI) kinematic imaging of star-bursting systems at *arbitrarily* large red-shift. Similarly, the radio

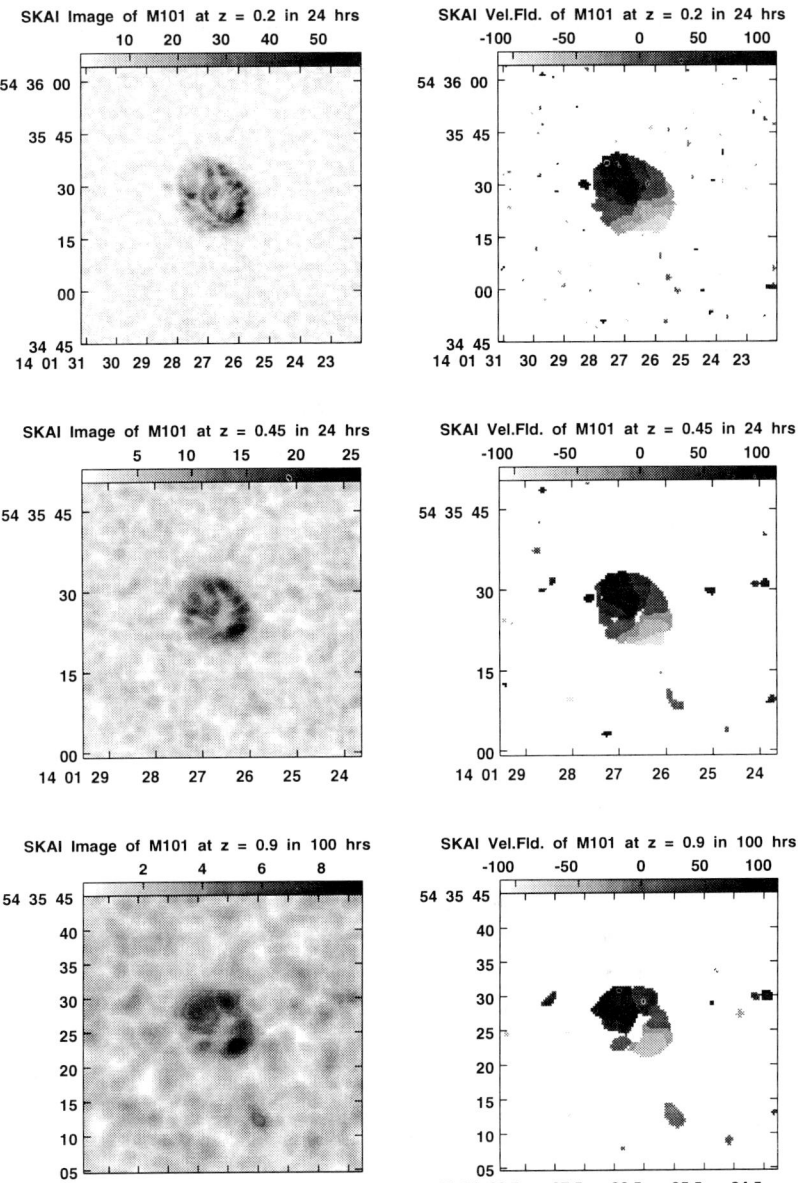

FIGURE 3. Simulated SKAI observations of M101 as it would appear at redshifts of 0.2, 0.45 and 0.9. Peak observed brightnesses are shown in the left hand panels and corresponding velocity fields on the right. The assumed integration time is indicated above each panel.

continuum emission, which is so intimately associated with massive star formation, could be imaged from galaxies even at the very earliest epochs, building on the current work reported by Kellermann in these proceedings.

2.1. *Serendipity*

Although it remains essential to consider the many branches of current research and to predict how they might be effected by a significant improvement in our capabilities, history has taught us that our imagination consistently falls short of the surprises that

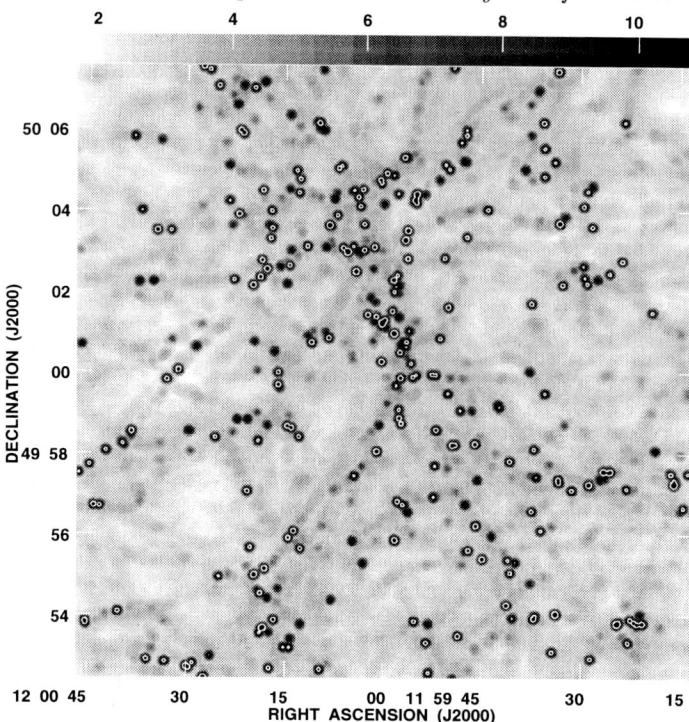

FIGURE 4. Simulated HI emission at z = 2 with SKAI detections overlaid. The linear grey-scale indicates the predicted peak brightness of HI emission in a 22.2/(1+z) Mpc cube and extends from $log(M_\odot/Beam) = 1.7 - 10.8$. The single white contour at $log(M_\odot/Beam) = 9.22$ is the 5σ SKAI detection level after a 1600 hour integration.

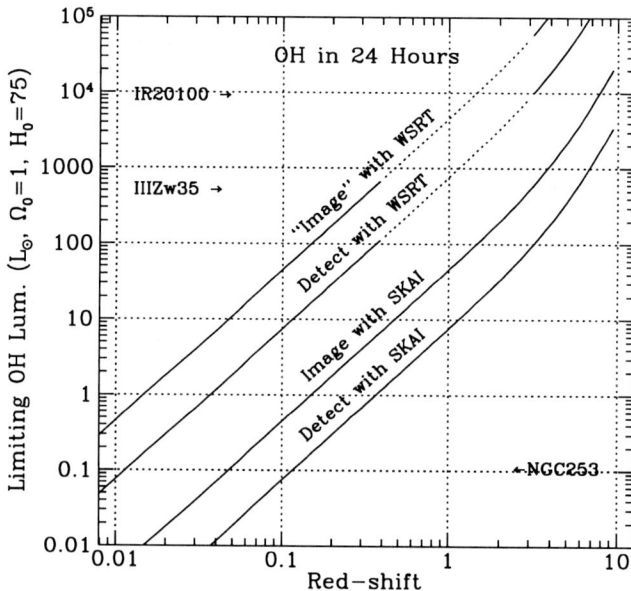

FIGURE 5. We show the detection (5σ in 50 km s^{-1}) and imaging (5σ in six channels of 50 km s^{-1}) limits for OH mega-maser emission as a function of redshift with current and planned instrumentation.

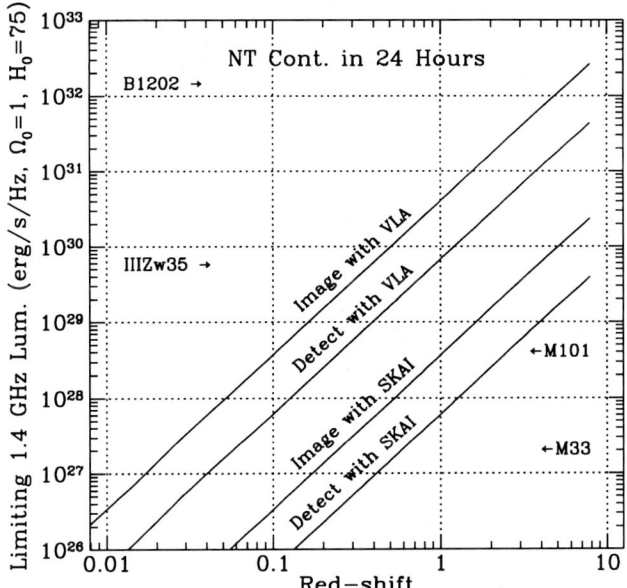

FIGURE 6. We show the limiting luminosity for a continuum observation at 1.4 GHz for detection (5σ) and imaging (30σ) of the non-thermal continuum associated with massive star formation as a function of redshift for current and planned instrumentation. The curves assume a power law spectrum of the form $S \propto \nu^{-0.7}$.

lie in store for us when a large step in instrumental capability is achieved. Much of the beauty of science lies in the fact that Nature has managed to provide us with an onion-skin parameter space, where unexpected discoveries accompany each new layer to which we penetrate. We must strive to make the most general possible improvements to our capabilities to maximize the cross-section for unexpected discoveries. A particular challenge lies in balancing the breadth of a new instrument against its depth in particular areas, while keeping within the constraints of a plausible facility cost.

3. Technical Specifications

Consideration of the many varied scientific drivers suggests the following basic technical specifications for the instrument:
- a primary frequency range of about 200 – 2000 MHz, to address the (red-shifted) spectral lines of HI and OH, with a strong desire for yet higher frequency coverage particularly for VLBI and stellar applications,
- a total collecting area of about 1 km^2, to achieve the desired sensitivity when employed with nearly sky-limited system temperatures,
- distribution over at least 32 elements, to achieve a sufficiently clean instantaneous synthesized beam and permit adequate modelling of time variable (interfering) sources
- a preference for forming these 32 elements from many smaller ones, so as to allow adaptive beam formation to excise interfering sources,
- better than about 1 Kelvin of brightness sensitivity for spectral line applications, implying a maximum array size of about 50 km
- spatial resolution of 0.1 to 1 arcsec for continuum applications, implying an array size of about 300 km

One way to satisfy the conflicting demands of spectral line and continuum applications

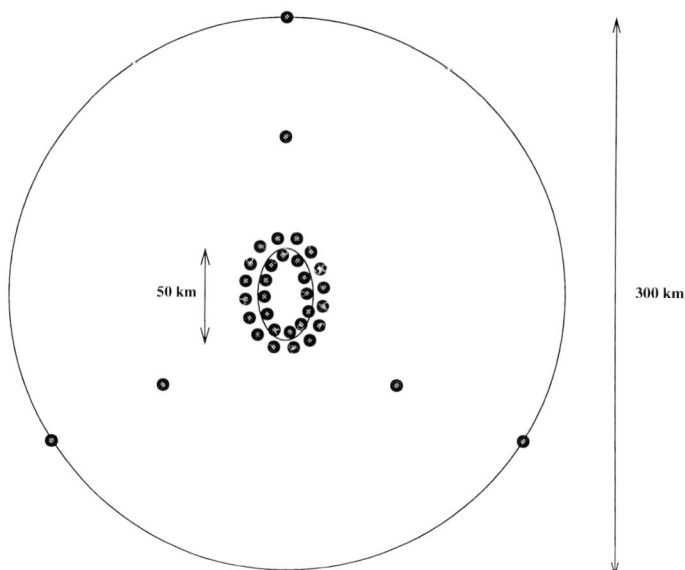

FIGURE 7. Schematic configuration of SKAI. Note that the unit telescopes are not depicted to scale, nor are the telescope locations accurately specified.

apparent above is to place a large fraction (say 80 percent) of the collecting area of the instrument within a region of about 50 km and to distribute the remaining fraction over a region of about 300-km diameter. In this way, the applications most limited by low surface brightnesses would retain most of their sensitivity, while other applications could utilize angular resolutions as high as about one tenth of an arcsecond.

A very schematic representation of what the SKAI might look like is given in Figure 7. The shaded circles indicate the position of the unit telescopes of the array. Note that the size of each circle has been greatly exaggerated to allow it to be seen on the scale of the illustration. A rather dense ring-like concentration of the telescopes over a region of about 50-km extent determines the beam size for which the brightness sensitivity is optimized (about 1 arcsecond at a frequency of 1420 MHz), while the additional elements, distributed over a 300-km extent, would make sub-arcsecond imaging possible over the entire frequency range of 200 – 2000 MHz. The thick annular distribution of the elements within 50 km is chosen to provide an approximately Gaussian naturally weighted synthesized beam. Although not indicated accurately in the Figure, care would obviously be taken to ensure good sensitivity to short projected baselines and the elimination of any "holes" in the spatial frequency coverage.

Other authors (Swarup, 1996) have suggested that it might be desirable to place as much as 50% of the total collecting area in one contiguous region, as has been done with the GMRT. Such a configuration would provide enhanced sensitivity to spatial scales of about 1 arcmin, while reducing the sensitivity to spatial scales of several arcsec. Trade-offs of this type will need to be made on the basis of our accumulated knowledge during the final design phase. In the mean time, the GMRT will have ample opportunity to demonstrate the utility of the course they have chosen to follow.

4. Telescope Concepts

Several possible element concepts for the SKAI are illustrated in Figure 8. At the heart of each of these concepts is a much greater reliance than ever before on mass-produced and highly-integrated receiver systems together with much more extensive digital electronics for beam formation. In the top panel we depict one conceivable extreme in a continuous range of possibilities. In this case the wavefront is detected by individual active elements comparable to a wavelength in size. Each of these is amplified, digitized and combined with the others to form an electronically scannable beam (or beams) with no moving parts whatsoever. The challenge in this case lies in achieving extremely low component and data-distribution costs since literally millions of active elements will be required. In the centre panel, some degree of field concentration is first achieved with the use of small reflectors before amplification, digitization and beam formation. In this case, active element number is reduced to some thousands and greater sky coverage at high sensitivity is also realized, although at the expense of the mechanical complexity of the drive and tracking system. In the lower panel we depict the other conceptual extreme, whereby a single large reflector is used for each of the unit telescopes. Extensive arrays of active elements would be employed in this concept to intercept the focal region of the spherical primary in order to efficiently illuminate the surface and allow multiple beams to be formed.

The adaptive beam formation technology which underlies all of these element concepts is extremely attractive for a number of reasons. Real-time beam formation with at least thousands if not millions of active elements provides a comparable number of degrees of freedom for tailoring the beam in a desired way. The basic properties of high gain in some direction and low side-lobe levels elsewhere are fairly obvious and traditional requirements. An additional possibility, which hasn't yet been applied in radio astronomy, is that of placing response *minima* in other desired directions, such as those of interfering sources. In addition, the way is naturally opened to exploit multiple observing beams on the sky to enhance the astronomical power of the instrument many-fold. These might be used to provide simultaneous instrumental calibration, support multiple, fully independent observing programs or enlarge the instantaneous field-of-view for wide-field applications. Finally, the great potential of adaptive beam formation has led to a strong commercial interest in this technology. This has opened the way to collaborative R&D efforts which are now beginning to take shape.

During the interval 1995–2000, a concerted effort at R&D for SKAI will be undertaken both within the NFRA and at collaborating institutes. The various concepts depicted in Figure 8 (and potentially new ones) will be worked out in sufficient detail to allow realistic cost estimates to be made. Proto-typing of cost effective technologies as an extension to the WSRT array is planned for the period 2001–2005. Assuming the successful completion of both technical preparations and funding arrangements, construction of the instrument is envisioned for the period 2005–2010. In this way we can look forward to opening the door to a new era of discovery.

REFERENCES

KATZ, N., WEINBERG, D.H., HERNQUIST, L. & MIRALDA-ESCUDÉ, J. 1996. *ApJ* **457**, L57

SWARUP, G. 1996. Array Configuration for a Radio Telescope of One Million Square Meter Area *URSI GA Abstracts*.

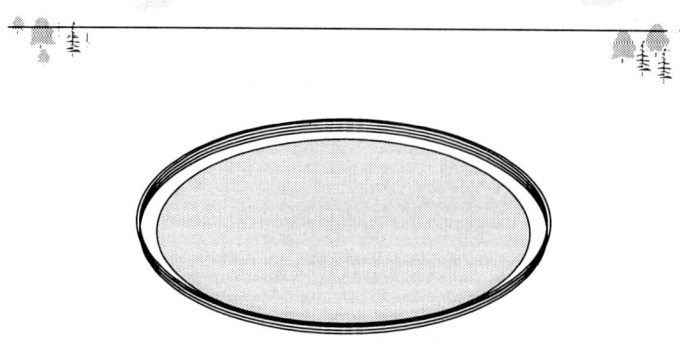

Adaptive Phased Array Element Concept

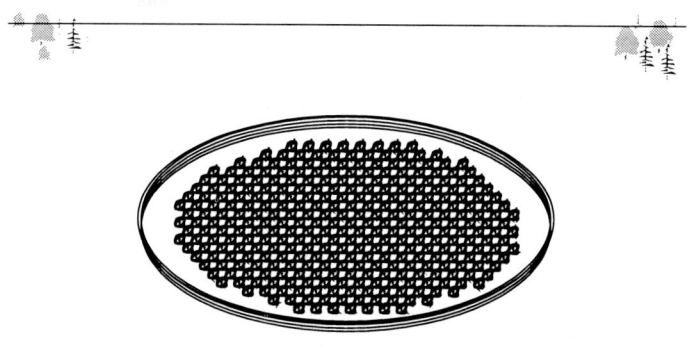

Adaptive Reflector Array Element Concept

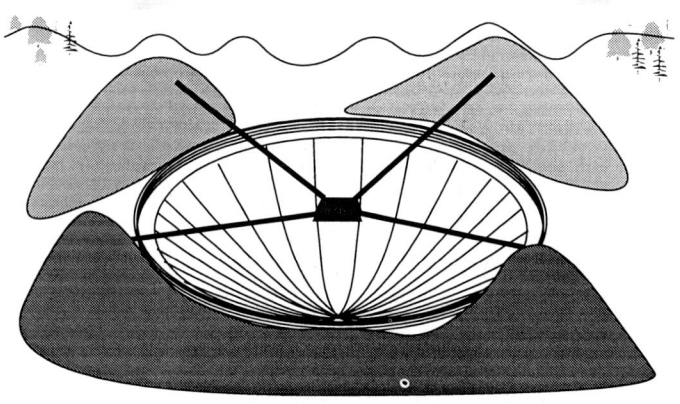

Passive Spherical Element Concept

← 300 meter →

FIGURE 8. Possible element concepts for the SKAI.

Alternative array configuration and antenna elements for the square kilometre array interferometer

By G. SWARUP

National Centre for Radio Astrophysics, TIFR
Pune University Campus, Pune 411 007, India.

A schematic design has been proposed by Braun (1995) for the Square Kilometre Array Interferometer (SKAI) consisting of 28 groups of antennas placed in a ring pattern of about 30 km × 50 km in extent, with 6 outlying antennas placed up to 150 km away from the centre. The possibility of developing a phased array for the individual antenna element has been discussed. In this paper, we discuss criteria for various antenna configurations and discuss the pros and cons of alternative choices of antenna elements, such as phased dipole arrays, spherical reflectors and low-cost parabolic dishes.

1. Introduction

Over the last few decades, radio-astronomical investigations have led to the discovery of many exceptional classes of objects. Radio astronomy has also given us valuable insights concerning the origin and evolution of the Universe. It has become increasingly clear that a radio telescope with a large collecting area of about one million m^2 operating at decimetre and metre wavelengths, which is an order of magnitude larger than any of the existing telescopes, is essential for elucidating some of the current puzzles of great astrophysical importance (Wilkinson 1990, Swarup 1991, Braun 1995).

A Large Telescope Working Group was set up by the International Union of Radio Science (URSI) in 1993 for developing a suitable design for a radio telescope with an area of about one million sq m. It is encouraging to note that the Dutch group has set up a small design group for investigating various concepts. A tentative conceptual design for a "Square Kilometre Array Interferometer (SKAI)" has been proposed by Braun (1995). In Sections 2 and 3, we summarize some of the scientific objectives of SKAI and the required technical specifications. Comments on a few alternative array configurations are given in Section 4. Choice of antenna elements for achieving a large collecting area over a wide bandwidth at an affordable cost is perhaps a formidable problem for which various possible solutions are discussed in Section 5. Conclusions are given in Section 6.

2. Scientific Objectives and Specifications of SKAI

SKAI is required to provide a very high sensitivity and arc or sub-arc second angular resolution over a wide frequency range in the decimetre and metre wave regions from about 150 to 1700 MHz. Although SKAI will be an extremely valuable instrument for studying a variety of astrophysical problems, some of the important scientific objectives are reviewed briefly. One of the most challenging problems is to study the epoch of galaxy formation and growth of large-scale structure in the Universe. This can be determined uniquely by searching for neutral hydrogen emission from proto-galaxies and proto-clusters prior to the formation of galaxies and quasars in the Universe. Searches for H I absorption of radio waves from high redshift radio galaxies and quasars will also provide valuable information about the high-z Universe. In particular, the proposed design specifications of SKAI should allow studies of H I distribution in an M101 type of galaxy with a linear resolution of about 2 kpc at redshifts, z, less than 0.2 and about 7 kpc at z ~ 1. It must be noted that studying the evolution of H I with cosmic epoch is very important as hydrogen is the basic building block of the Universe. SKAI will also be very valuable for understanding the origin and evolution of non-thermal emission seen in star-burst galaxies and galactic nuclei. SKAI will allow searching for pulsars, particularly of milli-second periods in our Galaxy and perhaps also in nearby galaxies. Studies of individual pulse profiles may elucidate pulsar emission mechanisms, and accurate timing studies of pulsars will allow probing of theories of gravitation and internal structure of neutron stars. Searches for relic radio galaxies, observations of radio stars with relativistic jets, large-scale redshift surveys, study of gravitational lensing, and radio emission from the Sun and planets are amongst a variety of phenomena for which SKAI will provide extremely valuable data.

3. Technical Specifications (The Wish List)

Besides providing a large effective collecting area of about 1×10^6 m^2 in the frequency range of about 150 to 1700 MHz, it is important that the system temperature of the SKAI receivers should be less than about 40 K in the frequency range of about 600–1400 MHz and less than 100 K at 327 MHz. It seems highly desirable that SKAI should operate at frequencies up to 150 MHz on the lower side in order to investigate the sources of re-ionization of the intergalactic medium up to a redshift of about 8. The operation up to 1720 MHz on the high side is important for studies of OH maser emission and for continuum polarization observations. The frequency coverage should be more or less continuous in the above frequency range with an instantaneous bandwidth of at least 100 MHz to be covered by about 1000–4000 spectral channels. A bandwidth of about 400 MHz in the phased array mode in the range 1000–1400 MHz will be very valuable for pulsar studies or for searching for spectral lines from the epoch of recombination at z ~ 1000. SKAI should have good polarization capacity. It is important that it has a built-in capability for rejecting man-made radio frequency interference. A wide field of view of the antennas would be useful. The antennas should be steerable over a wide declination range and capable of tracking in HA for several hours. The array configuration should allow adequate sensitivity for mapping both compact and extended features of weak radio sources. The antenna elements and the electronics should provide a high sensitivity at an affordable price and should require relatively low maintenance.

4. Array Configuration

The choice of array configuration depends upon the scientific objectives. Braun (1995) has proposed 28 elements placed in a 30 km × 50 km elliptical ring for a possible location of the array in Northern Europe. Six telescopes are to be placed in a Y array up to 150 km away from the centre. The central ring array will provide highest sensitivity for studying individual galaxies in H I with resolution of about 1 arcsec (~ 2 kpc), 1.7 arcsec (~ 7 kpc) and 4 arcsec (~ 14 kpc) at frequencies corresponding to redshifts of about 0.2, 1 and 3.5 respectively. Sensitivity of such an array will however be relatively poor for investigating protoclusters or a group of very young galaxies, whose overall extent may be of the order of a few arc minutes. In order to provide sufficient sensitivity for objects of low surface brightness, it may be preferable to place about one-third or half the elements in a compact array near one of the edges of the central part of the elliptical ring. It may be noted that observations with the D-array of the VLA have provided very valuable information about the low surface brightness features of radio sources.

5. Antenna Elements

The choice of suitable antenna elements for achieving large collecting area and wide bandwidth is quite formidable. The Dutch group has proposed the possibility of using about one million antenna elements, each consisting of about 1 m^2 "flat tile" with fractal dipole arrangement above a plane reflector and highly integrated electronics so that there are no moving parts. Highly integrated digital electronics is to be used for beam formation, interference rejection and tracking. This is a laudable objective but is only in a preliminary conceptual stage. The great advantage is that the tiles can be placed close to the ground so that their susceptibility to the man-made radio frequency interference (RFI) from terrestrial transmitters will be minimized. Also, multiple beams can be formed for tracking many sources at the same time. Further, it may be possible to minimise RFI by keeping the sidelobes of the array low in the directions of the terrestrial transmitters, even while the array is tracking a radio source. The proposed array will have over 100 million HEMT amplifiers and in addition at least 200 million active elements for other parts of the receiver systems. There are, however, many uncertain questions, a few of which are listed here : (a) Can a suitable flat tile array be produced economically with a bandwidth of about 10 or 12 : 1? (b) What is the likely system temperature? (c) What is the loss in sensitivity due to the limited declination range as a result of projection effects? (d) What are the likely gain variations during tracking of radio sources as a result of mutual coupling between individual elements?

An alternative is to use parabolic dishes. The advantages are as follows: (a) steerability over a wide declination range is quite practical and this also allows for better calibration capabilities than a flat tile array; (b) wide-band feeds with good polarization characteristics can be easily designed; (c) with the development of liquid nitrogen closed cycle refrigerators, it may be possible to obtain an overall system temperature of less than 30 K at 21 cm. Less complex electronics may imply better phase and gain stability leading to a large dynamic range of the array which is a must in order to optimally utilize its very high sensitivity. Another advantage of using dishes is that one could extend the frequency range of the dishes to 5 GHz, although with an antenna efficiency of only about 25% at 5 GHz for the sake of economy. A possible disadvantage of choosing parabolic dishes for SKAI is that a large number of dishes ($\sim$ 1800 dishes of 25 m or 7200 of 12-m size) may pose maintenance problems. It needs to be investigated whether any industrial components such as those in textile machineries or automobile industry which are economical due to mass production but also have a long life could be used for the drive system of the parabolic dishes.

Low cost fully steerable parabolic dishes of 45-m diameter have been developed for the GMRT project using a novel concept nicknamed SMART (Stretched Mesh Attached to Rope Trusses). We have recently developed a computer design for a 12-m dish which has a tonnage of only about 1.5 tonnes for the dish portion. The overall tonnage excluding the counter-weight is likely to be only 5 tonnes. The corresponding total weight of a 25-m dish is likely to be about 25 tonnes. It may be noted that the cost of a single 45-m diameter GMRT dish (Swarup et al. 1991; Swarup et al. 1996, present volume) at 1995 prices including civil, structural, mechanical and electrical parts, is estimated to be about US $ 400,000. If we take the cost to vary as d$^{2.5}$ where d is the diameter, the cost of an optimized 25-m dish with a wire mesh of 6 mm $\times$ 6 mm, 10 $\times$ 10 mm and 15 $\times$ 15 mm size may be estimated to be (Swarup 1991)

$$C = \left(\frac{25}{45}\right)^{2.5} \left(\frac{v}{140 km/h}\right) \left(\frac{S}{.07}\right) \times 400,000 \times q$$

where C \$ = cost of 25-m dish, v is the survival wind velocity (50 years), S is the solidity of the wire-mesh surface (we may use 0.38 mm stainless steel wire instead of 0.55 mm used for the GMRT dishes) and q = 0.8 is the mass production factor including possible improved design. Thus, the cost of 1800 dishes of 25-m diameter will be about \$ 135 million at 1995 prices. The total physical area of these will be about 880,000 m^2 and effective area of about 600,000 m^2. The total cost of the project including electronics and other overheads may be about US \$ 250 million. It would be desirable to locate such an international telescope at a site with low winds and low RFI environment. A location closer to the equator, such as in India, will give a large sky coverage.

The Chinese astronomers have proposed 32 stationary spherical reflector antennas of about 200 m diameter each (similar to that of the Arecibo Radio Telescope) for SKAI taking advantage of the unique terrain of Guizhou Province in South-Central China, which has one of the world's most extensive natural Karst formations of rough hemispherical shapes. The economics of this concept need to be worked out.

A possible compromise solution between the flat tile array and parabolic dishes is to consider the use of parabolic cylinders. If the cylinders are placed in an east-west direction, a large declination coverage is easily obtained mechanically. Steering is possible without much loss of sensitivity up to ± 3 hours in HA. Parabolic cylinders have also the advantages of low maintenance. The cost/m^2 area of a parabolic cylindrical antenna is likely to be appreciably lower, say 50 to 60%, compared to that of a parabolic dish. However, the design of a broad-band feed over a range of about 12:1 may be problematic. One solution is to place feeds at different frequencies on the four faces of a rotating cage along the focal line of the cylinders but this will lead to mechanical complexity. Offset parabolic cylinders can be easily fabricated, similar to the Bologna or Ooty radio telescope or the Japanese IPS array. The Molonglo synthesis radio telescope has successfully used an east-west symmetric parabolic cylinder. Bracewell et al. (1963) had in fact proposed use of a 'Venetian blind' configuration of east-west parabolic cylinders for a 'Future Large Radio Telescope' inspired by Jan Oort's call in 1961 for a square-kilometre array giving 1 arcmin resolution at 21 cm!

6. Conclusion

The proposed Square Kilometre Array Interferometer will be an extremely valuable instrument for studying a variety of astrophysical objects and particularly for elucidating several important big questions, such as the epochs of formation of the first structures in the Universe and also of the re-ionization of the intergalactic medium. The SKAI is likely to remain an un-surpassable instrument for decades to come. The cost is relatively modest. However, the challenging problem is to design and develop low-cost antenna elements with suitable electronics so that the SKAI has frequency agility over a wide bandwidth, large effective collecting area of the order of half to one million sq m, wide declination coverage and low maintenance. With many rapid advances in the fields of antennas and electronics a suitable innovative solution is likely to be found by a dedicated R&D team within a few years.

7. Acknowledgements

The author thanks Dr. D.J. Saikia for valuable comments.

REFERENCES

BRACEWELL, R.N., SWARUP, G. & SEEGER, C.L. 1962. *Nature*, **193**, 412

BRAUN, R. 1995 "The Square Kilometer Array Interferometer" to appear in *The Westerbork Synthesis Telescope, Its 25th Anniversary and Beyond*, Eds. E. Raimond & R. Genee (Kluwer, Dordrecht).

SWARUP, G. 1991. *Current Science*, **60**, 106

SWARUP, G., ANANTHAKRISHAN, S., KAPAHI, V.K., RAO, A.P., SUBRAHMANYA, C.R. & KULKARNI, V.K. 1991. *Current Science*, **60**, 95

SWARUP, G., ANANTHAKRISHAN, S., SUBRAHMANYA, C.R., RAO, A.P., KULKARNI, V.K. & KAPAHI, V.K. 1996; present volume.

WILKINSON, P.N. 1990. *Proc. URSI/IAU Colloquium No. 131 on "Radio Interferometry – Theory, Techniques and Applications"*, Socorro, USA, 8-12 October 1990

A proposed large radio telescope of new design

By T. H. LEGG

Herzberg Institute of Astrophysics, National Research Council, Ottawa, Canada

A new type of radio telescope is proposed which may make very large telescopes more affordable. The telescope has an almost flat primary reflector that is slightly adjustable in shape and made up of identical flat panels supported by the ground. A very long focal length imposes the unusual condition that the receiver be carried by an airborne vehicle such as a powered, helium-filled balloon. Residual errors in the controlled position of the vehicle are dealt with in several ways. The telescope has the potential of wide sky coverage, a diameter of several hundred metres, and a range from decimetre to short centimetre or, with smaller panels, mm wavelengths.

1. Introduction

As has been well demonstrated at this meeting, there would be great scientific value in building radio telescopes that are substantially more sensitive than existing instruments. Single, very large collecting areas would, by themselves, lack angular resolution and be confusion limited for many observations and therefore inappropriate. Instead, a number of telescope elements will have to be joined interferometrically to synthesize larger apertures. This requirement for aperture synthesis means in turn that long baseline tracks are needed and that the individual telescope elements must therefore be steerable over large areas of sky.

So far, fully-steerable radio telescopes have been built exclusively by supporting a paraboloidal reflector on some form of structure that can be rotated about two axes. With this type of telescope, the reflecting surface itself accounts for a small fraction of the cost and weight. The structure needed to support and move the reflector is the part that is expensive. A number of radio telescopes have been built in the past which avoid much of cost of the rotating structure (Blum et al, 1963; Gordon and Lalonde, 1961; Korolkov and Pariiskii, 1979; Kraus et al, 1961) but these have been limited in their sky coverage. A new type of radio telescope is proposed here which would not have this limitation and which may provide an economical means of obtaining very large collecting areas and sensitivities.

2. The Way It Would Work

The telescope reflector would be made up of a large number of inexpensive, identical square panels as illustrated in Fig. 1a. The panels are supported at their corners by linear actuators which are set directly on concrete footings in the ground. Both the panels and the overall reflector are flat, or almost flat. However, the reflector can be slightly altered in shape by adjustment of the actuators. This allows the reflector to take the form of an offset paraboloid that is appropriate to the position of the radio source being observed. Wide sky coverage is thereby obtained, though the collecting area is reduced by a foreshortening effect at large zenith angles (by a factor of two, for example, at a zenith angle of 60 degrees).

Because the reflector is everywhere close to the ground, and its panels all identical and flat, it can be constructed inexpensively. This arrangement implies that the focal

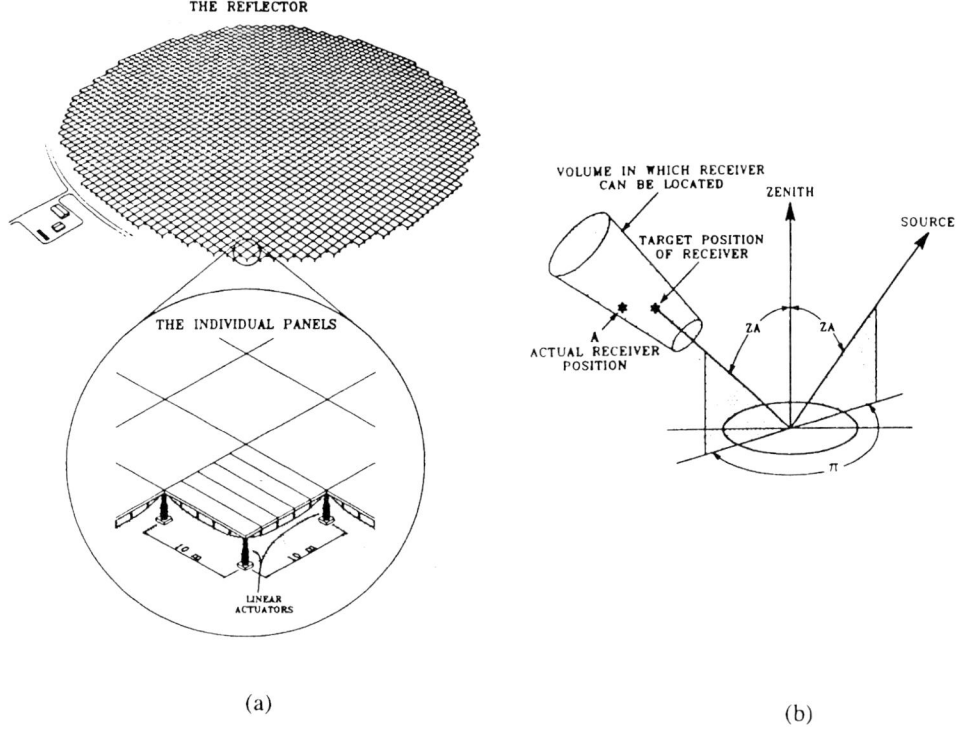

FIGURE 1. (a) (left) The reflector is made up of identical square, flat, panels that are supported by actuators set on footings in the ground. (b) (right) An airborne receiver is positioned as closely as possible to a target position. If the measured position deviates from the target position, to point A, the reflector would be slightly adjusted to move the source diffraction pattern to point A.

length of the reflector must be very long. As a result, one is faced with the very unusual condition that the telescope receiver be carried by some form of airborne vehicle. At first glance this condition may seem outlandish and unworkable. However, the basic problem of an airborne receiver is one of controlling its position and attitude so as to illuminate, focus and point the telescope correctly. This problem is substantially eased by having a reflector that, for entirely different reasons, has a surface that is adjustable with linear actuators.

The position of the airborne vehicle would be measured and controlled from the ground to bring the receiver feed as close as possible to a "target" position as shown in Fig.1b. This position is at the zenith angle of the radio source and at an azimuth 180 degrees from the source azimuth. If the measured position of the receiver deviates from the target position, however, the diffraction image of the radio source is moved so as to remain aligned with the receiver feed. This would be done by using the actuators to slightly tilt the entire surface in the required direction. The tilt would be small because the long focal length gives a large magnification of the surface motion in the motion of the diffraction image. The focal length of the reflector can also be adjusted with the actuators to keep the telescope accurately focussed despite a displacement of the receiver along the focal axis.

The airborne vehicle is the key component of the telescope. A powered, helium-filled balloon appears to offer the best characteristics in terms of reliability and low initial and operating cost. It is estimated that a spherical balloon, 10 m in diameter would have sufficient lifting capacity for the proposed telescope. There is a large saving in cost if, as for the present application, the balloon does not have to be certified to carry people. A spherical balloon has a known mass if it is in hydrostatic equilibrium. A sphere also has a known drag coefficient (about 0.2 for a diameter of 10 m). The acceleration of the balloon in sudden gust of wind can therefore be calculated, as can the power necessary to counteract a wind of a given velocity. To withstand a wind of 40 km/hour, for example, about 10 kW is needed with a propeller efficiency of 75%.

Small deviations of the receiver from the target position (say of the order of one or two metres) could be corrected by carrying a very large secondary reflector on the balloon (a design goal would be to carry the secondary inside the balloon). This would avoid moving the telescope reflector surface for small vehicle position errors. One possibility, for cm wavelengths, is to use a phased receiving array to feed the secondary reflector. The array would illuminate only the region of the secondary reflector that is covered by the diffraction image of the source. With a sudden change in vehicle position, the illuminated region would be moved electronically so as to remain in coincidence with the diffraction image, and thereby keep the telescope pointing constant.

A phased array also has the advantage that it can compensate for changes in geometry as a source is tracked. The primary reflector, as seen from the receiver, will be foreshortened in the vertical plane as the zenith angle increases. Depending on the path of the receiver, the horizontal angle subtended by the primary may also change. A phased array could be used to keep the illumination optimum despite these changes. This is not the case at 1.3 cm or mm wavelengths where receiving arrays are presently impractical. At these wavelengths a mechanically swivelled mirror would have to be used, instead of an array, to keep the secondary illumination coincident with the diffraction image. Also, the path of the airborne vehicle would have to be optimized to minimize the effects of the changing geometry. A study shows that if a mm wavelength feed of fixed pattern is optimized at a zenith angle of 45 degrees, the illumination efficiency can be kept to within about 8% of the optimum over zenith angle range from zero to 60 degrees.

Longer wavelength observations would be made using a smaller f/D ratio. This would be done by reducing the altitude of the vehicle and re-focussing the primary reflector. The cost of doing this is in the linear actuators. The long wavelength limit of a telescope is set by the need to bring the vehicle to an altitude where the diffraction image is no larger than the secondary reflector.

3. An Example

As an example of a telescope of the proposed type that might be suitable as an element of a synthesis array consider a telescope with the following characteristics:

Azimuth range:	360 deg	Panel side dimension:	10 m
Zenith angle range:	0 to 60 deg	Number of panels:	316
Reflector diameter:	200 m	Number of actuators:	346
Wavelength range:	1.3 cm to 21 cm	Actuator travel:	2.4m
Beamwidth:	0.25 to 20 arc min	Actuator accuracy:	1 mm
Focal length at zenith:	3000 to 900 m	Secondary diameter:	4 m
Sensitivity (at zenith):	about 5.2 K/Jansky	Balloon diameter:	10 m

A 200-m diameter may be close to an optimum for this type of telescope. Diameters up to about 500 m are possible but become progressively more difficult to feed at longer wavelengths. Most of the required actuator travel, and actuator cost, results from the need to reduce the altitude to 900 m at 21-cm wavelength in order to illuminate the primary reflector. The surface panels, 10 m on a side, can be flat for wavelengths of 3 cm or longer, but for 1.3 cm require a small curvature. Smaller panels are needed for mm wavelengths. A 100-m reflector at a 2.6-mm wavelength, for example, would require about 1500 flat panels 2.3 m on a side with 1550 actuators of 14-cm travel and 0.2-mm accuracy. A possibly important point is that flat, mm-wave panels can evidently be made inexpensively by stretching a thin sheet of metal between two machined supports. The reflector would evidently have good stability against wind and temperature changes (without a radome) as a result of its nearly uniform horizontal surface and small vertical dimension.

The general idea of the telescope seems to at least change the problem from one of making ever larger steel structures to a problem of controlling, or correcting for, the position of an airborne receiver. The scheme clearly needs to be proven, however, with a small prototype telescope.

REFERENCES

BLUM, E.J., A. BOISCHOT, AND J. LEQUEUX 1963. Radio Astronomy in France, *Proc. IRE Australia* **24**, 208

GORDON, W.E., AND L.M. LALONDE 1961. The Design and Capabilities of an Ionospheric Radar Probe, IRE Trans. Antennas Propag., AP-9, 17

KOROLKOV, D.V., AND YU. N. PARIISKII 1979. The Soviet RATAN-600 Radio Telescope, *Sky and Telescope* **57**, 324

KRAUS, J.D., R. T. NASH, AND H.C. KO Some Characteristics of the Ohio State University 360- foot Radio Telescope, *IRE Trans. Antennas Propag.* **AP-9**, 4-8

Further site survey for the next generation large radio telescope in Guizhou

By B. PENG[1], R. N. AN[1], Y. QIU[1], Y. NIE[2], B. ZHU[2], X. XU[1], AND R. STROM[3]

[1]Beijing Astronomical Observatory, Beijing 100080, PRC

[2]Institute of Remote Sensing Applications, Beijing 100101, PRC

[3]Netherlands Foundation for Research in Astronomy, Postbus 2, 7990AA, NL

We carried out a further site survey in a karst region in China for constructing the next generation large radio telescope (LT). One of the most suitable concepts for the realization of the LT is to build a passive spherical reflector. Data on more than 400 karst depressions in Pingtang and Puding counties of Guizhou province have been collected in a database using various selection criteria. Preliminary statistical results are presented in this paper, showing that even with only this limited sample it is still possible to find more than 30 depressions in each dimension.

1. Introduction

One way to realize the LT is to construct a passive spherical reflector of about 30 individual unit telescopes, each $\sim$300 m in diameter. The right terrain is critical to make such an instrumental concept happen. Valleys amid the hills of southwest China would be ideal for realizing the LT by very large spherical reflectors. We now refer to this effort as the Kilometre-Square Area Radio Synthesis Telescope (KARST) project.

Guizhou covers an area of approximately 170000 km^2 between latitudes of 24°35' to 29°09', and longitudes of 103°36' to 109°36'. 73% of the topography is made up of the karst landform. There are fewer than 3 days of snowfall every year. The relatively low latitude benefits the observations of Galactic objects, which is one of the major scientific drivers of the LT. The site survey started in May 1994: the geomorphologic features, distribution of the karst depressions, engineering geology, climate and social environment were studied (Nan et al. 1996). The entire planet already suffers from irradiation by space-based allocations at low frequencies. A critical consideration in choosing the best site is the interference environment. The preliminary results of interference monitoring there, which provide information about the frequency, strength and characteristics of the interfering signals, look quite promising (Peng et al. 1995), due to remoteness and the terrain shielding.

2. Statistical results

The karst landform development is subject to lithologic geological structure and hydrographic net characters. Subterranean cavities in limestone formation regions were formed due to heavy rainfall, and eventually collapsed leaving hemispherical depressions. The rock group of engineering geology consists of hard carbonate and hard psammite which can support a load of 250-650 t/m^2. In general, the architecture is stable in the south of Guizhou. Usually, there are no inhabitants in the karst depressions themselves.

Following the scientific objectives of the LT, the desired performance of 1-Kelvin sensitivity at 1-arcsecond resolution implies that most of the collecting area be distributed in a region of about 40-km diameter. A large number of depressions were surveyed with remote sensing (RS), geographical information system (GIS) and field observations. The

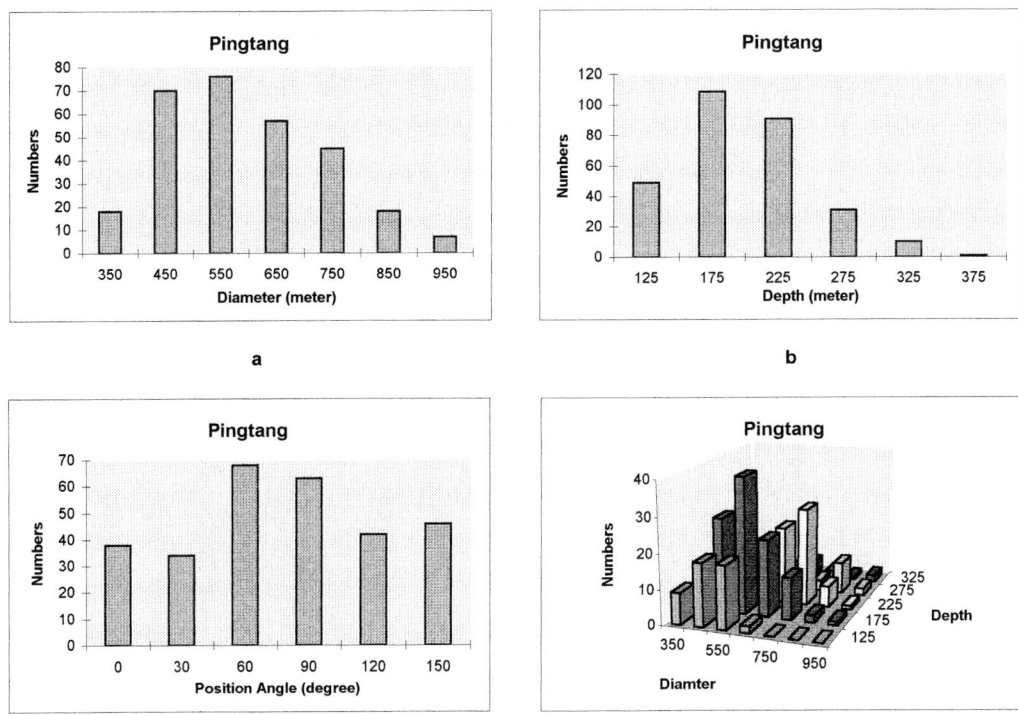

FIGURE 1. Distributions of the depressions in Pingtang county of Guizhou province: Numbers plotted against (a) diameter in metres, (b) depth in metres, (c) position angle in degrees: (d) diameter-depth.

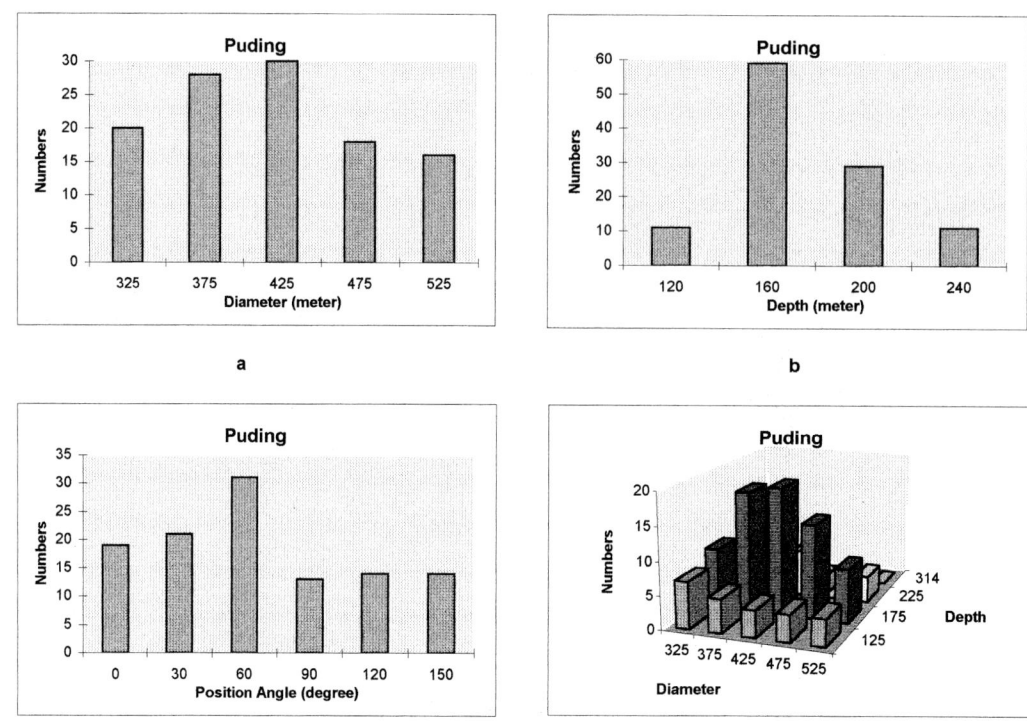

FIGURE 2. Distributions of the depressions in Puding county of Guizhou province: Numbers plotted against (a) diameter in metres, (b) depth in metres, (c) position angle in degrees: (d) diameter-depth.

distributions of the depressions in both Pingtang and Puding counties fit this requirement as shown in Fig. 1 (a, b, c and d) and Fig. 2 (a, b, c and d) respectively. These depressions usually take the shape of an anticone with diameters of 300–600 m at the top and depths of 100–250 m. Three-dimensional images of 15 depressions were produced with a high resolution of 5 m/pixel by using the DTM system. More than 400 depressions are included in a database, selected from elementary criteria as follows: diameter (D) should be larger than 300m, depth (d) greater than 150 m, peak number (n) surrounding the depression more than 3, and the ratio of major and minor axes of the depression smaller than 1.5. The PA in degrees refers to the position angle of the long axis of each selected depression. The u-v coverage was simulated by using the program HAZI in the Caltech VLBI package. The zenith angle limit is 35° for all antennas in our analysis.

3. Concluding remarks

In summary, remoteness together with the paucity of industrial development potentially benefit the radio environment. The on-the-spot investigation of a dozen depressions in Pingtang and Puding counties in Guizhou shows that a large number of suitable sites can easily be found to configure the KARST.

We wish to thank all the members of the Chinese KARST project team for their great efforts in the LT site survey, the Guizhou Provincial Science and Technology Commision and the KARST Comprehensive Research and Experimental Station in Guizhou for their practical help. This work is supported by the Chinese Academy of Sciences and the National Natural Science Foundation of China.

REFERENCES

NAN. R., NIE, Y., PENG, B., WU, S., YAN, Y., PIAO, T., KANG, L., TIAN, W., STROM, R., 1996. *Proc. of the LTWG-3 and W-SRT, Beijing*, in press

PENG, B., STROM, R., NAN, R., NIE, Y., 1995. *Astrophysics Reports* **26**, 68

Technical aspects for the Square Kilometer Array Interferometer (SKAI)

By A. van ARDENNE[1] AND F. SMITS[1]

[1]Netherlands Foundation for Research in Astronomy
P.O.Box 2, 7990AA Dwingeloo, The Netherlands

NFRA is exploring the possibilities of building a new generation of very large radiotelescopes by means of phased arrays. The main challenge is to provide sufficiently wideband, low–cost active receiving elements with the required sky coverage at acceptable investment cost of a few hundred MUS$. The areas of major concern to technical R&D are discussed. Also, trends and interested parties that can fruitfully contribute are identified.

1. Introduction

The total receiving aperture for a next generation of radiotelescopes is expected to be configured as 30 unit antennas of about 300-m equivalent diameter (Braun 1996) each. One way to realize this large collecting area is by means of a tile structure of flat receiving panels (Noordam et al. 1994). In contrast to most of the present generation of mechanically driven parabolic reflectors, the radiosource in this concept is followed electronically. Together with the appropriate electronic combination of received signals, the system as a whole can then be configured as an adaptive phased-array antenna. "Intermediate" concepts are possible in which the receiving function is physically separated from the electronic beamcontrol.

Other concepts have been discussed on different levels in a variety of international contexts and are studied elsewhere (see eg. other contributions to this workshop).

NFRA is focussing on a programme to explore the possibilities of the phased-array concept in more detail (van Ardenne 1996) primarily because it shows promise of an extremely versatile new generation of radiotelescopes. The technology of phased arrays is least mature for our applications and major challenges are to provide sufficiently wideband, low cost active receiving elements with the required sky coverage at acceptable investment cost of a few hundred MUS$. The realization of a practical observing instrument is therefore not trivial and more R&D is required to define an architecture that fits the expectations.

In this paper, we will briefly address this concept and the areas of major concern to technical R&D. Also trends and interested parties are identified that can fruitfully contribute to the definition and possible realization of such a new telescope.

2. Phased Array concept

The receiving surface of a planar phased array is realized using a flat structure of panels in each of which a number of flat receptor antenna elements is located. The relatively simple mechanical construction makes a low-cost light-weight construction possible. The electronic combination of received signals together with signal-processing techniques provides the potential of extreme flexibility. In this way, simultaneous reception from different directions is possible. Possibly, this "multibeaming" capability may result in new observing techniques extending beyond radioastronomy e.g. SETI. The adaptivity makes reconfiguration of the antenna beam(s) possible so as to minimize damage from sources of interference. This is relevant because of the increasing demand for frequency space from other developments e.g. telecommunications combined with the decade bandwidth desirable for **SKAI**. Emphasis on electronic techniques intrinsically profits from fast developments elsewhere e.g. for mobile communications [Beach et al (1994)] and microelectronics. This also implies a natural connection with interest groups and activities of colleagues at other knowledge institutes outside radioastronomy.

The concept of phased arrays has been in existence for tens of years primarily for military purposes. These are generally operating in receive and transmit mode in a hostile environment. This together with application-specific designs, makes the common implementations too expensive for our aims. Also in radioastronomy phased arrays are in use either by design or by use in special observing modes. Contrasting the much smaller bandwidth in these applications, we require a much wider band of approximately one decade between say 200 — 2000 MHz.

Adaptive arrays with subarrays per antenna unit (a "tile") are ideally performing "wave front sampling" without using traditional reflectors as the normal mode of operation. The signal-processing algorithms resulting in the equivalence to the common phased arrays are therefore more complex. Noordam et al. (1994) give an impression of the tile structure embedded in a larger part constituting one of the 30 unit telescopes. This "tile" telescope consists of 30,000 tiles of one square metre each. Each tile consists of a large number (say: 100 — 400) planar element antennae.

The total number of receiving elements is determined by the need to sample the wavefront with reasonable density at the highest frequency. This will result in not less than say, 10^8 elements. The aim therefore is simply stated, to build a synthesis instrument that can cope with these large number of inputs and transform them into calibrated maps of the sky. Because of this large number of receiving elements, the design of a tile, as the cost driving unit, is extremely important. Implementation for our purpose requires receiving elements with sufficient bandwidth and low-noise preamplifiers with integrated digital signal- and data-processing techniques both for (multi)beamforming and EMI-suppression with the appropriate signal distribution.

3. Some Technology Issues

For SKAI the primary driver is the collecting area while having a reasonable system temperature of say 50 — 100 K. Based on the (uncooled) system noise temperatures of state-of-the-art systems for the GMRT and for the new Westerbork system (operating from '97), it seems plausible to expect system temperatures of 80 K (200 MHz), 60 K (400 MHz) and 50 K (1-2 GHz) using GaAs transistors. One may use single very wideband receptors with fairly constant gain over frequency but this will lead to a decrease in collecting area versus wavelength squared. To get an acceptable area versus frequency, the total frequency band must therefore be broken into e.g. octave bands.

Estimates of the capacity of the central processing units indicate it must have the equivalent capability of a digital correlator with 10^7 complex channels. This is somewhere between 1 and 2 orders of magnitude larger than the new correlator back-end presently under construction for Westerbork and the Smithsonian Millimeter Array. To put this into perspective, the number of Gigabit operations per second of radioastronomical correlators over the last decades indicate a 1000 times increase over the last 20 years. The future requirement is about ten times beyond the PetaOperations/Second level but close to a straightforward extrapolation. Not too surprisingly, microelectronics developments both in cost and in power consumption show similar developments over these decades. These are positive indications that our aim is ambitious but realistic.

Estimates are based on the correlator performance of a basic system architecture (van Ardenne 1996). It is expected that digital beamforming, *i.e.* the complex weighting of each element-signal from individual antenna elements, takes place digitally. The weighting is analogous to the definition of an "array factor" resulting in the desirable beam on the sky.

The principle of digital beamforming in principle allows for a high level of functional integration at low cost. It is however likely that the SKAI concept will include complex weighting on both RF- and IF-level as well. This can be understood from the hierarchical method of beamforming within a tile ("intra-tile") and building up to levels between tiles ("inter-tile") and higher. The antenna elements within a tile are the basic receptors. These can be arranged according to some pattern; for example they can be fixed as in a dipole pair. For each beam, the signal-processing capability is comparable with existing correlators. This arrangement can be repeated at the highest levels resulting in increased processing requirements.

SKAI will operate in an electromagnetic environment heavily used by earth-based fixed and mobile infrastructures with an increasing demand by satellites for worldwide navigation (e.g. GPS) and communication purposes (e.g. Iridium) in a great orbital variety ("LEO", "MEO" and "GEO").

The matter of EMC/EMI has been and is brought forward by our community and others at different fora. It will have an inherent impact on **SKAI** system design in a variety of ways; for example, the need for sidelobe-cancelling techniques incorporating adaptive nulling and low-sidelobe antenna design while taking into account the high dynamic range, in particular in the first receiver stages. Furthermore, an advanced level of data massaging and algorithm development will be needed in order to discriminate the radioastronomical noise signature from the interferer.

To achieve the goals required for the Square Kilometer Array, the areas of major importance to technical R&D that we feel need to be addressed are summarized below. Other aspects e.g. political and organizational, are outside the scope of this contribution.

- Technological
  - Wideband and efficient array concept
  - Suitable system architecture and algorithms (EMI)
  - Low-cost electronics technologies and functional integration *e.g.* antenna, LNA and signal processing elements
  - Availability and integration of technologies *e.g.* microelectronics, materials, (fibre)-optics, manufacturing
  - Data manipulation and handling
  - Mechanical/structural, thermal properties
  - Interested parties
  - Investment costs
- Operational
  - Reliability/serviceability/maintainability
  - Data networks infrastructure/communications/EMI(local)
  - Operational costs

The matter of reliability is important due to the large number of elements and through its effect on the antenna performance, enters as a design aspect. The challenge therefore is to develop low-cost, low-power wideband and reliable antenna elements which can be integrated with receivers and signal-processing techniques with a high level of functional integration. These are not available and efforts are to be undertaken to make this approach successful for a large telescope operating at the beginning of the next century.

4. Strategies and Objectives

Clearly, the challenges of the phased array approach are considerable. Without a well defined R&D-programme, no conclusions can be drawn regarding the feasibility and the way the new telescope is to be realized. We envisage a programme that is set up in three main phases with a precursor activity. We will concentrate over a first 5-year period on three main areas of $R\&D$:

- **Wideband phased-array antenna concepts**.

The effective area of the antenna over the frequency band may result in several smaller band systems. Ease of manufacturing and low cost are crucial. Basic antenna elements based on planar (microstrip) antenna techniques in which mutual coupling effects are included, are therefore promising. The programme will initially concentrate on these. Alternative techniques e.g. using quasi-optics are considered also. A mixture of intermediate approaches and technologies may prove to be most interesting in the end but will be addressed later.

- **Uncooled, low-noise preamplifiers including integration aspects with both the antenna and the signal-processing functions.**

A high level of integration through mixed-mode (analog/digital) semiconductor technologies based on SiGe may prove to be very attractive. Low noise frontends based on GaAs cannot be excluded at this stage but it needs to be proven whether this is the optimum technology choice taking integration and cost aspects into account. Present results with SiGe heterojunction technologies are very promising at these frequencies. This technology combined with digital-processing functions are worth investigating.

- **Optimal signal processing techniques, algorithms and architectures.**

A formal description as is being developed in the context of the AIPS++ effort (the 'Measurement Equation'; Noordam 1996), is helpful. Efforts must be put into robust and efficient algorithms that can cope with EMI pollution e.g. from low orbiting satellites. The processing challenge can be inferred from the transformation of the 100M or so elements, to the making of reliable maps from say, $1-5$ beams.

Cooperation in other areas of interest, *e.g.* the area of quasi-optical techniques and RF-to-optics, is in the process of being set up.

The output of Phase 1 should result in the definition and selection of a technically realizable phased-array antenna concept with the constraints as mentioned before. Based on this, the system architecture and functionality can be defined and designs can be detailed. On the shorter term, we concentrate on starting a two year feasibility study as the first part of Phase 1. After that, an overview of architectures with pros and cons, available technologies and estimates of cost developments and prospects, are needed to define the further steps. The second 5 year phase is a (concept) demonstration project. NFRA aims at combining the result of Phase 2 with a North-South extension of the Westerbork array and the achievement of first scientific results. In the third and last 5 year phase, a full scale international project should be aimed at the gradual implementation of a full observational capability.

5. Technology programmes and cooperation

To expand the knowledge base and the possibility to increase communication with other parties, SKAI can profit from developments outside radioastronomy:

(a) Move from industries and research institutions with high technological profile from defence to civil applications.

(b) Rapid developments in mobile telecommunications regarding reliable, low cost, planar and integrated antennas, microelectronics, signal processing and functional levels of integration.

(c) Space-applications regarding multibeam antenna array technologies, materials and reliability.

(d) Manufacturability and implementation approaches from the consumer-electronics market.

Based on this, we aim at participating in (inter)national programmes to develop the required technologies in the various stages shown earlier. The establishment of an (inter)national platform with interested parties may be combined with programmes e.g. in the E.U. framework whose output may be relevant to the initial R&D. Although no E.U.-programme exists any more to support international concept studies relevant to LRTs, possibilities exit in connection with the new large technological programmes (e.g. ACTS) of the 4th-Framework.

6. Conclusions

A well defined RTD-programme is necessary to further advise on the the feasibility of a cost-effective system approach for the adaptive phased-array concept. This is most important in the area of wideband planar antennas with integrated low-noise receivers and some level of adaptive beamforming techniques. These must be implemented in low-power microelectronics technology using development results obtained in other areas e.g. wireless communications. Provided that the research and technology development programme is successful, a demonstration project may start 5 years from now.

REFERENCES

BRAUN, R. 1996. In *The Westerbork Observatory: Continuing Adventure in RadioAstronomy* (eds Raimond, E. & Genee, R.O.) Kluwer Dordrecht

NOORDAM, J.E. & VOUTE, J.L.L. 1994. *NFRA Internal Note*

ARDENNE, A. VAN 1996. *Proc. Third Workshop Sph. Radio Tel., Beijing*

BEACH, M.A. & XUE, H. & MCGEEHAM, J.P. 1994. In *Proc.Int.Conf.Ant.& Prop., Eindhoven(Neth.),p. 344*

1995 Spectrum Conservation. *Session 4B, Proc.Conf.Microwaves & RF, Oct. London*

WARNOCK, J. 1995 Silicon Bipolar Device Structures for Digital Applications: Trends and Future Directions. *IEEE-ED, pp 377-389, March 1995*

NOORDAM, J.E. 1996 The Measurement Equation of a generic radio telescope. *AIPS++ Implementation Note 185, NFRA*

ARGUS: a next-generation omnidirectional radio telescope

By ROBERT S. DIXON

The Ohio State University Radio Observatory, 2015 Neil Avenue, Columbus, Ohio 43210, USA

---

1. Overcoming the Legacy of Galileo

Ever since the invention of the first telescope by Galileo in 1609, the need to "point" it in the desired direction has always been assumed to be a fundamental requirement. But this assumption was based only on the technology used until now to build all almost all radiotelescopes. By marrying modern computing with an array of many small antennas, a new generation of telescopes can be built which look in all directions at the same time, just as if there were thousands or millions of conventional telescopes all in the same spot, each pointed in a different direction. The new telescope is called Argus, named after the mythological guard being that had 100 eyes and could watch in all directions at the same time.

A second assumed fundamental requirement that started with Galileo's telescope is that only one person can look through it at a time. This has evolved without conceptual change into today's method of assigning telescope time to individual observers. Existing radiotelescopes can typically only be used by one research project at a time. There is great competition to use the available telescope time, with some prospective users being turned away. Argus can conduct all desired research programs at the same time, with no mutual interference or compromise needed, making its scientific productivity much greater than comparable conventional telescopes. This capability is particularly important when seeking international funding and participation, since it makes wider usage possible.

A third assumption from Galileo's telescope is that precisely-made, moving components such as lenses or mirrors must be used. That has evolved into the the large dish radiotelescopes of today, which are ultimately limited by gravity, wind and differential temperature. Argus has none of these limitations. And as telescopes grew larger than Galileo's, they required massive yet accurate machinery to support and steer them. Argus has no machinery, no moving parts, no tight tolerances and little need for mechanical maintenance.

The radiotelescopes of today are typically large steel structures, one-of-a-kind with no mass production, with construction costs dominated by labour, which increases with time. An Argus telescope consists of a large number of very small antennas, connected together with a large number of small computers. Mass production of the pieces is automatic, and construction cost is dominated by the cost of computing, which decreases with time. Hence the cost of an Argus telescope must become less than a Galilean telescope at some point, even if its priceless all-seeing advantage is ignored.

The time has come to seriously consider a fundamentally different approach for radiotelescopes. Instead of large steel dish structures, a large number of small omnidirectional antennas can be used in an array to obtain much greater performance at ultimately lower cost. Such arrays are commonly called "phased" arrays, but that implies narrow bandwidth, so a more correct term for what is discussed here is a "timed" array.

2. Advantages of Argus Over a Dish-type Antenna

Compared to a conventional dish, an Argus timed array provides many advantages, including simultaneous high-gain omnidirectional sky coverage (no scanning), high sensitivity (arbitrarily long integration time), high resolution, variable beam size and shape, low and movable sidelobes, detection and tracking of transient and moving sources, adaptive and retroactive observations, interference rejection, higher aperture efficiency and fault tolerance.

Energy is falling on any radio telescope from all directions all the time, and the vast majority of it is ignored. The larger a dish antenna is, the worse it becomes in terms of using all the energy that falls on it. For example, the Ohio State radio telescope has an efficiency of about 10^{-4} percent, and Arecibo is about 10^{-5} percent. This is in contrast to Argus with nearly 100 percent efficiency. The sensitivity of an Argus array is the same as that of a dish having the same total effective collecting area and the same sensitivity receiver.

In terms of flexibility, an Argus array has a number of advantages. It can be easily expanded or changed in shape; its resolution can be chosen independently of its collecting area; and its resolution, beamshape, and sidelobes can be changed at will by software.

In terms of capability, an Argus array can do many things a dish cannot do, including observing multiple (all) objects simultaneously, tracking rapidly moving objects, detecting transient events in unknown directions, surveying the entire sky in a single integration period, observing adaptively in response to current results, and reobserving retroactively objects or events not recognized initially. The retroactive observations can be done by playing back the recorded data from the array elements, and if desired the beam and processing equipment can be reoptimized for the reobservation.

Argus has many advantages over a dish in terms of its ability to deal with radio frequency interference (RFI). The elements can be designed to have nulls at the horizon for rejection of terrestrial signals. The elements are on the ground, in contrast to the elevated feed of a dish, hence the signal strength of terrestrial signals is less. Small shield fences can be used around the elements or array if necessary for further rejection of terrestrial signals. The direction of any RFI signal is immediately known to Argus since one of its beams always points toward the RFI source, and it will be strongest in that beam. That beam will also provide a nearly noise-free version of the RFI which can be used to characterize and identify it and to blank it or cancel it in the rest of the beams. Diagnosis of RFI is immediate with no need to steer the telescope "off-source" to see if it goes away. Since each beam can be separately optimized, permanent nulls can be generated by each beam in the direction of known fixed RFI sources. Adaptive nulls can be generated in real time as needed to deal with transient RFI. Argus can also identify RFI sources by their distance, since it can simultaneously focus itself at all distances.

We have constructed and operated a prototype eight-element circular Argus array at 162 MHz (Bolinger 1988).

3. Argus Design Criteria

The elements of a general-purpose Argus array should have hemispherical coverage, aimed straight up. The best candidates are from the helix family. A multifilar contrawound conical helix can achieve these requirements.

The Argus array geometry should have approximately circular symmetry (for uniform beams), and not have uniform spacings (to avoid grating lobes). Placing the elements logarithmically spaced along the arms of a multiarm logarithmic spiral achieves these requirements.

The number of elements required in an Argus array to have performance comparable to a dish telescope depends upon the desired application and frequency. For example, if it is desired to observe only a single direction with maximum effective collecting area, then the number of elements for comparison with some existing dishes is:

Frequency (MHz)	Ohio State	Arecibo
50	180	1600
1500	180,000	1,600,000

But note that if an Arecibo-size Argus array were operated at 1500 MHz, it would also be looking in every other direction at the same time, and so would be equivalent to 1,600,000 Arecibos. If it is desired to do an all-sky survey in a given time, then the equivalent number of elements required is:

Frequency (MHz)	Ohio State	Arecibo
50	14	40
1500	420	1300

4. Argus Computing Architecture

The performance of an Argus array (as measured by its number of elements, number of beams, and bandwidth) is limited primarily by its computing power. Fortunately, available computing power is rapidly increasing and its price is falling. A small computer is used at each of the n elements, which does all computations that can be done on the data coming from that element. A different set of m small computers is used to perform the calculations for each of the m beams. In general, m is much greater than n, since the array is sparse (2).

All the element and beam computers may communicate via a token ring network. The element weightings used for beamforming are kept in lookup tables that are separate for each beam and can be rapidly changed as desired.

In addition to the element and beam computers, there is another much smaller group of small computers attached to the network, each dedicated to some special project. Examples of such projects include monitoring a pulsar, tracking a spacecraft, lunar occultation, identifying RFI, calibrating the system, etc. Each special project computer is free to use whatever data it wishes and make whatever calculations it wishes, with no interference with the main computers or with each other. Hence there is no limit to the number of special projects that can occur simultaneously. One particularly important special project is to record all the element data in a compressed form for later analysis. This makes it possible to reobserve an event that occurred long ago, but was not recognized at the time. The special projects computers can also be attached to the worldwide Internet, making it possible for anyone anywhere to control them and to obtain data from them.

Once an essentially noise-free image of the sky is obtained by long integration, a differential mode of operation can be used. In this mode, the telescope output displays only the differences between the "normal" sky and the current sky. This drastically reduces the amount of data to be displayed, and allows for immediate discovery of anything which has changed, moved, appeared, or disappeared. Such discoveries could automatically be announced immediately by one of the special projects computers to everyone around the world who chose to receive such announcements, via an Internet newsgroup or mailing list.

5. Converging Technologies

Many technologies are now advancing rapidly in directions relevant to Argus. The Institute of Electrical and Electronics Engineers (IEEE) recently published a special Communications issue (May 1995) on Software Radios. It describes how the cellular telephone industry expansion is forcing the development of radio technology that uses computers to process multiple complex communications signals instead of using traditional analogue filters and amplifiers. Q-Dot corporation in Colorado is now manufacturing CCD chips for dedicated electronic beamforming applications. Optical beamforming of microwave signals is an active area research area for military radar applications.

IEEE sponsored a conference on Mass Storage Systems in the Fall of 1995. The development of mass storage systems is being forced by the entertainment industry, who want to provide video-on-demand home access to every movie and television programme ever made. The amount of data produced by an Argus array having 10^3 elements, 10^5 beams, 10' resolution, 100s integration and 10^4 spectral points is about 1 terabyte/day. This is less than the data rate produced by a number of other scientific efforts today: NASA Earth Observation Satellite (8 TB/day), CERN (10 TB/day), Fermilab (3 TB/day). By the year 2000, optical tape cartridges will hold 1 TB each. Holographic storage systems are being developed by Optitek and GTE (Ross 1994) that are predicted to store a trillion bits in a cubic centimetre crystal of lithium niobate in three to five years. Holographic storage also allows massively parallel computations to take place within the storage itself, such as the addition or subtraction of entire images in a single step.

Protein-based computers are being developed at Syracuse University (Birge 1995) that can in principle be 1/50 the size and 1000× the speed of semiconductor-based computers. This computing technology was first recognized by the Russians, who have used it to build a yet-secret military radar processor. (Note that Argus requires computations similar to a radar.) Protein memory cubes can attain the same trillion bits per cubic centimetre as the crystals mentioned above. They must have very uniform composition, which is aided by low gravity manufacturing. Two space shuttle flights have already carried experiments to do this. Birge predicts that this technology will dominate computing within eight years.

Perhaps the ultimate in computation may emerge from the field of quantum computation. Seth Lloyd of MIT is pursuing the use of individual atoms as storage devices. A single electron energy level is used to represent a bit, which is flipped by laser radiation. A single grain of salt contains about 10^{18} atoms and could hence store that many bits. Because of internal instabilities and imperfections among the atoms, a quantum processor would have to spend about 99.9 percent of its time doing error detection and correction. Nevertheless the remaining speed is about 10^8 times faster than a Pentium processor. This technology is about 20 years from practical application.

6. The Big Picture

It is commonly believed that humankind is basically aware of everything that goes on around us in the universe. This may seem logical, given all the telescopes in operation around the earth. But the fact is that all telescopes combined see only a tiny fraction of the universe and frequency spectrum at any one time, and as larger telescopes are built, they see even less. In our quest for ever greater detail about the trees, we are ignoring the forest. There are undoubtedly transient events occurring all the time of which we are unaware; previous examples include pulsars and supernovae. We have no global view of our electromagnetic environment, encompassing both natural and manmade signals. We have an obligation to open our eyes widely and be aware of our surroundings so we can learn more about the universe and understand the big picture. Argus will make this possible.

REFERENCES

BOLINGER, JAMES. 1988. A Simultaneous Multi-beam Phased Array Using Digital Processing Techniques. Master's thesis. The Ohio State University.

BROWN, STEPHEN B. 1993. Radio Camera Arrays for Radio Astronomy. Master's thesis. The Ohio State University.

PHILIP E. ROSS, Forbes, Sept 26, 1994, p 170

ROBERT R. BIRGE, *Scientific American*, March 1995, p 90

Wired, March 1995, p124

Afterword: Towards high sensitivity and high resolution

By R. HANBURY BROWN

This conference celebrates the first radio observations made by Sir Bernard Lovell at Jodrell Bank 50 years ago. That's really quite a long time! Indeed 50 years is roughly half of the whole history of radio! If we had been holding this conference 50 years ago, we might have celebrated Marconi's first experiments with Hertzian waves in his father's garden near Bologna. In fact it is almost exactly 100 years since Marconi was granted a patent in this country for a system of wireless telegraphy based on electric waves; he would, I am sure, be very keen to discuss higher sensitivity.

As that handsome booklet on the Lovell Telescope tells us, Dr. Lovell returned to Manchester after the war to look for radar echoes from cosmic rays and to his surprise discovered a new and highly effective way of looking at meteors. In due course this work at Jodrell made major advances in meteor astronomy; not only did it add greatly to our knowledge of meteor streams, particularly daylight streams, but it also cleared up once and for all the question of interstellar meteors. Another early programme using radar was devoted to getting echoes from the moon; to mention only one result, it drew attention to the effects of Faraday rotation in the ionosphere.

I shall say no more about the work on radar astronomy. Most of you, I believe, are not concerned with radar, furthermore I don't know as much about it as I should. I was invited here to talk about the early days of radio astronomy at Jodrell. I shall define "early days" as being before the Lovell Telescope was commissioned in 1957 – the days before radio astronomy became Big Science.

In the pursuit of higher sensitivity for their work on meteors, in 1947 the group at Jodrell built with their own hands a very large paraboloidal reflector made out of scaffold poles and several miles of wire; it had a diameter of 218 feet and a focal length of 126 feet – a very enterprising and laborious thing to do! When I arrived two years later, they had already given up trying to detect cosmic rays with their big dish and were using it to look for what Grote Reber called "cosmic static". Almost ten years later Dr. Lovell gave me the chance. By the way I hope that any large antennas which you build in the future will turn out to be as useful for something unforeseen as that big dish proved to be.

Let me start by telling you a little about what we did with that 218-ft dish. In the summer of 1949 I took it over from Victor Hughes who was using it at 4.1m to record the "cosmic" radiation received in the zenithal strip. At that wavelength the half-power beam was about 4.5 degrees, but, having seen Reber's maps of the radio sky, I wanted something narrower. Given that the reflecting surface was formed by parallel wires 8 inches apart with a maximum departure from a true paraboloid of about 5 inches I chose a wavelength of 1.89m; at that wavelength the loss through the reflecting surface was rather high. in fact close to 37%, but the half-power beamwidth was 2 degrees.

In those days our major problem in measuring cosmic noise was not, as it is now, to get a lower noise factor or build a bigger antenna; it was to achieve adequate stability in the overall gain of the receiver. Anyway, I was joined by a PhD student, Cyril Hazard – a highly intelligent but fantastically untidy young man – and together we tackled this awful problem of stabilising the gain of the receiver. We went through all the obvious and tedious techniques – like the use of accumulators for the valve heaters, stabilised power

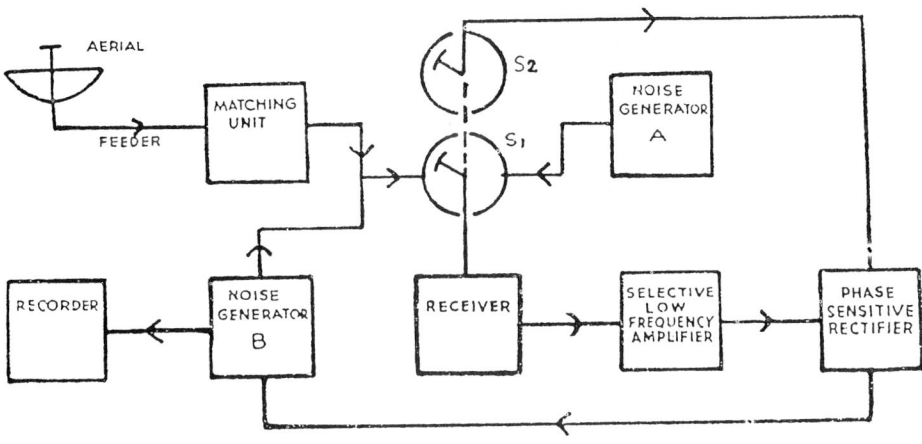

FIGURE 1. Block diagram of the receiver for the 218-ft

packs for high tension and so on, but ended up by borrowing a technique, it hurts me to say so, from Cambridge! For their work on solar radio noise Ryle and Vonberg had adapted, very ingeniously, a system used by Dicke et Princeton. All we had to do was adapt it for use with antennas with noise temperatures below room temperature – even in the winter at Jodrell it was warmer in the lab than the antenna temperature at 1.89m. As shown in the diagram (Fig. 1), we used a rotating switch to equalise the power from the antenna plus the noise from a controllable diode with the power received from a resistor at room temperature; in this system the receiver simply acts as a null detector. It worked like a charm, as you can see from this beautiful trace of the transit of the source in Cassiopeia through the antenna beam (Fig. 2).

Although it was exciting to look at the sky with the largest single telescope which had so far been used for radio astronomy, we soon got bored with looking at a 2 degree strip in the zenith and decided to try moving the beam by tilting the central mast. The mast was built in sections of steel tube which we were afraid of kinking and the maximum tilt which we dared to use was 17 degrees. It took at least two hours of hard and anxious work to move that mast through one beam width. I now know what it was like to point the "40-ft" telescope which Herschel built at the end of the 18th century or, worse still, to work on Lord Rosse's 72-inch reflector in the middle of the 19th century; to do astronomy in those days you needed to be in good physical shape.

Given a stable receiver we now set out to survey the whole sky between declinations +38 and +58. Rather a narrow band but even so it was enough to keep us busy for a few years. What we hoped to do was to relate this new radio view of the sky to the familiar view given by light.

Obviously a major part of the so-called background radiation was related to the structure

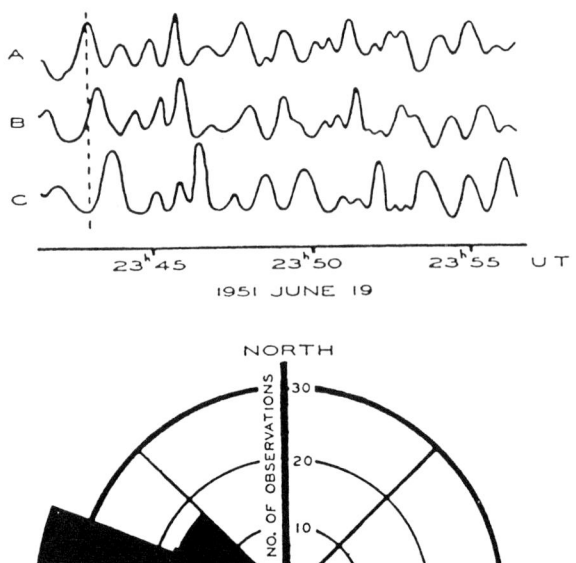

FIGURE 2. Trace of the transit of the source in Cassiopeia through the antenna beam of the 218 ft

of our Galaxy; it was clearly concentrated into the galactic plane but, even so, didn't correlate closely with the distribution of starlight. To be sure that was an interesting mystery, but not half as exciting as the bright point sources, the so-called "radio stars". What on earth, or rather, what in heaven, could they be?

In those days only three radio sources had been tentatively identified – Crab Nebula (NGC 1952), Virgo (NGC 4486), Centaurus (NGC 5128) – but to everyone's surprise most of the 100 or so known radio stars didn't line up with anything visible. Our survey of the sources in our strip showed that the stronger ones are concentrated into the galactic plane; whatever they are it was clear that they must be in our Galaxy. Although Mills in Australia had found the same thing, this concentration into the galactic plane was not confirmed by the Cambridge surveys made with an interferometer. The trouble was, as we found out later, that many of these strong sources had large angular sizes and so were resolved out by their interferometer – an interesting example of selective observation and of the importance of using a variety of techniques.

In our efforts to relate the radio and optical views of the sky we wondered if other galaxies also emitted radio waves. If so, a simple calculation suggested that we might be able to detect the nearest galaxy similar to our own, M31 in Andromeda. So in the autumn of 1950 we made 90 scans of the area around M31; it took us three months; all the work had to be done in the middle of the night to avoid ignition interference from traffic on a nearby road and 20 of the scans were ruined by charged rain falling on the primary feed. As you can see the isophotes plotted from these scans showed a radio source with a finite angular size in the position of the Andromeda Nebula with a central intensity of 50 Jy. We showed that the radio emission from M31 is similar to that from our own Galaxy and in doing so made the first crude map of an extragalactic source. Hazard and I went on to measure the radio emission from three other bright galaxies with the 218-ft dish and to survey some galactic clusters; but we didn't have the sensitivity to make a good job of it and we had to wait for the construction of the Lovell Telescope.

One interesting thing which we did with that big dish was to look for the supernova reported by Tycho Brahe in 1572. No visible remnant remains; nevertheless, during one of his brave, but discursive, weekly attempts to teach the Jodrell staff some astronomy Professor Kopal pointed out that the supernova of AD1054 (Crab Nebula) was known to be a strong radio source and so there might well be a radio source. We looked and found a quite strong radio source in the position of Tycho's star; as you know this identification was subsequently confirmed.

Nowadays our surveys with that 218-ft dish look pretty crude but to appreciate the difficulty of any early experiment one has to forget what came afterwards. One thing they certainly did was to confirm that a large, more sensitive and fully steerable paraboloid – the Lovell Telescope – would be a research tool worth building.

Let me now turn to another major part of the early work, the pursuit of high angular resolving power. Like all radio astronomers we wanted to solve the mystery of the so-called radio stars – were they, as most people believed, stars or nebulae in our own Galaxy or were they extragalactic? The obvious step was to measure their angular size.

As a first move we decided to measure the two brightest radio stars in the sky, in Cygnus and Cassiopeia; If, as was commonly believed, they really were stars then we might expect their angular sizes to be hundredths, if not thousandths, of a second of arc and to resolve them at metre waves our interferometer would have to work with baselines of thousands of miles. In those days highly stable oscillators had not yet been invented and there was no obvious way of making a conventional interferometer with such a long baseline. To solve this problem we developed a new form of interferometer, the intensity interferometer. Instead of trying to compare the relative phase and amplitude of the signals received at two widely separated antennas, we demodulated the two signals in independent receivers and then compared their intensity fluctuations in a narrow low-frequency band. For distances of up to a hundred miles or so we planned to transmit these low-frequency signals as modulation on a simple radio link; at greater distances, such as a thousand miles, we intended to make the comparison by recording them on tape.

In 1950 Roger Jennison and Das Gupta built a prototype intensity interferometer and tested it by measuring the apparent angular diameter of the Sun. The two receivers were tuned to 125MHz with a bandwidth of 200kHz; after demodulation in square-law rectifiers, the two low-frequency noise outputs, in a bandwidth of 1000Hz (1kHz-2kHz) were taken to a correlator. In 1951 they added a transportable aerial and radio link and used the instrument to measure the angular sizes and rough shape of Cygnus and Cassiopeia. Both sources were resolved with baselines of a few kilometres; their angular diameters were of the order of minutes of arc and to our disappointment neither of them was a star; Cygnus was found to be a double source. At about this time both sources were identified optically, Cassiopeia appeared to be a galactic nebulosity and Cygnus to be two galaxies in collision. We could easily have resolved both of them with a conventional interferometer and done the experiment in less than half the time. But every cloud has a silver lining; to interpret his results on Cygnus Jennison developed the method of phase closure which later proved to be of inestimable value to radio astronomy.

Talking of scintillation one of the successful programmes at Jodrell in the early days was the study of ionospheric scintillations. The intensity of the radio stars, observed at metre waves, was known to fluctuate and at first people thought it was a feature of the sources themselves. In 1949 Jodrell and Cambridge made simultaneous observations of

Figure 3. Group photograph of the staff at Jodrell Bank in 1951

the sources and found the fluctuations to be dissimilar at the two places, a result which suggested that they did not arise in the sources themselves. Subsequent research, led by Gordon Little and Alan Maxwell, showed that the fluctuations arise when the radio waves pass through the F region of the ionosphere, and that by observing their relative timing in spaced receivers it is possible to measure the direction and speed of the winds in the ionosphere.

In another programme seeking high angular resolution, we set out in 1952 to measure the angular size of the 23 sources in the field of view of the 218-ft dish. We made a simple interferometer by connecting the big dish and a small portable antenna together with a cable. Immediately we found that six of the strong sources close to the galactic plane had angular sizes of the order of 1 degree which suggested that they originated in extended nebulosities of low optical surface brightness. In an attempt to resolve the other sources we slowly extended the baseline, developing the necessary techniques of long baseline interferometry as we went, the control of fringe frequency, the use of supersonic delay cells and so on. But even at the maximum practicable length of our cable, about 500 wavelengths, we still couldn't resolve these other sources. Clearly there were at least two different types of source and to resolve the ones with small dimensions we were going to need a much longer baseline.

As a next step Henry Palmer and his colleagues replaced the cable with a radio link and in 1954 they began the long and tedious task of extending the baseline, spending a good deal of their time persuading suspicious farmers to accept the presence of the portable aerial on their land. To cut a long story short, by 1956 they had reached the highest pub in England, the Cat and Fiddle, at a distance of 20km from Jodrell; three of the sources were still unresolved and must therefore have angular sizes less than 12 seconds of arc. In those

FIGURE 4. Jodrell Bank from the air in 1949

days that was really something! It meant that these sources had brightness temperatures comparable with the remarkable source in Cygnus – could they, we wondered, be objects like Cygnus but further away?

Although these results were published in 1957 it was not worth using large telescopes to look for these small diameter sources until their positions had been measured much more precisely. In 1959 this was done at Cambridge and as a consequence Rudolf Minkowski, using the 200-inch telescope at Palomar, identified one of these three sources (3C295) as a galaxy with a visual magnitude of 20.9 and unusually strong emission lines like those in Cygnus; with a redshift of 0.461 it was the most distant object known in the universe at that time. This led on to the recognition of the class of radio galaxies, but that is outside my frame of the early days. Henry Palmer and his colleagues carried on this work with the Lovell Telescope and it was the forerunner of Merlin which is now one of the major programmes of Jodrell Bank.

Looking back on those early days I remember Jodrell Bank as a pleasant and exciting place to work – great mysteries like the radio stars were, and still are, particularly exciting. I spent 12 years at Jodrell and might well have stayed for another 20 had not Richard Twiss and I called on Jennison and Das Gupta one day in 1952 to watch the intensity interferometer working. By chance it was a day when the signals from the source were scintillating violently and yet, to our surprise, the correlation was unaffected. The penny dropped and we realised that by using the same principle we could make an optical interferometer which would be totally insensitive to atmospheric scintillation. But first we had to demonstrate that the same principle could be applied to light waves. This was necessary because in those days radio engineers didn't worry about photons but physicists did; furthermore the physicists told us in no uncertain terms that our system would not work with light waves because it demanded that the photons in two mutually coherent beams of light must arrive in time coincidence – manifestly, so they said, that was impossible.

To cut a long story short we demonstrated the correlation of photons in the laboratory and in 1955 I built an optical intensity interferometer out of two old army searchlights. In the winter of 1955/56 I measured the angular diameter of Sirius, the first measurement ever to be made of a main-sequence star. Whatever else it may be, Jodrell Bank is not an ideal site for optical astronomy; it took me four months and 60 nights of observing to get 18 hours of good observations on Sirius. However the instrument worked perfectly and, after a great deal of grubbing around for money, we built a full-scale stellar interferometer which we installed in a place with a better view of the sky, Narrabri Observatory in Australia. Towards the end of 1961 I went there for one year to install the new instrument, got trapped, and came back 28 years later, and that is why I am only qualified to talk about the early days of Jodrell.

There is, I need hardly say, a parallel story of those early days which can only be told by Sir Bernard. Like mine, that story starts in 1949 and ends in 1957 with the completion of the Lovell Telescope. Things don't always go smoothly in that story as you can tell from reading his book – The Story of Jodrell Bank – but it has a happy ending. As we all recognise he built, against considerable odds, a remarkably fine instrument which in subsequent years has proved of great value to astronomy. I am only too glad to help celebrate the 50th anniversary of his first observations at Jodrell Bank.

Author index

S. Ananthakrishnan 217
A. van Ardenne 282
R.W. Argyle 50
D.J. Axon 149
W.A. Baan 73
I. Bains 57
L.R. Ball 46
D.S. Balser 53,93
T.M. Bania 53,93
B. Bates 101
R. Beck 117
M. Bode 33
R.S. Booth 194
R. Braun 260
G. de Bruyn 233
M. Bryce 57
W.B. Burton 97
F. Camilo 14
Y. Chikada 256
R.J. Cohen 65
J.E. Conway 153
W.D. Cotton 157,165,161
D. Dallacasa 161
R.D. Davies 101
R.J. Davis 33,37,50
J.R. Dickel 61
R.S. Dixon 287
W. van Driel 229
V.K. Dubrovich 189
J.S. Dunlop 129,167
S. Eyres 33
L. Feretti 157
S.T. Garrington 50
E. Gerard 229
G. Giovannini 157
L.I. Gurvits 252
N.G. Hamilton 139
R. Hanbury Brown 292
D. Hartmann 97
V.A. Hughes 42
J.M. van der Hulst 106
D.R. Jiang 161
P.M.W. Kalberla 97
V.K. Kapahi 217
N. Kawaguchi 256
F.P. Keenan 101
S.N. Kemp 101
H.T. Kenny 37
H. Kiuchi 256
N. Kobayashi 256
M.J. Kukula 129,139
V.K. Kulkarni 217
S. Kwok 37
L. Lara 157
T.H. Legg 274

R.P. Linfield 252
H. Lloyd 33
E. Lüdke 165,161
M.R.W. Masheder 88
J. Meaburn 53
U. Mebold 97
G. Mellema 53
V. Migenes 46
Y.C. Minh 84
M. Miyoshi 256
L.V. Morrison 50
C.G. Mundell 143
Y. Murata 256
T. Muxlow 57,149
J. Nakajima 256
R. Nan 278
S.J. Newell 46
Y. Nie 278
R.J. Oliver 88
I. de Pater 2
A. Pedlar 53,129,139,149
B. Peng 278
R.A. Perley 242
J. Pezzani 229
Y. Qiu 278
A.P. Rao 217
W. Reich 225
E.I. Robson 207
R.S. Roger 101
R.T. Rood 53,93
M. Rowan-Robinson 177
H.S. Sanghera 165,161
R.T. Schilizzi 161
E.R. Seaquist 23
C.R. Shaw 101
F. Smits 282
R.E. Spencer 46
R.G. Strom 2,278
C.R. Subrahmanya 217
G. Swarup 217,269
F. van der Tak 2
A.R. Taylor 37
P. Thaddeus 88
P. Thomasson 53
A.K. Tzoumis 46
J.S. Ulvestad 252
T. Venturi 157
G. Westphalen 97
R. Wielebinski 6,211
K.A. Wills 149
P.N. Wilkinson 149
T.L. Wilson 53,93
S. Withington 203
X. Xu 278
B. Zhu 278

Object index

ρOph 28,29
0050-727 47
0103-762 47
0218+357 180
0540-697 47
0836+29 158
0921-630 47
0957+561 180
1146+596 154
1323-619 47
1353+055 154
1516-569 47
1642-455 47
1659-487 47
1722-363 47
1744-299 47,48
1934−638 48
1946+708 154-156
29 Dra 52
3C31 158
3C84 154,234
3C147 165-166,234
3C236 154-156
3C264 158
3C273 200
3C286 161
3C293 154
3C338 158-160
3C465 158
4C31.04 154-156
4C35.03 158
6C0140+326 167-175
6C1232+39 167
A665 181
A2218 181
A2670 109
Algol 51
Arp 220 73
B2 0902+34 171
BD+30 3639 57
Cen A 153,195
Cepheus A 42
Circinus X-1 46,47
Comet Shoemaker-Levy 2
CP1919 6
Crab 14
Cygnus A 130

Cyg X-3 46
E1821+643 135
F10214+4724 197
F568-1 107
GGTau 195
GRSJ1655-40 28,46
GRS1915+105 28,46
GX340+0 47
GX339-4 47
H1-36 41
H1413+117 197
HM Sge 37
HoII 110
Hydra A 154
IC10 110
IC289 54
IC342 125-126
IC4182 179
IKTau 70
IR13218+0552 75
IR14070+0525 75
IR20100-4156 75
IRC+10216 195
IRC+10420 71
Jupiter 2
LMCX-1 47
M13 101-104
M31 109,117-125,182
M32 182
M33 109-114,120-126,179
M51 109,124,153,195,211
M81 124-126.177-188
M82 149-152,211
M83 107,124
M87 153
M92 185
M96 179
M100 179
M101 109-111,263
MBM7 84
Mkn6 140
Mon OB1 88
NGC253 126,211
NGC300 179
NGC315 153,158
NGC891 112-113,120,211
NGC1052 76,154

NGC1068 76,130,153
NGC1566 123,124
NGC1808 195
NGC2276 120,123-126
NGC2484 158
NGC2639 76
NGC3079 73,76,211
NGC3242 53
NGC3556 120
NGC3627 211
NGC4151 141,143-148
NGC4254 125
NGC4258 76
NGC4261 153-154
NGC4496 179
NGC4536 179
NGC4565 117,211-212
NGC4631 211
NGC5055 124
NGC5253 179-180
NGC5462 111-114
NGC5775 119
NGC5907 211
NGC5929 140
NGC6543 54
NGC6720 54
NGC6946 110-112,117-126
NGC7009 54
NGC7027 58
NGC7662 54
NMLCyg 70
Nova Cyg 1992 33
Nova Cas 1993 33
PSR0045-7319 20
PSR0011+47 5
PSR0136+37 5
PSR0154+61 5
PSRJ0218+4232 18
PSR0329+54 5,6
PSR0355+54 5,6
PSRJ0437-4715 18
PSR0525+21 5
PSR0540+23 5,6
PSR0655+64 14
PSR0656+14 5
PSR0740-28 5
PSRJ0751+1807 17

PSR0820+02 14
PSR0823+26 5
PSR0912+02 5
PSR0942-13 8
PSR0950+02 5
PSRJ1012+5307 18
PSRJ1022+1001 18
PSR1112+50 5
PSR1133+16 5,6
PSR1227+25 5
PSR1257+12 14,20
PSR1502+66 5
PSRJ1518+4904 18
PSR1534+12 15,20
PSR1612+07 8
PSR1620-09 8
PSR1630+27 5
PSR1640-08 5
PSR1701-16 6
PSR1716-16 8
PSR1750-24 8
PSR1758-23 8
PSRJ1803-2712 20
PSR1821-11 8
PSR1849+00 8
PSR1855+09 14,15,20
PSR1913+16 14,20
PSR1929+10 6
PSR1937+21 14
PSR1953+29 14
PSR1957+20 15
PSRJ2019+2425 15
PSR 2020+28 6
PSR2021+51 6,7
RAqr 41
RTVir 68,69
Sco X-1 46
SMCX-3 47
SS433 46
TXCam 68
TXFS2226-184 76
UHer 68
V1016Cyg 41
V395Car 47
Virgo cluster 177-188
VLA1623 28
VXSgr 68

Subject index

Arecibo 6,14-18,20,73
Argus 287-290
ARISE 252-255
Astrometry 50-52,241
Australia Telescope 17,47,49,106
Baade-Wesselink method 177
CH 73
CO 42,44,84-87,88-91,122,196
    survey 88
Comets 2-5,232
Compact radio sources 153-156,165,161
Cosmology
    abundances 53-56
    distance scale 177-188
    luminosity functions 167-176
    H_o 29,177-188
    primordial spectrum 189
Deceleration parameter 186
Depolarization 63-64,165
Deuterium 53-56,189
Dominion RAO 84,101
Effelsberg 6,10,53-56,225-228
Evolution, chemical 53-56,93,189
Fe$_{II}$ 43
Free-free
    absorption, Seyferts 133,140-142
    absorption in M82 149-152
    emission 25
Galaxies
    active 129-138,139-142,143-148
    149-152,157-160,165-167,252,161
    bars jm1,127,143-148
    cold dust with IRAM 211-212
    disks 117-129,143-6,153-4
    evolution 167-176
    FR$_I$ 157
    jets 139,157,165,161
    magnetic fields 117,165,161
    Nançay telescope 231
    polarization 121-123,165,233-237,161
    radio emission 106-115,117-127,129-138, 139,143-148,149-152,157,165
    radio quiet 129-138
    Seyfert 131-134,139-142,143
        unified schemes 131-134,139-142,147
    starburst 149-152
        and SKAI 261
    unified schemes 158

Giant Metrewave Radio Telescope 217-224,266
H$_I$
    GMRT observations 223,269
    in galactic clouds 84-87,89,101-104
    in galaxies 97-100,106-116,153-156
    in Seyfert galaxies 143-148
    kinematics 106-108
    Nançay telescope 229
    in Perseus arm 89
    and SKAI 260-268,269-270
    against supernovae 29
    surveys 97-100
    and WSRT 233-237
Gravitational lenses 181,199,269
Green Bank 140-ft 16
GBT (Green Bank Telescope)
H$_{II}$
    abundances 53
    regions 88,93
H_2 43
Helium 53-56,93,189
HEMT 11
High latitude clouds 84-87
HIPPARCOS 50-52
HST 50,57-60,135,153-154,238-239
Hubble constant 28,177-188
IRAM 194,211-212
JCMT 207-210
Jodrell Bank 16-20,101,213
    Lovell telescope upgrade 213-216

K$_I$ 101
Leiden/Dwingeloo survey 97-100
Lovell Telescope see Jodrell Bank
Luminosity functions see Cosmology
Magellanic Clouds 32
Magnetic fields
    in galaxies 122-129
    Jupiter 2
    in masers 65-72
    in SNR 62

Masers
    H_2O 31,65-72,75,153
    OH 31,65-72,73,153
    SiO 65-72
    megamasers 73-82
    pumping 68,79

and space VLBI 252-255
MERLIN 33-36,37-40,50-52,57-60,65,131
  139,143-148,149-152,238-241
    phase-referencing 50-52,238
Millimetre astronomy
  CfA telescope 88
  with IRAM 211-212
  large mm-submm array 194-202
  large southern array 194-202
  millimetre array 194-202
  SCUBA 207-210
  sites 198-200
  submm spectral-line imaging 203
NaI 101,104
Nançay telescope 229
Novae see Stars, novae
Parkes 16-21
Planets 2
Planetary nebulae 53-56,57-60
  ^{3}He abundance 53-56
  interacting wind models 57
  comparison of radio and optical 57-60
Pulsars
  binary 14-22
  GMRT observations 223
  high radio frequencies 6-13
  millisecond 14-22,231
  and Nançay telescope 229
  polarization 10
  pulse shapes 7-13
  searches 14-22
  spectra 8-13
Quasars
  astrometry 50-52
  compact steep-spectrum 161-164,165-166
  luminosity functions 167-176
  microquasars 29
  radio-quiet 134-136
  SKAI 269
  WSRT observations 233-237

Radio galaxies 129,157-159,167-176
Reference frames 50-52

SCUBA 207-210,212
Seyfert galaxies see Galaxies, Seyfert
Shocks
  in Cepheus A 43

in extragalactic jets 162,163
in galaxies 143-148,124

in galactic cloud 87
in PNe 57
in supernovae 29-30,61
SKAI (Square kilometre array interferometer)
  Argus design 287-291
  and stars 23-32
  and supernovae 61-64
  technical 260-268,269-272,278-281,282-286
Stars
  binary 23,33-36,37-40,46-49
  bipolar outflow 42-45
  Herbig Ae, Be 28
  mm observations 195
  IR emission 43
  jets 42-45
  masers see Masers; stars, OH/IR
  Miras 23
  and Nançay telescope 229
  novae 25,33-36
    light curves 33-36
    outburst 33-36
  OH/IR 31,66
  observations with large telescopes 23-32
  pre-main sequence 28
  RSCVn 28,51
  solar-type 27
  supernovae 29-30,61,149-152,179
  symbiotic 25,37-40
    colliding winds model 37-40
    STB model 37-40
  Wolf-Rayet 25,26
  X-ray binaries 14,29,46-49
Sunyaev-Zel'dovich 181
Supernovae see Stars, supernovae
Telescopes see name of telescope
Tully-Fisher relation 177,229
VLA 3-4,33,37-40,42-45,56,
  57-60,66,68,106,129,149,157-160,242-251
VLBA 68,153-156,157-160,242-251,252-255
VLBI 49,50-52,65,68,71,153-156,157,
  165-168,161,252-255,256-259
  in space 252
WEASEL 203
WSRT 3-4,106,233-237